Geophysical Monograph Series

Geophysical Monograph 228

Flood Damage Survey and Assessment

New Insights from Research and Practice

Daniela Molinari
Scira Menoni
Francesco Ballio

Editors

This Work is a co-publication of the American Geophysical Union and John Wiley and Sons, Inc.

WILEY

Published under the aegis of the AGU Publications Committee
Brooks Hanson, Director of Publications
Robert van der Hilst, Chair, Publications Committee

For details about the American Geophysical Union visit us at www.agu.org.

Wiley Global Headquarters
111 River Street, Hoboken, NJ 07030, USA

For details of our global editorial offices, customer services, and more information about Wiley products visit us at www.wiley.com.

Library of Congress Cataloging-in-Publication Data is available.

ISBN: 978-1-119-21792-3

Cover image: Flash flood that occurred in the Umbria Region in the 2013 flood in the municipality of Scheggia e Pascelupo (GRID-Politecnico di Milano, 2013). Insert 1: Wet documents in a firm flooded in the municipality of Marsciano, in Umbria in the 2012 flood (GRID-Politecnico di Milano, 2012). Insert 2: Debris flow that occurred on November 2013 in Umbria in the municipality of Gualdo Tadino (GRID-Politecnico di Milano, 2013).
Cover design by Wiley

Printed in the United States of America.

Set in 10/12pt Times New Roman by SPi Global, Pondicherry, India

10 9 8 7 6 5 4 3 2 1

CONTENTS

CONTRIBUTORS

Andrea Ajmar
Geodatabase Administrator
Information Technology for Humanitarian Assistance,
Cooperation and Action (ITHACA)
Turin, Italy

Shahrzad Amouzad
PhD Student
School of Architecture
Oxford Brookes University
Oxford, UK

Danilo Ardagna
Associate Professor
Department of Electronics, Information and
Bioengineering
Politecnico di Milano
Milan, Italy

Funda Atun
Research Fellow
Department of Architecture and Urban Studies
Politecnico di Milano
Milan, Italy

Francesco Ballio
Full Professor
Department of Civil and Environmental Engineering
Politecnico di Milano
Milan, Italy

Tatiana Bedrina
Researcher
CIMA Research Foundation
Savona, Italy

Nicola Berni
Head of Functional Centre
Umbria Region Civil Protection Authority
Foligno (PG), Italy

Piero Boccardo
Associate Professor
Interuniversity Department of Science, Design and
Land Policies
Politecnico di Torino
Turin, Italy

Marco Broglia
Scientific Officer
European Commission Joint Research Centre
Ispra, Italy

Maria Brovelli
Full Professor
Geomatics Laboratory
Department of Civil and Environmental
Engineering
Politecnico di Milano
Milan, Italy

Christina Corbane
Scientific and Technical Project Officer
European Commission Joint Research Centre
Ispra, Italy

Tom De Groeve
Acting Head of the Global Security and Crisis
Management Unit
European Commission Joint Research Centre
Ispra, Italy

Martin Dolan
Research Fellow
School of Architecture
Oxford Brookes University
Oxford, UK

Tiernan Doyle
Project Coordinator
VOAD and Resilience Networks
BoCo Strong
Boulder, Colorado, USA

Daniele Ehrlich
Senior Staff Member
European Commission Joint Research Centre
Ispra, Italy

Melanie Gall
Research Assistant Professor
Department of Geography
University of South Carolina
Columbia, South Carolina, USA

Fabio Giulio-Tonolo
Geomatics Expert and Remote Sensing Specialist
Information Technology for Humanitarian Assistance,
Cooperation and Action (ITHACA)
Turin, Italy

Yetta Gurtner
Researcher
Centre for Disaster Studies
James Cook University
Townsville, Queensland, Australia

Sören-Nils Haubrock
Director
Beyond Concepts GmbH
Osnabrück, Germany

Adriana Keating
Research Scholar
International Institute for Applied Systems Analysis
Laxenburg, Austria

David King
Associate Professor and Director
Centre for Disaster Studies
Centre for Tropical Urban and Regional Planning
School of Earth and Environmental Sciences
James Cook University
Townsville, Queensland, Australia

Heidi Kreibich
Senior Scientist and Head of WG Flood Risk and
Climate Adaption
Section Hydrology
GFZ German Research Centre for Geosciences
Potsdam, Germany

Jan Kucera
Scientific Officer
European Commission Joint Research Centre
Ispra, Italy

Jessica Lamond
Associate Professor
Architecture and the Built Environment
University of the West of England
Bristol, UK

Karen MacClune
Senior Staff Scientist
Institute for Social and Environmental
Transition-International
Boulder, Colorado, USA

Marco Massabò
Project Leader
CIMA Research Foundation
Savona, Italy

Mirjana Mazuran
Research Assistant
Department of Electronics, Information and
Bioengineering
Politecnico di Milano
Milan, Italy

Scira Menoni
Full Professor
Department of Architecture and Urban Studies
Politecnico di Milano
Milano, Italy

Guido Minucci
Research Fellow
Department of Architecture and Urban Studies
Politecnico di Milano
Milan, Italy

Daniela Molinari
Researcher
Department of Civil and Environmental
Engineering
Politecnico di Milano
Milan, Italy

Brendan Moon
Chief Executive Officer
Queensland Reconstruction Authority
Brisbane, Australia

Meike Müller
Geoecologist
Deutsche Rückversicherung NatCat-Center
Düsseldorf, Germany

Carolina Arias Munoz
PhD Student
Department of Civil and Environmental Engineering
Politecnico di Milano
Milan, Italy

Ray Ogden
Professor and Associate Dean
School of Architecture
Oxford Brookes University
Oxford, UK

Shadrock Roberts
Director
Ushahidi
Resilience Network Initiative
Athens, Georgia, USA & Nairobi, Kenya

Roberto Rudari
Research Director
CIMA Research Foundation
Savona, Italy

Kai Schröter
Researcher and Project Manager
Section Hydrology
GFZ German Research Centre for Geosciences
Potsdam, Germany

Julio Serje
Program Manager/Senior Software Engineer
United Nations Office for Disaster Risk Reduction
Geneva, Switzerland

Michael Szoenyi
Flood Resilience Program Lead
Zurich Insurance Group
Zürich, Switzerland

Annegret Thieken
Professor
Institute of Earth and Environmental
Science
University of Potsdam
Potsdam, Germany

Kanmani Venkateswaran
Research Associate
Institute for Social and Environmental
Transition-International
Boulder, Colorado, USA

Nicholas Walliman
Senior Lecturer
School of Architecture
Oxford Brookes University
Oxford, UK

Annett Wania
Scientific Officer
European Commission Joint Research
Centre
Ispra, Italy

PREFACE

In this book, state of the art methods and procedures for post-flood damage data collection and analysis are discussed, suggesting also best practices that may guide the reader toward the improvement of the quality and comparability of data and analyses across time and geographic areas.

The fact that better data are needed is a common plea put forward by researchers in many areas of investigation, including risk analysis. The call for better data on natural hazards impacts is certainly not new and has been on the agenda for a long time. So why bother now? Today the novelty stands at multiple levels to justify the proposal of such a thorough reflection proposed to the reader.

First, not only scientists are concerned about lack of data. It also has become a strategic issue for a variety of stakeholders, pertaining both to private and public sectors, who hold responsibility in different ways for disaster risk management. This explains why in the book contributions from a variety of actors can be found, ranging from institutions working at different spatial scales, to reinsurers, to practitioners. The reasons are varied and reflect specific interests and the mission of each actor. For governments, public administrations, and national and international organizations, the need to be able to compare events across time and space has become a prominent factor as the number and the extent of disasters have been constantly increasing over the last years putting at risk lives, public investments, and economic development. To fully appreciate the root causes of such an increase, there is the need first to be able to rely on the data related to the most obvious indicators, such as the number of victims, lost assets, and damages to items and systems.

Different studies suggest that such an analysis of trends over time and across geographic areas is not really possible given the low quality of available databases and the lack of agreed upon standards that are used to collect data when a disaster strikes and afterward. In front of the evidence of increased impacts and associated costs of repair and lost revenue, particularly in times of financial crisis, the need of programming investments in mitigation becomes key, in order to achieve the best results in terms of avoided damage at sustainable costs. However, such appreciation clearly requires that the background information on which such evaluations of potential investments is done be reliable at least at a minimal level, which apparently is not the case as for now.

It would be a mistake, however, to think that such concerns take into account only public bodies. Private organizations at large would greatly benefit from an enhanced capacity to estimate and prepare for damage before an event strikes. Insurance companies have relied until now on the large amount of data that is available in their databases. However, such data are very partial, of varied quality, depending significantly on the skills and time devoted to surveys by experts appointed to set the claims after an event. Such data are useful for identifying key variables benefitting from a very large number of surveyed values, but the data cannot account for extraordinary situations (linked for example to catastrophic events or whenever cascading effects are implied) or to appreciate the interaction of factors in very complex environments. As urbanized areas have grown exponentially over the last few decades so has the complexity of disasters. A variety of interdependent and tightly interconnected systems (including social, economic, built up, natural) have created the starting point for unanticipated damage that can be very costly. Gaining a finer understanding of how a variety of initial conditions in different environments produce larger and more complex ways to solve problems is becoming an issue also for insurance and reinsurance companies. In the meanwhile, studies [*Rose and Huyck*, 2016] have shown that the cost of collecting new and more data is fairly repaid by the possibility of better appraising how the emergency context affects businesses and what the factors are that provoke the highest impact on businesses' capacity to recover quickly.

The reasons for a growing interest by a variety of actors for enhanced disaster damage data that we have just discussed explain why now different initiatives at the national and international levels, such as the Working Group established by the European Union (EU) Commission, or the Sendai Framework for Action, have raised interest on the topic. At the heart of the reasons for such interest is certainly the recognition that data and information are the bricks of knowledge. It is not just a matter of accounting to better program resources to be allocated for disaster management or to evaluate trends of losses to identify the potential impact of climate or social changes leading to different patterns in the natural and the built environments. It is also an issue of identifying and selecting the most effective mitigation measures while gaining a better perspective on what the factors are of the risk function, hazard, exposure, and vulnerability, that have contributed most to the final outcome in terms of losses.

On the other hand, enhanced knowledge of natural hazards accomplished in the last decades is key for identifying what the most useful data are to collect. In addition, identifying the crucially missing information is key. Without both, a better understanding of how risk factors play in each context and better modeling capacity for forecasting damage before the event occurs will not be achieved.

The book is organized in five parts. Part I comprises two chapters that lead the reader into the international debate on loss data needs, discussing loss data requirements defined by the Sendai Framework and the main initiatives to meet such requirements.

Part II starts with a comprehensive overview of loss data storage at the global level, highlighting limits, strengths, and needs of available databases in order to accomplish the Sendai Framework requirements (Chapter 3). Then, the focus shifts to the national level with a critical discussion of flood loss databases in the United States of America (Chapter 4) and the German HOWAS21 database (Chapter 5), presented as a best practice of loss databases tailored to risk modeling needs.

Part III focuses on best practices of damage data collection, at both the meso and the local scale. As for the former, the experience gained in Germany after the Elbe flood in 2002 is analyzed (Chapter 7). In this instance, computer-aided telephone interviews were carried out to "survey" observed damage at residential buildings and firms. Such practice is now a standard in Germany after every flood event and could be considered for replication in other countries. As for the local scale, the survey experience gained in the Umbria region (Central Italy), after the 2012 flood, is discussed in Chapter 6. Such experience brought the development of a procedure for damage data collection, at the individual affected item scale, to be implemented every time a flood occurs in the region. The procedure has been designed to meet several user needs (i.e., emergency management, damage compensation, disaster forensic, and risk modeling) and includes specific forms for damage surveys.

Chapter 8 presents a comprehensive overview of the surveys carried out at the Centre For Disaster Studies Research (at James Cook University) on the occasion of 13 floods in Australia. Such an experience can be seen as a best practice situation to address issues that contribute to mitigation as well as to understand community

experience in a disaster. The main results from the study are described in terms of communities' vulnerability and resilience. In Chapter 9, that closes the third section, the main advantages and limits of crowdsourcing as a reliable and complementary source of loss data are discussed. In detail, the authors, who are practitioners working within humanitarian organizations and community-based flood relief organizations, describe their own experience by presenting several case studies. The latter constitute the basis for illustrating the value of crowdsourcing but reflect also on how to ensure its effective integration into disaster response.

Part IV supplies examples of data analysis, of how collected and stored data can be used to support multiple objectives for which data are collected. Following *De Groeve et al.* [2013], objectives can be synthetically indicated as accounting, forensic analysis, needs assessment, and improved risk modeling capacity.

The first contribution to this section deals with the Post-Event Review Capability (PERC) methodology (Chapter 10). The methodology has been designed as part of the Zurich Insurance's resilience alliance as a process to evaluate what happened before, during, and after a disaster, to identify the critical gaps and successes in the overall disaster risk management system, and to present actionable recommendations. Then, the use of damage data to develop complete event scenarios after flood events is discussed, providing an application to an Italian case study (Chapter 11). Chapter 12 presents the experience gained by the Queensland Reconstruction Authority in Australia after the 2010–2011 floods. The chapter highlights how the knowledge of observed impacts allowed the definition of the most suitable strategies to build a more resilient Queensland. The final contribution to this section (Chapter 13) supplies insights on the use of collected and stored data to carry out a forensic investigation of flood damage at the industrial sector. In particular, the chapter discusses how disaster forensics can be used to understand damage cause and mechanisms and then to define proper risk mitigation measures.

The last section (Part V) includes best practices on the use of Information and Communication Technology (ICT) supporting data collection, storage, and analysis. Chapter 14 focuses on the use of satellite data to survey and assess damage at the global scale. In particular, the Copernicus Emergency Management Service (EMS) is described making reference to some case studies. Chapter 15 describes tools developed within the Italian project Poli-RISPOSTA for data collection and analysis at the local scale. Such tools consist of mobile applications for data survey, spatial databases for the storage of data, and a web-GIS application for data analysis and representation.

Conclusions close the volume and include recommendations, guidelines, and best practices starting from the experiences described in the book.

REFERENCES

De Groeve, T., K. Poljansek, and D. Ehrlich (2013), Recording Disasters Losses: Recommendation for a European Approach, JRC Scientific and Policy Report.

Rose, A., and C. Huyck (2016), Improving catastrophe modeling for business interruption and insurance needs, Risk Analysis, doi:10.1111/risa.12550.

Daniela Molinari
Department of Civil and Environmental Engineering
Politecnico di Milano, Milan, Italy

Scira Menoni
Department of Architecture and Urban Studies
Politecnico di Milano, Milan, Italy

Francesco Ballio
Department of Civil and Environmental Engineering
Politecnico di Milano, Milan, Italy

ACKNOWLEDGMENTS

The editors would like to thank all the authors of the chapters for their valuable contributions to the book, as well as the reviewers of the various chapters for their critiques and suggestions that surely contributed to the improvement of the book, for this final version. The editors would also like to thank AGU-Wiley for fostering and supporting the realization of the book and, in particular, Dr. Rituparna Bose, Mary Grace Hammond, Vishnu Narayanan, Peggy Hazelwood and Shiji Sreejish for their help in the entire production phase. Finally, the editors acknowledge that the main ideas behind the design of this book come from activities carried out within the EU expert working group on disaster damage and loss data at the Joint Research Centre (JRC) and the research projects Poli-RISPOSTA (stRumentI per la protezione civile a Supporto delle POpolazioni nel poST Alluvione), which was funded by the Poli-SOCIAL funding scheme of Politecnico di Milano, IDEA (Improving Damage assessments to Enhance cost–benefit Analyses), a EU prevention and preparedness project in civil protection and marine pollution, funded by DG-ECHO, G.A.N. ECHO/SUB/2014/694469 and EDUCEN - European Disasters in Urban centres: a Culture Expert Network (3C [Cities, Cultures, Catastrophes]) funded by EU Horizon 2020, C.N. 653874, in which the editors have been actively involved.

Part I
Introduction

1

Overview of the United Nations Global Loss Data Collection Initiative

Julio Serje

ABSTRACT

The Year 2015 was marked by the emergence of three international agreements: The Sendai Framework for Disaster Risk Reduction, the 2030 Agenda for Sustainable Development, and in the Intergovernmental Panel on Climate Change (IPCC) Conference of the Parties (COP) 2015, a global legally binding agreement on Climate Change now known as the Paris Agreement.

All of these frameworks explicitly recognize the importance and usefulness of collecting and analyzing loss data in their corresponding implementations. The Sendai Framework, in particular, calls for the collection of data about disaster of all scales. It also calls for the collection of data about man-made, technological, environmental, and other hazards, with an emphasis on climate-related risks.

Most importantly, the Sendai Framework sets out seven targets, of which four relate to losses: mortality, people affected, economic loss, and damages to infrastructure. This implies that the coverage of national disaster loss data sets will have to be expanded to be global so that countries can report on these targets. This development represents a unique opportunity to build a bottom-up constructed global disaster loss database.

Many actors have collected national loss data for many years. For over a decade, the United Nations (UN) system has supported and promoted the construction of national disaster databases based on the Disaster Information Management System (DesInventar) methodology and software tools. Additionally, a number of countries have been collecting data with proprietary specifications and different levels of resolution. These include several countries that collect data at a localized level, for example, European countries where data are associated with compensation mechanisms.

DesInventar-based national data sets also cover small disasters, breaking down event data by municipality aggregates and using a rich set of indicators, which contain those that will be required to report against the Sendai Framework. The number of indicators implies bigger efforts may be required to build or retrofit and sustain these databases, which in addition can provide a clearer picture of damage trends and patterns at subnational scales and contribute to a better understanding of risk.

There are, however, methodological, conceptual, and practical challenges associated with a relatively localized data collection. These challenges may range from discrepancies in the perception of what an "event" is, to difficulties in the integration of multiple data sources, to the additional effort required to disaggregate information collected otherwise and the challenge of the economic valuation of the damage aggregates using a consistent and homogeneous methodology.

United Nations Office for Disaster Risk Reduction,
Geneva, Switzerland

Flood Damage Survey and Assessment: New Insights from Research and Practice, Geophysical Monograph 228,
First Edition. Edited by Daniela Molinari, Scira Menoni, and Francesco Ballio.
© 2017 American Geophysical Union. Published 2017 by John Wiley & Sons, Inc.

Despite these challenges, the 2015 edition of the Global Assessment Report on Disaster Risk Reduction (GAR) by the UN features analyses using a consolidated, homogenized, and standardized data set covering 82 countries and several states in India, which includes a uniform economic valuation of damage. The United Nations Office for Disaster Risk Reduction (United Nations International Strategy for Disaster Reduction [UNISDR]) has been using this data set as a proof of concept of what a global database could look like. The UN Initiative, which started in 2005 when only 15 countries had these data sets, has continued to approach 100 countries in 2015. It will continue with renewed enthusiasm in the next few years, with the target of global coverage by 2020, as stated by the Sendai Framework.

1.1. DISASTER RISK REDUCTION: A FRAMEWORK FOR ACTION

The concept and practice of reducing disaster losses and risk through systematic efforts to analyze and reduce the causal factors of disasters and therefore reduce its impacts is known today as Disaster Risk Reduction (DRR). Reducing exposure to hazards, lessening vulnerability of people and property, wise management of land and the environment, and improving preparedness and early warning for adverse events are all examples of disaster risk reduction [*UNISDR, 2009a*].

Progress in reducing risk has been undeniable over the past decades. However, global models suggest that the risk of economic losses is rising as a result of a series of factors, including increases in exposure and vulnerability, exacerbation of hazards because of climate change, and the rapidly increasing value of the assets that are exposed to major hazards [*UNISDR, 2015a*]. In addition, a large proportion of losses continue to be associated with small and recurring disaster events that severely damage critical public infrastructure, housing, and production, which are key pillars of growth and development in low- and middle-income countries.

The long road of international agreements that started with the declaration of 1990–1999 as the International Decade for Natural Disaster Reduction (IDNDR) [*UNISDR, 1999a*], and which produced the Yokohama Strategy and Plan of Action, and the subsequent Hyogo Framework for Action, has shown the international continuous concern about the growing impacts of disasters.

1.2. THE SENDAI AND OTHER FRAMEWORKS OF 2015

On 18 March 2015, representatives from 187 United Nations Member States gathered in Sendai, Japan for the Third World Conference on Disaster Risk Reduction and adopted the Sendai Framework for Disaster Risk Reduction (SFDRR) (*UNISDR, 2015*). Later in the same year, the 2030 Agenda for Sustainable Development was also adopted, and to finalize a golden year in international agreements, countries participating in the Paris COP 21 reached for the first time a global legally binding agreement on climate change, now known as the Paris Agreement.

The international community made a big effort to align these three processes as much as possible. In its first page, the Paris Agreement welcomes "the adoption of United Nations General Assembly resolution A/RES/70/1, 'Transforming our world: the 2030 Agenda for Sustainable Development,' in particular its goal 13, the adoption of the Addis Ababa Action Agenda of the third International Conference on Financing for Development and the adoption of the Sendai Framework for Disaster Risk Reduction" [*United Nations Framework Convention on Climate Change (UNFCCC), 2015*].

The Sendai Framework, the first of these to be adopted, sets "the substantial reduction of disaster risk and losses in lives, livelihoods and health and in the economic, physical, social, cultural and environmental assets of persons, businesses, communities and countries" as its main outcome. It also sets as its only goal to "prevent the creation of new risks and to reduce existing ones through different measures and thus strengthen resilience."

The 2030 Agenda for Sustainable Development embeds within its goals and targets all of the targets set by the Sendai Framework. Goal 11 Target 5 in particular comprises three of the seven targets of the Sendai Framework, all of them aiming at the reduction of human and economic losses [*UN, 2015*]. Targets in other goals, such as Goal 13 addressing climate change, also address similar challenges as those identified by SFDRR.

The Paris Agreement, in its Article 7 on adaptation, sets a global goal to increase adaptive capacity, strengthen resilience, and reduce vulnerability. This is the first time there is a formal agreement on a global adaptation goal. Article 8 on loss and damage (one of the problematic issues that delayed negotiations) includes reducing risk of losses and damages, early warning systems, emergency preparedness, and comprehensive risk assessment and management, all of which are aligned with the Sendai Framework Priorities for Action and Targets [*UNFCCC, 2015*].

1.3. THE SENDAI FRAMEWORK AND LOSS DATA COLLECTION

The Sendai Framework is structured around one main outcome and one goal, four priorities for action, seven targets and has a much wider scope than its predecessor, the Hyogo Framework for Action.

Priority 1. "Understanding disaster risk" states that disaster risk management should be based on a thorough understanding of disaster risk and losses in all its dimensions of vulnerability, capacity, exposure of persons and assets, hazard characteristics, and the environment. Such knowledge can be used for risk assessment, prevention, mitigation, preparedness, and response.

Priority 2, "Strengthening disaster risk governance to manage disaster risk" recommends clear vision, plans, competence, guidance, and coordination within and across sectors, as well as participation of relevant stakeholders and fostering collaboration and partnership across mechanisms and institutions for the implementation of instruments relevant to disaster risk reduction and sustainable development.

Priority 3, "Investing in disaster risk reduction for resilience" suggests public and private investment in disaster risk prevention and reduction through structural and non-structural measures, which are essential to enhance the economic, social, health, and cultural resilience of persons, communities, countries, and their assets, as well as the environment.

Priority 4, "Enhancing disaster preparedness for effective response and to 'Build Back Better' in recovery, rehabilitation, and reconstruction" recognizes there is a need to strengthen disaster preparedness and ensure capacities are in place for effective response and recovery at all levels. The recovery, rehabilitation, and reconstruction phases are critical opportunities to build back better than before and opportunities to integrate disaster risk reduction into development.

Both the Sendai Framework for reducing disaster risk and its predecessor, the Hyogo Framework for Action, explicitly recognize the importance and usefulness of collecting loss data as one of the actions that will help countries to increase the knowledge about the risks they face. In particular, the Sendai Framework Priority 1, "Understanding disaster risk," suggests among other activities the following:

"(d) Systematically evaluate, record, share and publicly account for disaster losses and understand the economic, social, health, education, environmental and cultural heritage impacts, as appropriate, in the context of event-specific hazard-exposure and vulnerability information;

(e) Make non-sensitive hazard exposure, vulnerability, risk, disaster and loss-disaggregated information freely available and accessible, as appropriate";

The text of the Framework calls for its application to *disasters of all scales* and, as opposed to the Hyogo framework, it requests countries to address and therefore collect data about hazards that are not only considered of "natural" origin:

"15. This Framework will apply to the risk of small-scale and large-scale, frequent and infrequent, sudden and slow-onset disasters caused by natural or man-made hazards, as well as related environmental, technological and biological hazards and risks".

To support the assessment of global progress in achieving the outcome and goal of the framework, seven global targets were agreed upon. Most importantly, out of these seven targets, four are related to losses and impacts.

These targets will be measured at the global level and will be complemented by work of the Open Ended Intergovernmental Working Group (OEIWG), tasked with the responsibility of developing appropriate indicators, with all the details and precise definitions that will be required, and defining the rules regarding how those indicators will be used to compute the targets [*UNISDR*, 2015]. The seven global targets, in summary form, follow:

(a) Substantially reduce relative (per capita) global disaster mortality.

(b) Substantially reduce the relative number of affected people globally.

(c) Reduce direct disaster economic loss in relation to global gross domestic product (GDP).

(d) Substantially reduce disaster damage to critical infrastructure and disruption of basic services, among them health and educational facilities.

(e) Substantially increase the number of countries with national and local disaster risk reduction strategies by 2020.

(f) Substantially enhance international cooperation to developing countries.

(g) Substantially increase the availability of and access to multi-hazard early warning systems and disaster risk information and assessments.

There are several consequences to the wider scope of the framework, the explicit recommendations of Priority Action 1 on loss data collection and, in particular, to the fact that Targets (a) to (d) are based on loss indicators. One is that countries are strongly encouraged to systematically account for disaster losses and impacts for a wide spectrum of disaster scales and a large set of hazards. This accounting must take into account an expectedly large number of loss indicators defined by the OEIWG, including human, infrastructure, and economic indicators. This set of indicators will allow, on one hand, the monitoring of the outcomes of the framework, reduction of losses, and the progress in achieving the targets, and on the other hand, it will allow improvement of the understanding of risk and the impacts of disasters in member states.

The work of the OEIWG has defined a relatively manageable but still numerous and complex set of indicators to

Table 1.1 Set of Indicators Agreed Upon by the OEIWG in Geneva.

Target A: Substantially reduce global disaster mortality by 2030, aiming to lower average per 100,000 global mortality between 2020 and 2030 compared to 2005 to 2015.	
A-1	Number of deaths and missing persons attributed to disasters per 100,000 population. (This indicator should be computed based on indicators A-2, A-3, and population figures.)
A-2	Number of deaths attributed to disasters per 100,000 population.
A-3	Number of missing persons attributed to disasters per 100,000 population.
Target B: Substantially reduce the number of affected people globally by 2030 with the aim of lowering the average global figure per 100,000 between 2020 and 2030 compared to 2005 to 2015.	
B-1	Number of directly affected people attributed to disasters per 100,000 population. (This indicator should be computed based on indicators B-2 to B-6 and population figures.)
B-2	Number of injured or ill people attributed to disasters per 100,000 population.
B-3	Number of people whose damaged dwellings were attributed to disasters.
B-4	Number of people whose destroyed dwellings were attributed to disasters.
B-5	Number of people whose livelihoods were disrupted or destroyed, attributed to disasters.
Target C: Reduce direct disaster economic loss in relation to global gross domestic product (GDP) by 2030.	
C-1	Direct economic loss due to hazardous events in relation to global gross domestic product. (This indicator should be computed based on indicators C-2 to C-6 and GDP figures.)
C-2	Direct agricultural loss attributed to disasters. *Agriculture is understood to include the crops, livestock, fisheries, apiculture, aquaculture, and forest sectors as well as associated facilities and infrastructure.*
C-3	Direct economic loss to all other damaged or destroyed productive assets attributed to disasters. *Productive assets would be disaggregated by economic sector, including services, according to standard international classifications. Countries would report against those economic sectors relevant to their economies. This would be described in the associated metadata.*
C-4	Direct economic loss in the housing sector attributed to disasters. *Data would be disaggregated according to damaged and destroyed dwellings.*
C-5	Direct economic loss resulting from damaged or destroyed critical infrastructure attributed to disasters. *The decision regarding those elements of critical infrastructure to be included in the calculation will be left to the member states and described in the accompanying metadata. Protective infrastructure and green infrastructure should be included where relevant.*
C-6	Direct economic loss to cultural heritage damaged or destroyed attributed to disasters.
Target D: Substantially reduce disaster damage to critical infrastructure and disruption of basic services, among them health and educational facilities, including developing their resilience by 2030.	
D-1	Damage to critical infrastructure attributed to disasters. (This index should be computed based on indicators D-2 to D-5.)
D-2	Number of destroyed or damaged health facilities attributed to disasters.
D-3	Number of destroyed or damaged educational facilities attributed to disasters.
D-4	Number of other destroyed or damaged critical infrastructure units and facilities attributed to disasters. *The decision regarding those elements of critical infrastructure to be included in the calculation will be left to the member states and described in the accompanying metadata. Protective infrastructure and green infrastructure should be included where relevant.*
D-5	Number of disruptions to basic services attributed to disasters. (This indicator should be computed based on indicators D-6 to D-8.)
D-6	Number of disruptions to educational services attributed to disasters.
D-7	Number of disruptions to health services attributed to disasters.
D-8	Number of disruptions to other basic services attributed to disasters. *The decision regarding those elements of basic services to be included in the calculation will be left to the member states and described in the accompanying metadata.*

measure these targets [*UNISDR*, 2015b]. Among the indicators considered, several are oriented to capture human losses, including those required to measure mortality and people affected, concepts that require very precise definitions and therefore precise indicators. A larger number of indicators will be required to measure direct economic losses and damages to critical infrastructure referred in Targets (c) and (d). At the time of writing this text, the OEIWG has put forward more than 20 indicators for consideration by the member states [*UNISDR*, 2015b], indicators that are deemed the minimum necessary for these measurements.

Systematically accounting for losses translates, in technological terms, to the creation of national disaster loss databases that are capable of recording the large number of loss indicators for disasters, at all scales, in a disaggregated manner, which is in agreement with the spirit of Priority Action 1 of the framework (see above). Priority 1 recommendations go even further, suggesting that these databases and information should be publicly accessible.

Table 1.1 compiles the set of indicators that have been agreed upon by the OEIWG in Geneva in the Third Session held in November 2016. This list of indicators is available in the United Nations General Assembly Resolution A/71/644.

1.4. WHERE WE ARE: BASIC PRINCIPLES OF THE UNITED NATIONS INITIATIVE

Although there are a few global disaster loss databases such as the Emergency Events Database (EM-DAT) [*Centre for Research on the Epidemiology of Disasters (CRED)*, 2011], NatCat from Munich Re, Sigma from SwissRe, and others, it is important to note that any reporting process to the Sendai Framework monitoring system has to be based on *officially endorsed data*, ideally collected and authenticated by national governments. These data should comply with the requirements of the framework, that is, it should address small- and large-scale disasters, slow and rapid onset events, it should cover a large number of hazards, including technological and man-made hazards, and most importantly, it should record a larger number of indicators not currently available in these global loss databases. Furthermore, if the recommendations of the framework are to be applied, databases should be built gathering disaggregated data that have to be usable at a subnational scale. Data should be disaggregated, at the minimum, by hazard, by event, and at a certain level of geography. For internal purposes, countries are encouraged to pursue even higher levels of disaggregation, for example, by recording human impacts in a gender-sensitive way or to collect data at asset level.

All of these minimum requirements imply that current national disaster databases will have to be expanded to reach global coverage once consolidated. Additionally, many existing databases and loss data collection systems will have to be retrofitted so that data sets contain all of the required indicators and comply with disaggregation requirements (see Chapter 3).

From the UN perspective, this situation represents a unique opportunity to build a bottom-up constructed global disaster loss database, allowing the process of global consolidation of data required to assess the progress in achieving the targets.

1.4.1. A Bottom-up Approach to Build a Global Database

The building of a global scale disaster loss database is not just the provision of a mechanism to measure Sendai Targets. Robust, official, systematic, and homogeneous measurements of losses will be a major contribution to the implementation of the Sendai Framework, and in general to disaster risk reduction, climate change adaptation, and sustainable development strategies.

National disaster loss databases will increase the capacity of countries to understand their risks and will provide a solid evidence base upon which to help countries to assess and address their disaster losses and impacts, particularly those associated with climate and weather-related hazards.

More specifically, loss databases will significantly improve the understanding of how disasters and risks affect the most vulnerable, and the databases could be used to better understand how climate variability impacts are trending and their true magnitude.

In those countries where no loss data are collected, or where information is kept only as paper archives, the UN has been proposing the use of a common simple but effective tool that implements the minimum requirements for the Sendai Framework. This effort, its challenges and achievements, and its future will be described in detail in the following sections.

In summary, this UN initiative has been implementing national disaster loss databases that comply with the following requirements:

• Data are collected for every hazardous event that has any type and level of damage registered, therefore, allowing the collection of information for disaster on all scales. Damage registered can be either quantitative (a number) or qualitative (a yes/no marker or a textual description of the damage).

• For each hazardous event, a set of indicators that is very similar, if not the same, as those discussed in the OEIWG for Sendai Targets are collected and recorded. Each indicator collected has precise definitions and even recommendations on data collection issues and problems [*UNISDR*, 2011b].

• For each hazardous event, the main and triggering hazards (from a local perspective) are recorded. The list of hazards used in the initiative is also standardized as much as possible; the IRDR[1]-suggested definitions of perils [*IRDR*, 2014] have been adopted by the initiative.

• For each hazardous event, summary loss indicators are collected and recorded separately for each of the geographic units affected; geographic units are in general equivalent to a municipality. It is important to note that collecting loss data at asset level has not been encouraged (but neither discouraged) given its level of complexity and the repercussions on data privacy, legal, and financial liabilities and other factors.

The initiative has been using the "DesInventar" free open source software and methodology [*UNISDR*, 2011b]. In addition to implementing the above criteria for data collection and storage, the software tools provide basic analysis and reporting tools without which the data collection itself would not be as valuable.

It has to be recognized though that several other countries follow different approaches to collect data. The recent studies of the Joint Research Centre (JRC) Working Group [*JRC*, 2013; *JRC*, 2014; *JRC*, 2015] show that within the European continent there are disparities in the types of data indicators, thresholds, hazards, and resolution of the data collected (which may range from building or asset level to national aggregates), and in those mechanisms that trigger data collection. In particular, it has been found that a number of European countries collect data at building/asset level for purposes of compensation, be it from official funds [the case of Spain, for example, *Defensa Civil Española*, 2014] or from insurance policies [the case of France, for example, *Observatoire*, 2015].

In these cases, the United Nations, in collaboration with countries, intends to build automated interfaces to consolidate the information up to a level equivalent to municipality. Such data sets will be aligned and compatible with the products obtained in the rest of the world, in a common resolution. Most importantly, data aggregates will avoid privacy and data protection problems that could prevent the data from being made publicly available.

Active work is also happening in Europe to standardize and adopt similar hazard/peril classifications as the IRDR and to ensure the consolidation process will render the set

of indicators proposed and defined by the OEIWG (see Chapter 2 in this book).

Despite the initial expectations that rich-information countries could easily comply with all of the requirements for Sendai Framework monitoring, it has been seen that not all databases in developed countries contain all of the indicators required. The Sheldus database, for example, in the United States (US) [*Cutter et al.*, 2005; Chapter 4 in this book] only contains a subset of the indicators proposed, and a similar situation has been found in some European countries. For instance, no indicators are collected around critical infrastructure or people affected in many of these databases. However, it is expected that the amount of digital data, the diversity of data sources, and the abundance of resources will result in a coherent integration of all the information required for monitoring the framework.

The final consolidated global data set will be, therefore, a feasible possibility within a few years from now, because it must be finished by 2020 in accordance with Sendai Framework requirements. See Box 1.1 for sample output of consolidated data for South American countries.

UNISDR already has been conducting consolidation exercises with data from a growing number of countries to build the data sets used for analysis posted in the Global Assessment Report (GAR). The data set started with 12 countries in the 2009 edition of GAR, then 21 in the 2011 edition of GAR, followed by 56 in the 2013 edition of GAR, and with the latest edition of GAR in 2015 featuring a consolidated data set containing data for 82 countries and 2 Indian states [*UNISDR*, 2015c].

This data set, of more than half a million records, was used for several research activities and as a proof of concept of the possibilities of consolidation of relatively homogeneous data sets. As documented in Annex II of the GAR 2015 Report, this consolidation was successful although it faced several challenges and some manual work.

Most of the problems faced were related to homologation of hazards, not only because of differences due to the particular context of the participating countries, but also because of linguistic and translation issues. Another area in which a careful examination of the data was required is quality control because some of the raw data still contained rogue or invalid values that had to be removed from the main body of data.

1.4.2. Economic Assessment of Direct Losses– United Nations Methodology

A major challenge faced while building the proof-of-concept data set was the lack of consistent, homogeneous, and documented evaluations of economic loss assessments of the impacts of disasters. As documented in several studies [*Dilley et al.*, 2013], all disaster loss databases register economic losses in a very poor manner.

---

[1]*Integrated Research on Disaster Risk (IRDR) is a decade-long research programme co-sponsored by the International Council for Science (ICSU), the International Social Science Council (ISSC), and the United Nations International Strategy for Disaster Reduction (UNISDR).*

Box 1.1 Sample output of consolidated data for 10 countries in South America.

Integration of data across boundaries is a feasible exercise if the data sources are compatible not only in format but also conceptually. This map shows the spatial distribution of the frequency of disasters associated with extreme precipitation at the second administrative level (municipality). Data from Brazil exists, and it is expected to become publicly available in the near future. Similar data sources exist for practically all countries in Central America and North America, meaning that for the first time, a continental view of the historical distribution, trends, and patterns of disasters can be readily obtained.

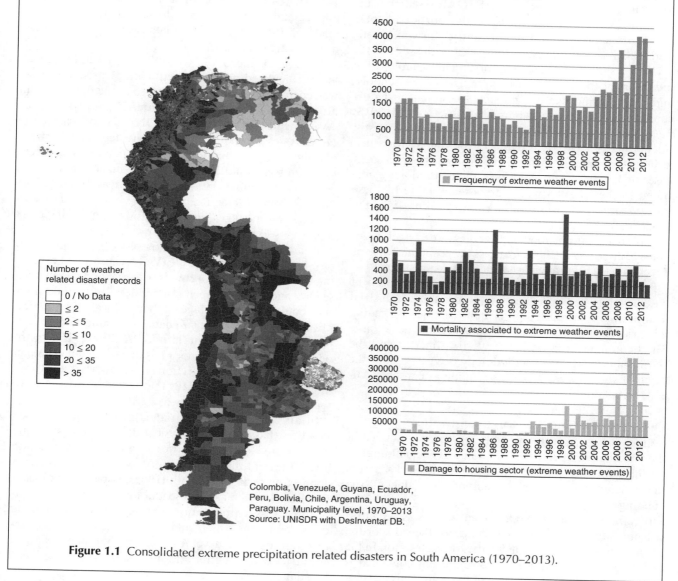

Number of weather related disaster records

- 0 / No Data
- ≤ 2
- 2 ≤ 5
- 5 ≤ 10
- 10 ≤ 20
- 20 ≤ 35
- > 35

Frequency of extreme weather events

Mortality associated to extreme weather events

Damage to housing sector (extreme weather events)

Colombia, Venezuela, Guyana, Ecuador, Peru, Bolivia, Chile, Argentina, Uruguay, Paraguay. Municipality level, 1970–2013
Source: UNISDR with DesInventar DB.

Figure 1.1 Consolidated extreme precipitation related disasters in South America (1970–2013).

A good manifestation of this issue is the extremely low coverage of data on economic losses, a problem that is common to most disaster loss databases, with the possible exception of insurance databases, where insured losses are operational assets and total losses are inferred using indexes such as market penetration. The well-known EM-DAT (see Chapter 3 in this book) only contains 25% of records with an economic assessment figure. Existing national databases contain 20% or even fewer records with dollar figures. Additionally, in all of these cases, national and global, methodologies and parameters used to estimate the economic loss are undocumented, if not unknown, and at the minimum, are not homogeneous or inconsistent given the disparity of the actors, contexts, and the circumstances in which the measurements were taken.

Target (c) of the Sendai Framework puts additional pressure on the requirements to collect loss data by requesting countries to assess "direct economic loss" defined as the value of the assets lost as consequence of disasters (loss of stock, in economic terms).

By applying a systematic and relatively simple approach to calculate direct economic loss, the GAR research team found it was possible to estimate a large portion of total direct losses recorded in the 82 countries for which data were available in the consolidated GAR data set of 2015.

Using a simple and consistent pricing methodology for indicators of losses in houses, roads, agriculture, schools, and health facilities, it was possible to estimate a significant part of total direct economic loss [*GAR*, 2011, 2013; *Velásquez et al.*, 2014]. However, this estimation still doesn't take into account damages to other sectors such as industrial and commercial, and costly infrastructure in cases of large disasters. However, the methodology proposed to the OEIWG will address many more of these missing sectors and will address known weaknesses of the GAR methodology.

In particular, the methodology addresses each sector separately, proposing methods to assess the economic value of direct damage using a replacement value methodology.

For all of the sectors that refer to built environment (i.e., housing, health, education, commercial, industrial facilities), the methodology is quite simple, estimating the price using the value of construction as a base. The Economic Commission for Latin America and the Caribbean (ECLAC) methodology suggests that the value of the physical damage to buildings can be calculated based on the following:

- the size of the building
- the price per square meter of construction
- the damage to furniture and equipment contained in the building (as a percent of the value of the building)
- the associated infrastructure (utility networks, access roads, landscaping, as a percent of the value of the building)

In turn, the values of the equipment and associated infrastructure are estimated as a percentage of the value of the construction, a percentage that varies on each sector. In the case of houses, for example, the equipment contained is suggested to be 25% of the value of the house; this percentage is much higher in health and industrial sectors.

For transportation infrastructures, the methodology uses rehabilitation costs per lineal meter, extracted from common projects in the sector.

Agricultural damage is estimated as a proxy value calculated based on the output of the crops. The underlying principle is that direct losses (seeds, fertilizers, pesticides, labor, and other costs that comprise what farmers invest in their crops) can be estimated as a percentage of the expected yield of crops.

It may be possible in the future to better estimate direct and total losses, based on conclusions from rigorous economic assessment of disasters conducted by the UN using the economic assessment methodology developed by ECLAC and the World Bank, which showed that direct losses represent statistically between 50 and 80% of total losses with this percentage higher in geological events [*ECLAC*, 2012]. In a subsequent phase, wider impact and macroeconomic losses could also be estimated if the quality of the data is high and adequate methods are developed.

Annex II of the GAR Report 2015 showed that direct losses calculated with this methodology are statistically well correlated and are usually close to the figures evaluated by UN-ECLAC, World Bank Damage and Loss Assessments (DaLA) and UN-PDNA (Post-Disaster and Needs Assessments). The report suggested that by extrapolating the figures found in these 82 countries, real economic losses could be significantly higher than losses reported by global data sources such as EM-DAT or NatCat from Munich Re, also taking into account losses in other sectors such as industrial and commercial sectors that were still to be accounted for.

To address some of the weaknesses of this methodology, the Secretariat of the OEIWG has proposed extending the loss indicators to cover industrial and commercial sectors and has developed a more detailed methodology that could take advantage of better local construction prices and asset average size data, to produce more accurate economic assessments [*UNISDR*, 2015e]. This methodology also opens the door to using very detailed data in countries where data collection is done at asset level or at intermediate levels of details that would greatly improve the accuracy of the assessment.

In all cases, the Secretariat is proposing, as a best practice, that all of the physical damage indicators are collected and kept by countries as important information asset. Physical damage indicators will allow the future connection of loss data with risk assessments or disaster forensics. It will make the Sendai Framework assessment of direct losses more transparent, and will allow, among other things, the incremental improvement of the assessment as countries develop better methodologies and as countries collect better and more comprehensive baseline data.

1.5. WHERE DO WE GO? EXPERIENCE FROM THE PAST INDICATES CHALLENGES FOR THE FUTURE

In 2008, when the first Global Assessment Report was being prepared to be launched in one year, approximately 15 countries were found to be using the DesInventar methodology. Most of the countries were in Latin America and, more incipiently, in several of the countries that were affected by the tsunami of December 2004. A first consolidated data set was assembled, aiming to look deeper into the real extent and importance of small and medium disasters. A sample of data from 12 countries was used to define, for the first time in numerical terms, the concepts of "Extensive" and "Intensive" risk. It was estimated that the number of countries with national disaster loss databases by 2008 was less than 30 [Global Risk Identification Program (*GRIP*), 2008], from which 90% were using the DesInventar methodology.

Since then, the number of countries covered by a DesInventar standardized electronic system for loss data collection has increased to over 90, under concerted efforts of the UN mainly represented by United Nations Development Programme (UNDP) and UNISDR, and other organizations including the European community and the World Bank. As stated before in this chapter, the GAR edition for 2015 contained a consolidated data set for 82 countries and 2 Indian states, and another set of countries joined the initiative during 2015, which is now approaching 100 countries in total.

Building more than 60 new data sets in a period of seven years has resulted in a wealth of experience and an important data asset.

1.5.1. Challenges and Achievements of National Databases

The next few sections of this chapter summarize the achievements, but especially the challenges, that countries and the UN system have faced while building a large number of disaster loss databases in the past decade.

Is important to underscore that the majority of this work has been done in developing countries, some of which are even classified as Least Developed Countries (LDC), and in many Small Island Developing States (SIDS), which are, of course, the focus of the development and humanitarian work of the UN. Only recently, the initiative has welcomed countries from the developed world, where a very different set of challenges occur.

Achievements of the initiative can be seen at national and global levels. The contribution of the group of Latin American countries that started the initiative under the umbrella of LA RED (LA RED de Estudios Sociales en Prevención de Desastres en America Latina[2]) has to be recognized as a pioneer work that brought to Sendai and other frameworks important ideas and hypothesis about the nature and significance of small and medium disasters, among other things.

It would be difficult to condense all of the achievements and products outcome of the initiative within countries in a few paragraphs. A few examples of loss accounting systems that are truly institutionalized and embedded into the national risk reduction mechanisms are the cases of Sri Lanka,[3] Indonesia,[4] Turkey,[5] Ethiopia, Cambodia, and Panama, among many other countries.

The Secretariat of the Pacific, a regional intergovernmental body, has developed and maintains Pacific Damage and Loss (PDALO), a data set covering 22 SIDS, many of which have very little capacity to maintain the system by themselves. Analysis of their data has been issued as documents in the Pacific Disaster Network, and loss data analysis is used as one of the inputs for the Pacific Catastrophe Risk Assessment and Financing Initiative (PCRAFI) system [South Pacific A*SPC/SOPAC*, 2014].

There are many examples of disaster loss data usage for policy analysis. Good examples are the applications in Latin American countries, where governments have adopted policy recommendations based on the impacts of the El Niño phenomenon [*LA RED/ENSO*, 2007]. In Tunisia, Niger, Mali, and several other African countries, disaster loss databases are providing, for the first time, evidence-based results of risks historically faced by these countries, which in some cases challenges the current perception of risks of governments. For example, in Mali, the impact of insect infestations was confirmed to have similar or greater impacts than floods.

More and more, loss data are used as input, calibration, validation, and complement of risk assessments and as linking data with climate change processes. Lebanon has recently produced a flood risk assessment that contains historical mapping and measures of impact of past flood disasters.

Data from the initiative have been crucial in shaping the current discourse of UNISDR in risk reduction. Four consecutive editions of the Global Assessment Report have strong basis, reflected in entire chapters and annexes devoted to the topic, on the findings arisen from the analysis of individual and consolidated data sets.

The ongoing work of JRC aimed at producing a recommendation for loss data collection to Member States [*JRC*, 2013, 2014, 2015] has gathered, perfected, and adopted many of the ideas, best practices, and lessons

---

[2]*The Network for Social Studies on Disaster Prevention in Latin America. See http://www.la-red.org*

[3]*See www.desinventar.lk. System includes subnational profiles for districts, public awareness, and education sections and publications.*

[4]*See http://dibi.bnbp.gov.in. Data Informasi Bencana Indonesia (DIBI) system is decentralized, with provincial subsystems. The data are linked, and the open source software has been reused for a poverty eradication project system and other applications.*

[5]*See https://tuaatest.afad.gov.tr/map.jsp. The Turkish system is coupled with a DRR knowledge base system.*

learned by the UN system and UNISDR over the years (see Chapter 2 in this book). UNISDR and UNDP contributions were provided by the Secretariat during the entire process of development of the recommendations.

Nevertheless, the most important reflection of the global impact of the initiative is its influence on the targets of the Sendai Framework and its indicators and their presence in other international agreements. The current set of indicators being discussed by OEIWG are an almost perfect match with the indicators and definitions collected for more than a decade with DesInventar. UNISDR intends to continue with the initiative, now with the renewed goal of turning it into the formal reporting mechanism for Targets (a) to (d) of Sendai, which are replicated in several of the indicators of the Sustainable Development Goals (SDG), to continue its use as prescribed by the Paris Agreement and, as usual, as a crucial step for countries to better understand their risks.

1.5.2. Challenges in Developing Countries: Scarcity of Data, Quality Control, and Sustainability

When starting the work of the UN system in developing countries, the first activity conducted is usually a capacity-building exercise where the basic concepts associated with risk reduction in general and those related to disaster loss databases in particular are explained. The software tools are introduced to a usually large number of stakeholders. These stakeholders range from emergency management bodies, which are usually the "hosting agency" of the initiatives, to data providers such as line ministries and to end users in planning and finance sectors.

The ease of use of the software tools used for data collection and analysis gives, at first, a false sense of the task being simple, but when actual research and data collection starts, the list of challenges is enormous. UN-supported databases are built in two clear phases: a first stage in which historical research is conducted during which loss data is obtained for a certain number of preceding years, normally 20 to 30 years, and a second stage during which the resulting data set is kept up to date by means of near real-time data collection.

Both stages face the following common challenges and difficulties:

Unavailability of information: The first common challenge that researchers face is simply the unavailability of data. No information is systematically collected or, if eventually collected, it is not properly stored. Many institutions that deal with this type of information only keep paper files, which after some time are discarded or destroyed, because there is no awareness of their importance.

Disaggregation: National data sets must cover disasters at all scales. There is a tendency in the humanitarian world to aggregate the total impact of a disaster in order to provide consolidated figures that are required for

planning the emergency response. For small or medium disasters, which do not cross the borders of the target geographical unit (similar to a municipality) this is not a problem. For events that cover multiple areas, breaking down event data by geographic unit implies greater efforts in building national databases, which then can provide a clearer picture of damage trends and patterns at subnational scales.

Conceptual issues: There are methodological, conceptual, and practical challenges associated with a relatively localized data collection, ranging from discrepancies in the *local perception* of what an "event" is, the associated hazards, to the *local perception* of its date, duration, and other problems that jeopardize the integration of information from different sources. A meteorological event that causes landslides in one municipality and in another municipality causes flash floods over certain periods of time that could span over weeks are not easy to connect as being part of the same hazardous phenomena.

Integration of information: In most cases, the backbone of the disaster loss data collection in each country is the agency in charge of emergency management (the 'hosting agency'), which has access, coordinates operations, and produces many of the loss indicators that are typically related to human losses and those related to shelter, food security, and health. In theory, information from other agencies should feed and complement this initial picture, for example, with sectoral data coming from utilities management agencies (roads, water, sanitation, communications, etc.). Integrating these data tends to be a rather difficult task for reasons already stated above, but it is difficult also because of data sharing problems between agencies, which in many cases work as functional isolated silos.

Capacity of institutions: Especially in very low income countries, there is a lack of, or low, capacity among those engaged in data production. This is unfortunately the case for many emergency services where personnel deployed to the field have basic search and rescue training but in many cases lack the knowledge and technology required to properly assess physical damage to structures and other tasks required to obtain highest quality loss indicators.

Vision of disasters as catastrophes: Despite clear indications in both the Hyogo Framework and the Sendai Framework, there is still a tendency to disregard the data or even consider disasters as many damaging small hazardous events. Reflecting on a process that has also happened internationally and making extensive risk information available are the best triggers and justifications of specific risk reduction and risk management for extensive risk.

Disparities among countries: Several aspects make data collection on each country specific for its context. Legal regulations can impose restrictions and change definitions of loss indicators. For example, the concept of a 'missing person' can be legally established in different ways depending on each country's law, and the same can happen for the definition of hazards. Another important consideration is the concept of what a 'municipality' can be. Sizes and definitions of administrative divisions vary dramatically in each country making the choice of a type of administrative unit a sensitive issue.

Disaggregation by gender/age/other: There is a lot of pressure and work from different groups in the international community (including specific mentions in all of the three frameworks mentioned: Sendai, SDGs, and Paris) to introduce gender-specific approaches to disaster management, and this is also reflected in requirements for data collection. The experience in building disaster loss databases tells that collecting (and later analysing and using) this information requires a huge effort in changing the mindset of those deployed in the field to attend the emergencies. It is important to realize that most of the information on affectation to human lives is collected for humanitarian purposes (for example, per family or household), and only in a few cases is the information actually required in disaggregated form for operational reasons. Specialized agencies in sectors like education and health may be the key to produce and record this information.

Public access: Unfortunately, loss data are seen in many contexts as a legal or a financial liability, not to mention a political liability. Access to mortality and health information can be extremely regulated, especially in developed countries where there is strict data privacy legislation. For example, in the US, it is almost impossible to publish a record of a hazardous event with only one victim, because then the victim could be identified and associated with the disaster [*Cutter*, 2005]. The existence of compensation mechanisms is one of the reasons access to emergency loss data can be restricted, because it could give arguments to petitioners to request compensation. Political and national security liabilities are also noteworthy; good examples of this are the several countries in the Middle East that joined the initiative but have chosen not to share their information based on national security considerations.

The young institutionalism of African countries: One of the main targets of the UN initiative is the African continent. A big challenge here is the incipient status of institutions, particularly in countries that only became independent in the past half century. Many of these young institutions do not have the accumulated knowledge (and archives) that are required to conduct historical research, and their capacities are still to be highly reinforced.

Economic valuation: The economic valuation of the damage using a consistent and homogeneous methodology is a common problem for a high number of countries. Although today there are sophisticated methodologies

for assessing the economic value of losses, there is no simplified methodology that can be applied in the myriad small-scale disasters that do not justify the deployment of specialized engineering valuation teams. The proposed UNISDR simplified methodology and its implementation in the software tool DesInventar aims at filling this gap.

Sustainability: Probably one of the main challenges in the past for disaster loss data sets is an issue with many faces that have been addressed by the UN using different but systematic approaches [*UNDP*, 2009]. Ownership from part of the government, resources needed for the operation of the system, continuity of political priorities and operational procedures, staff turnaround, continuous need for capacity building, unawareness of the results and applications, among others, are some of the challenges that are to be overcome when ensuring the sustainability of national disaster loss databases.

1.5.3. Developed Countries: Challenges in Information-Rich Environments

Several developed countries have mature and publicly available disaster loss databases, among them the US, Canada, Australia, Spain, and Slovenia. Looking closely at some of these data sets, it is immediately apparent that not all of them will be sufficient to respond to the demands of the Sendai monitoring system. Others, like the Spanish data set, are fully ready to be used and even have an interface with DesInventar, which allowed the data set to be integrated into the GAR 2015 consolidated loss database.

During 2014 and 2015, JRC conducted a series of workshops and produced three documents on the topic of loss data for European countries [*De Groeve et al.*, 2014, 2015]. Progress in the European continent is evident, but paradoxically the richness in data may play against the goals of an integrated system to collect loss data.

During the development of these workshops, a number of countries also provided sample data trying to test the comprehensiveness of the data sets and, from the UNISDR point of view, to look at the possible compliance of the data sets vis-à-vis the minimum requirements for the Sendai Framework targets (see Chapter 2 in this book).

Data from the European continent will be critical to assemble the puzzle of a global data set. Given the progress and actual practices found in member states, it is foreseen that European governments and those countries that have advanced loss databases, will have to build automated interfaces in collaboration with UNISDR to produce the desired output as requested by OEIWG.

However, the process is not without difficulties and challenges, although these will be very different from those faced by developing economies.

The apparent excess of data may play against the goals of sharing and integrating information. The more data sources that are available, the more chances there are to find discrepancies in aspects such as conceptualization, definition, electronic formats, and glossaries, among other things. Excess of information will not necessarily be an advantage, because it will come with different problems associated with integration. One of them will be the old and well-known problem of data sharing, which may be less patent in developed systems but which will be surely found.

Privacy and data protection issues have to be carefully managed so that no citizen rights are violated but at the same time, the aggregated figures required for the monitoring must be obtained. Compensation systems may also be seen as potentially introducing not only biases in the inventories of the impacts but in its economic valuation, and could also possibly contain political, national security, legal, and financial liabilities that may limit the collection of data and its dissemination and use.

1.6. CONCLUSIONS

The UN Global Loss Data Collection initiative has already achieved what is probably its most important goal that was set when it was conceived more than a decade ago. It has permeated the awareness of practitioners, researchers, and academics. Most importantly, it has made the public and governments more aware of the importance of considering the impacts of disasters at all scales in the process of national disaster risk management.

The Sendai Framework, explicitly stating it applies to "the risk of small-scale and large-scale, frequent and infrequent, sudden and slow-onset disasters caused by natural or man-made hazards, as well as related environmental, technological and biological hazards" is a recognition, to a large extent, of the work and conclusions that have been conceived and produced based on the evidence collected by the initiative, and published insistently in the GARs.

Building loss databases has been so far considered an optional element of the battery of tools and instruments that governments at all levels can use in their path to better understanding risks. Having three international frameworks (i.e., the Sendai Framework, the Agenda for Sustainable Development, and the Paris Agreement) consistently encouraging the construction of these artifacts as part of their implementation and the monitoring mechanisms is a game changer. Now loss data collection is becoming an almost mandatory instrument to be implemented globally by all member states.

The road ahead, however, is still partially unclear and rough. Much capacity building, provision of implementation means, and good institutional mechanisms have to be put in place for the monitoring systems to attain global coverage and to become sustainable during the next 15 years, the minimum period specified in the frameworks. Improved methodologies to collect, measure, and assess the economic impact of disasters will enhance the accuracy and usability of these systems.

It is a historical responsibility for national governments to start, revamp, or continue the systematic collection of loss data as part of the process of contributing to a better understanding of our risks, the impact of climate change, and in general, the process of sustainable development. As citizens of the world, all of us who contribute to the topic will be, at the end of the day, contributing to make this a better planet for us and for generations to come.

REFERENCES

Cardona, O.D., M. G. Ordaz, M. C. Marulanda, and A. H. Barbat (2008), *Estimation of probabilistic seismic losses and the public economic resilience—an approach for a macroeconomic impact evaluation.*

CIMNE, EAI, INGENIAR, ITEC (2013a), *Probabilistic modeling of natural risks at the global level: global risk model. Background paper prepared for the 2013 global assessment report on disaster risk reduction*, UNISDR, Geneva, Switzerland. http://www.preventionweb.net/gar.

CIMNE, EAI, INGENIAR, ITEC (2013b), *Probabilistic modelling of natural risks at the global level: the hybrid loss exceedance curve. Background paper prepared for the 2013 global assessment report on disaster risk reduction*, UNISDR, Geneva, Switzerland. http://www.preventionweb.net/gar.

Compass International Inc. (2012), *Global construction cost and reference yearbook 2012.*

CRED (2011), EM-DAT *The OFDA/CRED international disaster database*—www.emdat.net. Universite Catholique de Louvain, Brussels, Belgium. http://emdat.be/. Visited February 2016.

Cutter, Susan L., and Christopher T. Emrich (2005), "Are Natural Hazards and Disaster Losses in the U.S. Increasing?" *Eos, Transactions American Geophysical Union 86* (41): 381. doi:10.1029/2005EO410001.

Defensa Civil de España, Dirección General de Protección Civil y Emergencias (2014), Catalogo Nacional de Inundaciones Historicas. Accessed at http://www.proteccioncivil.es.

Dilley, M., and V. Grasso (2013), A comparative review of country-level and regional disaster loss and damage databases. Bureau for Crisis Prevention and Recovery, United Nations Development Programme.

ECLAC (2003), *Manual para la estimación de los efectos socioeconómicos de los desastres naturales* (report LC/MEX/G.5). CEPAL, Banco Mundial, Mexico DF.

ECLAC (2012), *Valoración de daños y pérdidas: Ola invernal en Colombia* 2010–2011. ECLAC, IDB, Bogota.

FAO (United Nations Food and Agriculture Organization) (2012), Post Disaster Damage, Loss and Needs Assessment in

Agriculture. This document can be accessed online in: http://www.fao.org/docrep/015/an544e/an544e00.pdf.

European Commission Joint Research Center (2013), Recording Disaster Losses: Recommendations for a European approach. Available at: http://publications.jrc.ec.europa.eu/repository/bitstream/JRC83743/lbna26111enn.pdf.

European Commission Joint Research Center, (2014), Current Status and Best Practices for Disaster Loss Data Recording in EU Member States. A comprehensive overview of current practice in the EU Member States. Available at: http://drr.jrc.ec.europa.eu/Portals/0/Loss/JRC%20SOTA%20Loss%20Report_11182014.pdf.

European Commission Joint Research Center (2015), Guidance for Recording and Sharing Disaster Damage and Loss Data: Towards the development of operational indicators to translate the Sendai Framework into action. Available at: http://drr.jrc.ec.europa.eu/Portals/0/Loss/JRC_guidelines_loss_data_recording_v10.pdf.

Global Risk Identification Programme GRIP (2008), Disaster Loss Data Standards. Methodology and tools. United Nations Development Program, Geneva. Available at http://www.gripweb.org/gripweb/sites/default/files/methodologies_tools/Disaster%20database%20standards_black.pdf.

Integrated Research on Disaster Risk (2014), Peril Classification and Hazard Glossary. Available at http://www.irdrinternational.org/2014/03/28/irdr-peril-classification-and-hazard-glossary.

LA RED/ENSO (2007) ENSO What? LA RED Guide to getting radical with ENSO Risks, lead author, Gustavi Wilches-Chaux. LA RED de Estudios Sociales en Prevención de Desastres en America Latina. Available at http://www.la-red.org/public/libros/2007/quENOSpasa/Qu-ENOS_pasa_ENG_ENSO_What.pdf.

Observatoire National des Risques Naturels (2015), last vistited at http://www.onrn.fr/.

OSSO *Desinventar.org—DesInventar Project*. Corporación OSSO, Cali, Colombia. http://desinventar.org/en/.

SPC SOPAC (2013), Pacific Damage and Loss (PDaLo) Regional Disaster Impact Report, Samantha Cook lead author. Available at http://www.pacificdisaster.net/pdnadmin/data/original/SPC_SOPAC_2013_PDalo_Regionalreport.pdf.

United Nations Development Programme (2009), Guidelines and Lessons for Establishing and Institutionalizing Disaster Loss Databases. Risk Knowledge Fundamentals. UNDP Regional Centre, Bangkok. Available at http://www.undp.org/content/dam/undp/library/crisis%20prevention/disaster/asia_pacific/updated%20Guidelines%20and%20Lessons%20for%20Estabilishing%20and%20Institutionalizing%20Disaster%20Loss%20Databases.pdf.

United Nations Development Programme (2013), *A comparative review of country-level and regional disaster loss and damage databases*. Bureau for Crisis Prevention and Recovery. New York. Available at http://www.undp.org/content/dam/undp/library/crisis%20prevention/disaster/asia_pacific/lossanddamagedatabase.pdf.

United Nations Economic Commission for Latin America and the Caribbean (2012), *Handbook for Disaster Assessment*. Santiago, Chile. This document can be accessed online in: http://repositorio.cepal.org/bitstream/handle/11362/36823/S2013817_en.pdf?sequence=1.

United Nations Framework Convention on Climate Change (2015), Adoption of the Paris Agreement, Conference of the Parties Twenty-first session Paris, 30 November to 11 December 2015. Accessible at https://unfccc.int/resource/docs/2015/cop21/eng/l09r01.pdf.

UNISDR (The United Nations Office for Disaster Risk Reduction) (1999), International Decade for Natural Disaster Reduction (IDNDR) programme forum 1999 proceedings. United Nations International Strategy for Disaster Reduction, Geneva. Accessible at https://www.unisdr.org/we/inform/publications/31468.

UNISDR (The United Nations Office for Disaster Risk Reduction) (2009a), *Terminology on Disaster Risk Reduction*. UNISDR, Geneva, Switzerland. This document can be accessed online at: http://www.unisdr.org/files/7817_UNISDRTerminologyEnglish.pdf.

UNISDR (The United Nations Office for Disaster Risk Reduction) (2009b), *GAR 2009: Global assessment report on disaster risk reduction: risk and poverty in a changing climate*. United Nations International Strategy for Disaster Reduction, Geneva. Available at http://www.preventionweb.net/english/hyogo/gar/2015/en/home/previous-gar.html.

UNISDR (The United Nations Office for Disaster Risk Reduction) (2011a), *GAR 2011: Global Assessment Report on disaster risk reduction: revealing risk, redefining development*. United Nations International Strategy for Disaster Reduction, Geneva. Available at http://www.preventionweb.net/english/hyogo/gar/2015/en/home/previous-gar.html.

UNISDR (The United Nations Office for Disaster Risk Reduction) (2011b) *Desinventar.net database global disaster inventory*. United Nations International Strategy for Disaster Reduction, Geneva.Available at http://www.desinventar.net.

UNISDR (The United Nations Office for Disaster Risk Reduction) (2013a), *GAR 2013: Global Assessment Report on disaster risk reduction: from shared risk to shared value; the business case for disaster risk reduction*. United Nations International Strategy for Disaster Reduction, Geneva. This document can be accessed online: http://www.preventionweb.net/english/hyogo/gar/.

UNISDR (The United Nations Office for Disaster Risk Reduction) (2013b), *GAR 2013 ANNEX II: Loss Data and Extensive/Intensive Risk Analysis*. United Nations International Strategy for Disaster Reduction, Geneva. This document can be accessed online: http://www.preventionweb.net/english/hyogo/gar/2013/en/gar-pdf/Annex_2.pdf.

UNISDR (The United Nations Office for Disaster Risk Reduction) (2015), Sendai Framework for Disaster Risk Reduction 2015–2030. *United Nations International Strategy for Disaster Reduction, Geneva*. Available online: http://www.unisdr.org/we/inform/publications/43291.

UNISDR (The United Nations Office for Disaster Risk Reduction) (2015a), *GAR 2015: Global Assessment Report on disaster risk reduction: Making development sustainable: The future of disaster risk management*. United Nations International Strategy for Disaster Reduction, Geneva. This document can be accessed online: http://www.preventionweb.net/english/hyogo/gar/.

UNISDR (The United Nations Office for Disaster Risk Reduction) (2015b), *Indicators to Monitor Global Targets of* the Sendai Framework for Disaster Risk Reduction 2015–2030: A Technical Review. Background paper presented for the open-ended intergovernmental expert working group on indicators and terminology relating to disaster risk reduction. Geneva, Switzerland. This document can be accessed online: http://www.preventionweb.net/files/45466_indicators paperaugust2015final.pdf.

UNISDR (The United Nations Office for Disaster Risk Reduction) (2015c), Proposed Updated Terminology on Disaster Risk Reduction: A Technical Review. Background paper presented for the open-ended intergovernmental expert working group on indicators and terminology relating to disaster risk reduction. Geneva, Switzerland. This document can be accessed online: http://www.preventionweb.net/files/45462_backgoundpaperonterminologyaugust20.pdf.

UNISDR (The United Nations Office for Disaster Risk Reduction) (2015d), GAR 2015 ANNEX II: Loss Data and Extensive Risk Analysis. United Nations International Strategy for Disaster Reduction, Geneva. This document can be accessed online: http://www.preventionweb.net/english/hyogo/gar/2015/en/gar-pdf/Annex2-Loss_Data_and_Extensive_Risk_Analysis.pdf.

UNISDR (The United Nations Office for Disaster Risk Reduction) (2015e), Concept note on Methodology to Estimate Direct Economic Losses from Hazardous Events to Measure the Achievement of Target C of the Sendai Framework for Disaster Risk Reduction: A Technical Review United Nations International Strategy for Disaster Reduction, Geneva. This document can be accessed online: http://www.preventionweb.net/documents/framework/Concept%20Paper%20-%20Direct%20Economic%20Loss%20Indicator%20methodology%2011%20November%202015.pdf.

Velásquez, C. A., O. D. Cardona, M. G. Mora, L. E. Yamin, M. L. Carreño, and A. H. Barbat (2014), "Hybrid loss exceedance curve (HLEC) for disaster risk assessment." Nat Hazards (2014) 72:455–479. DOI 10.1007/s11069-013-1017-z.

2

Technical Recommendations for Standardizing Loss Data

Daniele Ehrlich, Christina Corbane, and Tom De Groeve

ABSTRACT

The Sendai Framework for Disaster Risk Reduction calls for standardized loss data for risk knowledge generation and for monitoring four of its seven global targets. Current global and national loss databases do not always contain the necessary data that allows assessing the indicators proposed for monitoring the progress towards the global targets of the Sendai Framework. This chapter briefly analyses the requirements and proposes a conceptual loss database model that would accommodate the requested information. It then describes the technical challenges related to the implementation of the proposed model and provides recommendations on the way forward.

2.1. INTRODUCTION

With enactment of the Sendai Framework for Disaster Risk Reduction (Sendai Framework), countries have committed to report on global targets and to reduce disaster risk [*United Nations Office for Disaster Risk Reduction (UNISDR)*, 2015]. The implementation and monitoring of the Sendai Framework requires the collection of quantitative loss data for monitoring four of its seven global targets (see Chapter 1 in this book). That milestone policy framework implicitly identifies standardized loss data as an indispensable tool for disaster risk reduction decision-making. In addition, five targets proposed for monitoring the Sustainable Development Goals (SDG) also pertain to disaster risk reduction (DRR) and are, therefore, areas of potential synergy between the two policy frameworks [*Aitsi-Selmi et al.*, 2015].

The Sendai Framework is not only about monitoring targets. Its main aim is to understand risk through the generation of risk knowledge (see Chapter 1 in this book). Standardized loss data is a part of explicit risk knowledge

generation [*Weichselgertner and Piegeon*, 2015]. It helps in identifying loss trends, evidencing the risk drivers and forecasting future losses through probabilistic risk assessments [*UNISDR*, 2015b].

However, the available loss data varies greatly due to the legacy of past unstandardized loss collection and recording that follows different traditions and are still in use today as summarized below. Recording loss data due to hazard impacts on people, societal assets, and the environment is not new. The catastrophic natural events that affect the natural and built environment have always fascinated people. They stimulate the feeling of awe to the force of nature. The suffering that ensues, engraved in the minds of those that survived, and the consequences that the society of that time had to face, was in itself a reason of reporting long before the introduction of science. Without a scientific method to rely on, those early records were based on qualitative description of damage and suffering. That anecdotal and descriptive legacy of recording loss information, which is still in use today, falls short in precision, consistency, and reference to the characteristics of the damaged assets and, thus, is difficult to use for knowledge generation as advocated by the Sendai Framework.

European Commission Joint Research Centre, Ispra, Italy

Flood Damage Survey and Assessment: New Insights from Research and Practice, Geophysical Monograph 228,
First Edition. Edited by Daniela Molinari, Scira Menoni, and Francesco Ballio.
© 2017 American Geophysical Union. Published 2017 by John Wiley & Sons, Inc.

The scientific inquiries on hazards have prompted new ways of accounting for losses. Seismology attempted to relate the shaking of intensity of events with the damage. Volcanology, meteorology, geology, and hydrology, each within their own discipline, have equally attempted to associate hazardous event intensity with the damage by producing intensity damage functions related to physical assets from which vulnerabilities can be derived. The good practices established in a single hazard loss recording [i.e., *Salvati et al.,* 2010; *De Groeve et al.,* 2013] need to be extended to multi-hazard recording procedures. In fact, only multi-hazard loss recording can account for the cascading effect and the multi-hazard attribution that are indispensable in planning for risk reduction.

Loss data are also collected for policy-related objectives. Examples include compensation, reparation, declaration of the state of emergency, and soliciting donor's aid for reconstruction purposes [*De Groeve et al.,* 2013]. Those objectives do not require standardization because often the data collection is for singular events. However, the generation of risk knowledge as it is advocated in this chapter requires established systematic processes for recording loss data and for collecting data on exposure and vulnerability that can be compared in space and time.

At the time of writing, that standardization in data recording is hampered by the diversity of mandates and institutions involved. Nationwide loss data are typically collected by governmental departments with different aims or legislations that include compensation or that are used to fund risk reduction measures [*De Groeve et al.,* 2014]. Global loss data are generated and maintained by re-insurance companies for commercial purposes and by academic institutions for research purposes. The records in these databases are, by and large, not comparable because they are collected and recorded differently [*Margottini et al.,* 2011; *Wirtz,* 2014], and thus, they are not suitable for spatial and temporal comparisons [*Cutter and Gall,* 2015].

The degree of coverage and completeness varies globally. High-income countries typically report better than do low-income countries and not only for the availability of resources [*United Nations Development Programme (UNDP),* 2013]. The private sector is more effective in collecting loss data in some high-income countries where insurance penetration is higher compared to that of low-income countries that have typically lower insurance penetration for the combined effect of lower value assets and inability to pay premiums. Often loss data, even if available, are not accessible, neither from the private nor the public sector [*De Groeve et al.,* 2013]. The standardization of loss data has been advocated for some time [*UNDP,* 2008] and is starting to be addressed. The UNISDR has promoted mainstreaming of loss data collection practices by expanding the Disaster Information Management System (DesInventar) methodology to the many countries that still require capacity building [*UNISDR,* 2015]. (See Chapter 1 in this book). The DesInventar procedure and database may be adapted to better capture the variables required to build indicators for the Sendai Framework. Programs such as the DATA Working Group of the Integrated Research on Disaster Reduction (IRDR-DATA) have produced the "Peril Classification and Hazard Glossary" that is one attempt to provide a standard attribution of hazards worldwide [*IRDR,* 2014]. That classification and peril glossary is now tested within the main global loss data holders, within several regional organizations including the European Union (EU) [*De Groeve et al.,* 2015]. IRDR has also addressed human and economic losses analysis [*IRDR-DATA,* 2015] contributing to generating discussion on the potential loss indicators for measuring progress toward the targets of the Sendai Framework. However, implementation gaps still persist between the reporting requirements and the practices in loss data collection and recording.

This chapter analyzes requirements for loss databases to be used for international reporting and for knowledge generation. It describes a conceptual loss database model and the challenges related to its implementation in an operational setting. It also reports on an attempt to derive indicators for the Sendai Framework from 14 loss databases available within the EU. In conclusion, this chapter provides recommendations for future loss data recording to be used to address disaster risk reduction and that is required for international reporting

2.2. REQUIREMENTS FOR LOSS DATABASES

The technical requirements for establishing a loss database, as discussed in this chapter, should be part of a larger process of data capture for generating risk knowledge. This process should be initiated by the policy makers in governmental institutions that are mandated to assure security over their national territory.

The policy makers should be among the main recipients of the knowledge generated through the loss database. In fact, the incremental process of loss data collection/gathering and structuring/standardization in a database will provide an assessment on the risks, and it will possibly help in evaluating the risk reduction process and its successes. Implementers, practitioners, and scientists will also be important users of the risk knowledge generated but mainly in support of the ultimate objective of risk reduction managed by the decision makers. The disaster risk assessment and risk reduction processes involve at least three steps that each involve different stakeholders (Figure 2.1); the decision makers that define the requirements for data that aim to generate

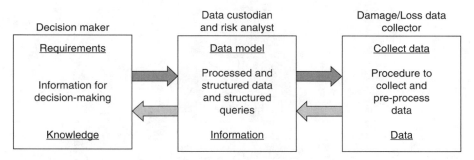

Figure 2.1 The requirements determine the data model that in turn determines the data to be collected [Re-drawn from *De Groeve et al., 2014*].

risk knowledge; and the database custodian that gathers the loss data, and then structures and populates the database to be able to answer the requests from decision makers. The database custodian, based on the requirements from decision makers and from practitioners that he serves, will define the technical specifications for the data that need to be collected (Figure 2.1) or gathered from available databases. The discussion on the governance of the loss database will be addressed in section 2.3.4.

Policy requirements would be based on questions including the following: a) What hazards are generating losses and what is the degree of damage? b) What assets get damaged and where is damage occurring? c) What hazard is affecting which sector of the society most? d) Which disaster types affect economic losses most [*De Groeve et al., 2014*]? The disaster risk knowledge developed will provide insights to be used to identify issues including the hazard that can be reduced (prevention), mitigated, or to be prepared for; where is damage most likely going to occur in the future; and what the expected future losses will be. These requirements can be addressed with a multi-hazard loss database since the damage or loss records pertain to the assets, not to the hazard. The attribution of the damage to one or more hazards (peril) will assure the proper loss accounting.

In the Sendai Framework reporting context, the decision maker (left box in Figure 2.1) provides the requirements, for example, the four global targets. The custodian and risk analysts will then interpret the discussion from the inter-governmental working group on indicators and build the indicators to be used to monitor the four global targets (center box in Figure 2.1). The custodians will define the technical specification for the data that need to be collected from existing loss databases and in the future through data collection protocols (right box in Figure 2.1). Beyond the four global targets, the Sendai Framework also calls for the generation of risk knowledge as elaborated in the next section.

2.2.1. Risk Knowledge

Risk knowledge can be generated through the process of loss accounting, forensic analysis, and risk modeling that rely on loss data combined with exposure, vulnerability, and hazard data (Figure 2.2). The datum standardization is enabled by a data processing infrastructure (available by the data custodian or risk modeler), typically within a geographical information system environment that enables the indispensable structuring of the datum in space and time. It also provides the modeling environment.

2.2.1.1. Accounting. Loss accounting is the principal motivation for recording the impact of hazards, and it aims to document the trends in time also providing insights to policy makers of the effectiveness of the DRR measures. High quality loss data with a good temporal and spatial resolution may be used to establish the historical baseline needed for monitoring the level of impact on a community or country. In fact, loss accounting is being considered as the backbone for setting the baseline (i.e., a decade of national observations on mortality and economic loss data) and measuring the progress toward the agreed targets within the Sendai Framework.

Loss accounting is a complex process and requires a definition of what assets need to be considered. It is a common practice to subdivide losses in direct, based on damage of physical measurable objects, and indirect losses, those based on the interruption of services or flows [*De Groeve et al., 2013*]. This chapter focuses mostly on direct losses as defined in *Meyer* [2013]. Indirect losses are addressed even if there is an awareness that indirect loss accounting is not sufficiently developed to be integrated systematically in loss databases. In fact, there is no consensus on the definition on indirect loss and how to report [*Molinari et al., 2014*]. Also, indirect loss reporting is still, by and large, a modeling exercise [*Carrera et al., 2015*].

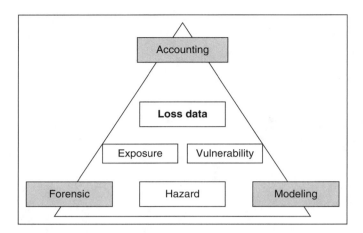

Figure 2.2 The three application areas that generate risk knowledge, loss accounting, forensic analysis, and modeling, rely on loss, exposure, vulnerability and hazard data sets that ideally would be structured in a common spatially referenced database.

Loss data, direct loss, is ideally collected at fine scale, for example, at the asset level. That fine scale level of accounting is implemented in a number of European Countries [*De Groeve et al.*, 2013] including Slovenia (see Chapter 3 in this book). A system of aggregation can then provide municipal, subnational, and national summaries that could then feed national accounting or the indicators system of the Sendai Framework. The fine-scale collection of loss data would guarantee the accountability and reliability of the loss record provided at the international level.

Loss data alone provide trends that can be understood only if the complementary exposure and vulnerability information is provided. The exposure provides the reference against which the losses can be measured. In fact, proper loss accounting would be based on loss estimates normalized by the exposed assets and attributed to a given hazard. For example, reports of 100 destroyed buildings have a different relevance if the assets impacted by the hazardous event were 1,000 or 10,000 buildings. Similarly, structural information on the building stock, often referred as structural vulnerability, provides the information on the structural fragility that is used to model losses of possible future hazardous events.

A number of loss assessment tools for national or international loss assessments are available. For example, the Damage and Loss Assessment (DaLA) methodology developed by the Economic Commission for Latin America and the Caribbean (ECLAC) [*Global Facility for Disaster Reduction and Recovery*, 2010] is used to generate damage and needs assessment to inform donors' conferences in the aftermath of major disasters for countries that have requested international assistance [*World Bank*, 2010]. The DaLA methodology builds on loss data collection, recording, and analysis for the purpose of identi-

fying root causes of disasters and determining recovery and reconstruction needs. Over the last 40 years, ECLAC has conducted specific loss assessments in a systematic manner generating historical evidence.

At the national level, loss data collection methodologies may be multi-hazard focused (Federal Emergency Management Agency [FEMA][i]), and methodologies specific to flooding are also in use [*Molinari et al.*, 2014].

2.2.1.2. Forensic.

Fine-scale disaster loss data recording generates crucial and unique evidence for disaster forensics. This allows identifying loss drivers by measuring the relative contribution of exposure, vulnerability, underlying risk factors, coping capacity, mitigation, and response to the disaster, which provides the lessons learnt to improve disaster and risk management. Disaster forensics collected for individual events is critical evidence for evaluating the effectiveness of specific disaster prevention measures and disaster prevention policy as a whole. Disaster forensics use loss data for a qualitative and quantitative hind cast modeling aimed at explaining the unfolding of disasters and thus highlighting the root causes (see Chapter 13 in this book). Forensic analysis uses loss data in combination with hazard intensity measures to derive empirical vulnerabilities.

2.2.1.3. Modeling.

Disaster models aim to address questions such as "What can go wrong? How likely is it that this will happen? If it does, what are the consequences?" [*Kaplan and Garrick*, 1981; *Kirchsteiger*, 1999]. The questions may be addressed using a combination of two modeling approaches: probabilistic and deterministic. Deterministic modeling aims to identify what can go wrong and how bad, based on a point event and is typically

---

[i] *https://emilms.fema.gov/IS559/lesson6/Toolkit.pdf*

geographically constrained. It is used to size response mechanisms and allows identification of shortcomings in protection measures at local to city level. Deterministic models are simple to understand because they simply project potential future losses due to individual events and as such, are also easier to convey to the wider audience [*Corbane et al., 2015*].

Probabilistic models also aim to inform on future losses. At the asset level, they provide information aiming to identify maximum loads that a structure can withstand. The assessments are conducted also at the national level for use by government financial institutions to address future financial losses. Probabilistic models at the global level inform the international and donor community. In fact, the worst disasters have not happened yet as modeled in global assessments through probabilistic risk models [*UNISDR, 2015*].

Disaster risk models typically comprise three main modules: hazard, vulnerability (exposure and vulnerability) and loss. The latter combines the hazard module and the exposure module to calculate different risk metrics, such as annual expected loss (AEL) and probable maximum losses (PML) for various return periods. The AEL and PML are used to complement historical analysis and are particularly useful for decision makers in assessing the probability of losses and the maximum loss that can result from major future events.

2.3. THE LOSS DATA MODEL

A data model is the description of the classes together with the definition of the data fields as well as relationships among the classes. It determines the logical structure of a database, and in which format data can be stored, organized, and manipulated. This section outlines the elements of the data model that are important and that should be included in national data models (Figure 2.3).

It starts from a disaster event, identified unambiguously (likely with an event identifier). A hazard event identification number modeled after the GLobal IDEntifier (GLIDE) number proposed by the Asian Disaster Reduction Center (ADRC) should be adopted [*De Groeve et al., 2015*]. It would allow for an unambiguous linking of loss records associated with the same disaster event and would enable interoperability among different loss databases. There may be several versions of loss records associated with the event, for example, through updates and corrections (where data becomes available), temporal versions to capture event dynamics (evolution of losses), or estimates of different organizations.

For each version, three sets of indicators of disaster losses (hazard event identification, the affected elements, and the damage and the loss indicators) can be recorded after the occurrence of a disaster, as well as metadata and quality assurance information. Metadata contains information such as entry date, author, validation status, and information on the methodologies used for assessing the damage and estimating the human and economic losses. The affected element may correspond to a house, a municipality, a province, or a country. The set of the affected elements is a subset of all exposed elements (elements at risk) located in the affected area. The affected area may be assessed either using the location of the affected elements or, at a coarser scale, by locating the municipalities or administrative units that include affected elements. Within the EU, the data specifications

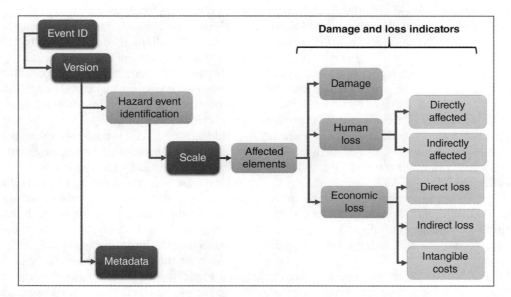

Figure 2.3 Conceptual loss data model that could be used to share loss data.

for the affected elements are described under the "Exposed Element" feature in the Infrastructure for Spatial Information in Europe (INSPIRE) Natural Risk Zones Data Specification [*INSPIRE Thematic Working Group Natural Risk Zones,* 2013]. A country may choose to record damage and loss data at a given scale and then aggregate at coarser scales (e.g., the municipality level may be obtained by aggregating losses recorded at the asset level or it may be assessed directly). The scale at which damage and loss data are recorded influences directly the quality of aggregated losses. Collecting data at the asset level will decrease uncertainty of loss indicators and increase the transparency of economic losses caused by a disaster.

2.3.1. Human Losses

A number of human loss frameworks have been proposed. *De Groeve et al.* [2014] described a possible model for harmonization of human loss indicators that is reported below. Other frameworks exist, for example, the one proposed in the IRDR Guidelines on Measuring Losses from Disasters: Human and Economic Impact Indicators [*IRDR,* 2015] that breaks down the loss indicators based on three priority levels, those that are required and those that may be collected if resources are available.

The different human indicator frameworks agree in principle, with modification related to slight changes in terminology and hierarchy and combination of indicators. Typically, the mandatory fields include the following indicators [*De Groeve et al.,* 2014]:

• The term "death" corresponds to the number of people who died during the disaster or sometime after as a direct result of the disaster. Missing corresponds to the number of persons whose whereabouts since the disaster are unknown. It includes people presumed dead without physical evidence. Data on deceased and missing persons are mutually exclusive.

• Directly affected people are a subset of exposed people (people living in the affected area that are thereby subject to potential losses) that suffered either impacts on their livelihood immediately after the disaster, or people who were displaced, isolated, and had impaired impacts on their physical integrity such as injured people. This definition is consistent with other interpretations such as the one proposed by EM-DAT [*Guha-Sapir et al.,* 2016].

• Indirectly affected people correspond to people in the affected country that suffer indirect effects of the disaster and can be within or outside the affected area, which divides them into secondary and tertiary level of indirectly affected people.

Of all of these indicators, the last two are arguably the most challenging for definition, measurement, and interpretation [*Guha-Sapir and Hoyois,* 2015]. The human loss indicators should be desirably defined according to the following four principles:

• Precise: Human loss indicators must have clear and preferably mutually exclusive definitions (one person is counted only once).

• Comprehensive: Human loss indicators must cover all affected people and every affected person be accounted for.

• Measurable: Human loss indicators are measured by public, private, or media organizations, or the indicators can be assessed in the field under current emergency management practices.

• Practical: Human loss indicators must match existing practices, there should be a one-to-one match with fields in existing databases, and the required changes are kept to a minimum.

2.3.2. Damage

This set of indicators corresponds to the total or partial destruction of physical assets existing in the affected area. They represent a summary of the damage in cases where aggregates are generated. The intention of these indicators is to provide a minimum set of physical damage indicators in the form of aggregated damage figures at spatial units above the asset level (i.e., municipality, region, country). These indicators may be used for validation and calibration of economic loss assessments and are useful as part of risk assessment and disaster forensic processes. The data will also ensure computability with the global targets for disaster risk reduction set in the Sendai Framework.

The minimum recommended measurement units required to monitor the Sendai global targets include the following damage indicators:

• House destroyed: the number of household units leveled, buried, collapsed, or damaged to the extent that they are no longer habitable/repairable

• Houses damaged: the number of household units with minor damage, not structural or architectural, which may continue being lived in, although they may require some repair or cleaning

• Education centers: the number of schools, kindergartens, colleges, universities, or training centers destroyed or directly damaged by the disaster

• Health facilities: the number of health centers, clinics, or local and regional hospitals destroyed and directly or indirectly affected (damaged or destroyed) by the hazardous event

A second set of indicators, in part already adopted in some databases (i.e., DesInventar), would include the following:

• Government buildings: the number of governmental and administrative buildings directly damaged or destroyed by the hazardous event belonging to national, regional, or local government

• Industrial facilities: the number of manufacturing and industrial facilities directly affected (damaged or destroyed)

• Commercial facilities: the number of individual commercial establishments (i.e., individual stores or warehouses damaged or destroyed

• Transportation: the length in kilometers of damaged/destroyed roads and railways, the number of damaged/destroyed bridges, airports, or marine ports

Damaged assets pertaining to the agricultural sector should also be recorded. For example, the total area of cultivated or pastoral land or woods destroyed or affected may be reported as hectares of lost production. The agricultural sector may also include the loss of livestock as the number of four-legged animals lost regardless of the type of event (i.e., flood, drought, epidemic).

2.3.3. Economic Loss Indicator

Several works attempted to define economic loss indicators related to disasters [Centre for Research on the Epidemiology of Disasters (*CRED*), 2011; *Meyer et al.,* 2013]. The definitions proposed here are based on the works of the EU loss working group [*De Groeve et al.,* 2014, 2015]. The economic losses represent market-based negative economic impact of a disaster. These consist of direct losses, indirect losses, and intangible costs, as follow:

• Direct losses relate to the monetary value of physical damage to capital and tangible wealth assets. Direct losses also may be measured in terms of flows of foregone production. Indirect losses include lower output from damaged or destroyed assets and infrastructure and loss of earnings due to damage to transport infrastructure such as roads and ports, including business interruption.

• Indirect loss may also include costs associated with the use of more expensive inputs following the destruction of cheaper sources of supply.

• Intangible costs relate to costs that accrue to assets without an obvious market price, which is difficult to depict in monetary terms.

Several frameworks exist for the assessment of economic losses: the Damage And Loss Assessment methodology [*The International Bank for Reconstruction and Development/The World Bank,* 2010], the Organization for Economic Co-operation and Development (OECD) Framework for Accounting National Risk Management Expenditures And Losses of Disasters [*OECD,* 2014], and the *IRDR Guidelines on Measuring Losses from Disasters* [2015]. These frameworks can hence be considered for defining the fields pertaining to economic losses.

2.3.4. Governance

The loss data collection and data recording process must identify roles and responsibilities. The decision maker provides the objectives, legislation, and tasks and that will identify three actors: the database coordinator, the curator, and the quality manager.

Damage and loss data recording need to be coordinated. The coordinator must ensure the application of a coherent methodology and foster the sharing of good practice. The coordinator is typically a government-appointed body that assures the application of the methodology nationwide. It also assures the aggregation and verification of the figures from local to national level. The coordinator supervises the loss and risk modeling database. He/she identifies the data that need to be collected and provides the specification to the field data collection teams on what data to collect. The coordinator provides guidance and designs protocols to transforming available data sets into useful data and information for use in databases. He/she verifies that the information is properly updated at regular yearly intervals and replies to users requests. Finally, the coordinator maintains the information management system. He/she is responsible for training personnel to process the data and assures the link with the existing database.

The curator should be the technical arm of the coordinator and is responsible for processing the collected data for input in the database. The curator calculates the codified values of database fields and identifies unclear or missing values that must be investigated.

The coordinator would convert the data in the proper units as defined by the entry methodology, use external references for the validation and verification process, and apply or even provide links to background information related to outside sources of information when not part of the loss database itself.

The quality manager ensures the disaster damage and loss data are recorded according to the principle of precision, comprehensiveness, comparability, and transparency [*De Groeve et al.,* 2014]. Precision refers to the correct use of terminologies and the consistency of the loss indicator. Comprehensiveness ensures that the database fields are populated. Comparability shows that each event is accompanied by a unique identifier to allow for the comparison of disaster impacts among the same hazard type, among different hazard types, across countries, across sectors, and through time. Transparency assures that damage and loss data have a geographical location, are accompanied with temporal information, and are associated with an uncertainty value.

2.4. TECHNICAL CHALLENGES

Loss data recording and the generation of risk knowledge rely on clarification of terminology, adherence to scale and measurement, hazard attribution, and comprehensive reporting. The issues are elaborated below. The section also reports on an attempt to generate indicators for the Sendai Framework based on 14 European loss data sets.

2.4.1. Terminology

Disaster management is multi-disciplinary. It is based on an understanding of societal functioning and by combining expertise from the social sciences, the physical sciences, as well as the decision makers and the practitioners. The communication requires a common language and a set of common definitions that are not completely available. The shortcomings are mostly felt in the definitions of indicators. The UNISDR and a number of agencies and programs are addressing terminology. Partial results of the preliminary discussion are available from [*De Groeve et al., 2015; IRDR-DATA, 2015*], and a DRR terminology was proposed by the Intergovernmental Informal Working Group [*UNODRR, 2015*].

2.4.2. Granularity and Scope

The available data quantifying losses from past disasters show that the granularity and the scope are two important dimensions of loss data. The granularity expresses the measurement scale [*Wu and Li, 2009*]. It provides an indication on the detail of loss data recorded. Scope, referred to also as geographical scale [*Wu and Li, 2009*], relates to the extent of the hazardous event.

The loss database also will have a scope, which is the geographical space covered, and granularity, which is the smallest spatial unit that can be accommodated within the database. The scope and granularity of loss data must be consistent with that of the loss database entries.

The geographical extent of loss databases is typically at the national level. Once loss data are available at a given reporting unit, it may then be aggregated at the next geographical level (i.e., municipality) and further on at the regional and national level.

The ideal loss database has national scope and local granularity. This is the case, for example, when the information is collected based on census information, or when citizens report their insurance claims. These databases would report the property (or asset) with physical location, size, and value and the losses from a hazardous event.

Most of the international databases do not record disasters at the local or municipal level, and as such, fail to provide geographically consistent reporting. Many of the local databases are geographically fragmented, often managed at a sub-national level, and thus, are uncoordinated. Therefore, the data cannot be aggregated to provide national statistics or for global reporting.

2.4.3. Intensive and Extensive Losses

Loss data collection should be independent from the declaration of a disaster that it is often the event that prompts the loss data recording process. All losses attributed to natural hazards should be recorded [*UNISDR, 2015*]. That alone will allow capture of the phenomenon in its entirety. Many current loss database records are based on disaster definitions that imply thresholds, which results in the underreporting of extensive disasters. This practice introduces bias to the comprehension of the phenomenon and thus also the possibility to act to assess risk and to take the proper risk reduction measures [*UNISDR, 2015*].

2.4.4. Hazard Classification

A disaster damage and loss is event-based and is required to be linked to a specific hazardous event. The event should be uniquely identified (spatially and temporally), classified to provide basic summary statistics (e.g., aggregation by peril type, year), and recorded by severity level to relate to the probability of occurrence for calculation of average annual losses. Hazard event identification allows attributing the losses to a hazard/peril. The attribution assumes a peril classification.

International and continental attempts to attribute losses are under way. Some, such as the European INSIPRE initiative, are encoded in legislation [*INSPIRE Thematic Working Group Natural Risk Zones, 2013*]. Others, such as the IRDR-DATA working group, are the effort of practitioners at the international level who have advanced the discussion on international hazard/peril classification [*IRDR Data Working Group, 2014*]. Standardization of hazard/perils at the international level remains one of the challenges of international loss reporting.

2.4.5. Identification, Start/End Date

Damage and loss indicators, which comprise damage, human, and economic losses, are the core of the disaster loss database. They describe the level of damage on individual assets or on a number of damaged/destroyed assets covering several dimensions to thoroughly recording the impact of the disasters. The degree of detail of damage depends on the availability of quantitative information in the area affected. Therefore, the damage and loss indicator is not only a name of data field with the value and the physical unit, but it is also accompanied by metadata including the time of recording/updating, the source and uncertainty as well as information on the assessment methodology. The unit should be standardized, for example, the unit for affected population should be persons.

Data in other units (families, households) should be converted to number of persons, with an associated uncertainty estimate.

Definitions of the fields, the format of their codified value (as well as a variety of information about the codified value that also needs to be collected, managed, and shared to assure their quality) should follow standard definitions to provide comparability and consistency.

2.5. CHALLENGES FOR REPORTING IN THE EUROPEAN UNION

In the EU, the member states and the European Commission worked together on the establishment of guidelines for recording and sharing disaster damage and loss data as a first step toward the development of operational indicators to translate the Sendai Framework into action [*EU Expert Working Group on Disaster Loss Data,* 2015].

The guidelines outline the elements of the disaster damage and loss data model that are important and that should be reflected in national data models. They attempt to address the technical requirements presented in section 2.2 by building on the loss data model described in section 2.3.

Figure 2.4 provides a synthesis on a preliminary test for loss reporting for the Sendai Framework. It is based on the compilation of a common loss database that included 14 European Loss data as input. The experiment aimed at testing the usefulness of the EU guidance for Recording and Sharing Disaster Damage and Loss Data [*EU Expert Group,* 2015], and at identifying areas of improvement in reporting and challenges for expanding the model to non-EU countries.

The 14 loss databases have shared aggregated statistics following the minimum requirements set out in the EU guidance and using a common reporting sheet.

All loss databases shared disaster loss data at the national level. The granularity of reporting occurs at the third level (municipality) of the Nomenclature of Territorial Units for Statistics (NUTS), the Standard Subdivision that was designed by Eurostat for statistical reporting within Europe.

There are a number of disparities in reporting. For example, some data sets are mono-hazard and others are multi-hazard [*De Groeve et al.,* 2014]. The recorded loss indicators vary between the countries. Some countries report human losses, and others focus more on economic losses. The time frame covered also varies between the

National loss database	Scope (hazard types covered) Complete / Partial / Missing					Hazard event identification			Affected people			Affected property	
	Geophysical	Hydro-meteorological	Climatological	Biological	Technological	National unit	Subnational units	Hazard event ID	Age	Sex	Location of residence	Sector	Loss ownership
01	•	•		?	SEVESO	•	•	•					
02	•	•		?	SEVESO	•	•	•	•	•	•	•	•
03	EQ,LS	FL,ST		?	SEVESO, ARIA	•	•	•				•	Insured
04		FL		?	SEVESO, ZEMA	•	•	•	•	•	•	•	•
05			WF	?	SEVESO	•	•	•			•	•	•
06				?	SEVESO	•	•	•					
07	LS	FL	WF?	?	SEVESO	•	•	•				•	
08	•	•	•	?	SEVESO, SADO	•	•	•	•	•	•	•	•
09	•	•	•	•	SEVESO, GIES	•	•	•				•	•
10				?	SEVESO	•	•	•				•	•
11		FL		?	SEVESO	•	•	•			•	•	•
12	•	•	•	?	SEVESO	•	•	•				•	•
13	?	•	?	?	SEVESO	•	•	•				?	Mainly insured losses
14	•	•	•	?	SEVESO	•	•	•				?	?

Figure 2.4 Comparison of hazard reporting and main loss indicators for 14 loss data sets originating from European Union countries.

countries and ranges between 47 years in the longest reporting loss database and 8 years in the shortest reporting loss database with 1966 being the oldest recording date and 2014 the most recent.

Some countries reported loss data beyond the minimum requirements (e.g., by disaggregating loss indicators in terms of insured and non-insured losses). Only one country provided a quality indicator for each recorded. This is a variable that is considered indispensable to assess the reliability of the reported data and the assessments.

This preliminary analysis of the 14 loss databases shows that the lack of homogeneity relates mostly to time span, the type of variables recorded, and the hazards to which the loss is attributed.

Future reporting, even with an existing data reporting framework, will need to address the following challenges when combining data from different databases: differences of objectives, variety of practices of loss data collection, and different mandates and legislations that enforce some aspects of recording as opposed to others. Most of the data recording procedures cannot be radically modified. However, minor modifications may be introduced to allow for sharing loss data at a level where most loss databases can provide data.

In some countries, only a fraction of economic losses, the insured losses, are assessed. The role of the private sector is essential because it is the main entity collecting loss data in some countries. A minimum requirement could be to report by loss ownership (private/public).

The loss data in Europe covers essentially natural hazards with some exceptions, and the scope of Sendai includes natural, technological, and biological hazards. Gathering loss data for different types of hazards would require collaboration among different institutions at the national level and the appointment of a responsible institution for compiling the loss data in a common database.

The implementation and the testing of the EU guidance by other countries would be useful for assessing the status of disaster loss in the EU, evaluating the progress toward more coherent loss indicators, and the possibility of calculating the required indicators to be reported in the context of the Sendai Framework. Figure 2.5 shows the status of loss data in the EU compared to what is required by the proposed Sendai indicators.

The EU experience in sharing disaster loss data in an open-data policy context following a common framework is the first step toward improving accountability, transparency, and governance, which are the key principles of the Sendai Framework. The next step is to ensure that the best data are made available. Improving the quality of disaster loss data (especially economic losses) is a priority. This can be achieved through the implementation of quality control and validation pro-

cedures. The European Commission together with an expert working group from EU member states has developed an approach for handling uncertainty in a transparent way and is available as Annex 2 in *De Groeve et al.* [2014].

2.6. RECOMMENDATIONS FOR BEST PRACTICES IN LOSS DATA RECORDING

Standards and good practices for disaster loss data recording are a moving target. Constraints given by political frameworks have been negotiated as part of the post-2015 development agenda, in the Sendai Framework, the SDGs, and the climate change framework. Nevertheless, from the analysis of strengths and gaps in existing loss databases and technical discussions at the national, regional, and global level, some recommendations emerge (Chapter 3 and Chapter 4 in this book).

First, it is recommended to record disaster losses along with uncertainty data. Loss data are often only partially known or estimated using a variety of methodologies. Only when recording enough information on the uncertainty associated with the estimates can the data be used for risk knowledge generation. Pedigree methods [*De Groeve et al.*, 2014; Chapter 5 in this book] or explicit recording of uncertainty intervals are two possible ways of implementing this, but other methods can be devised.

Second, it is recommended to define the type of the owner (individuals, business, government, non-governmental organizations) when recording loss data. This allows for providing statistics on losses in the public sector, the industry sector, and the private sector. Separate from the owner type of the building, the losses of a particular building are typically borne partially by the insurance industry, partially by the owner, and partially by public funds (e.g., disaster compensation funds). The loss owner, those that bear the losses (individuals, business, government, non-governmental organizations and insurance companies) should be recorded. In case not all losses are recorded (e.g., only insured losses), it is recommended to develop a method for estimating the total losses across all loss-bearing entities (e.g., applying a coefficient factor on insured losses).

Third, it is recommended to record not only the results of economic loss assessments but also the way the estimates have been produced, including a well-documented method/model, auxiliary data used, and assumptions made in the assessment in the form of metadata. The costs of planning and implementation of risk prevention measures are not considered here because they relate to risk management expenditures rather than to disaster losses. The reporting of economic losses at

Member State	A-2 Number of deaths	A-3 Number of missing	B-2 Injured or ill	B-3a- Number of evacuated	B-3b- Number of relocated	B-4- Number of people whose houses were damaged	B-5- Number of people whose houses were destroyed	B-6- Number of people who received food relief aid	C-2- Direct agricultural loss	C-3- Direct economic loss due to industrial facilities damaged or destroyed	C-4- Direct economic loss due to commercial facilities damaged or destroyed	C-5- Direct economic loss due to houses damaged	C-6- Direct economic loss due to houses destroyed	C-7- Direct economic loss due to damage to critical infrastructure caused by hazardous events	D-2- Number of health facilities destroyed or damaged	D-3- Number of educational facilities destroyed or damaged	D-4- Number of transportation infrastructures destroyed or damaged	D-5- Number of time basic services have been disrupted
Suggested global indicators																		
Target A: A1-Number of deaths and missing due to hazardous events per 100,000 (A-2+A-3)		*Target B:* B1- Number of affected people per 100,000 (sum of B-2 to B-6)							*Target C:* C1- Direct economic loss due to hazardous events in relation to global gross domestic product (sum of C-2 to C-7)						*Target D:* D1- Damage to critical infrastructure due to hazardous events (sum of D-2 to D-5)			
01						•						•						
02						•						•						
03	•	•	•	•		•			•		•	•		•	•	•	•	
04						•			•	•	•	•		•			•	
05	•			•		•	•		•			•		•	•	•	•	
06									•									
07	•	•	•	B-1		•						•					•	
08	•			•		•						•						
09	•	•	•	•		•			•			•		•	•		•	
10	•					•			•		•	•		•	•	•	•	
11	•		•			•						•		•	•	•	•	
12	•	•	•	•		•						•		•	•		•	•
13	•	•		B-1		•												
14																		

Figure 2.5 The indicators that are considered for developing composite indicators used in Sendai recording and the corresponding data available from European loss databases.

international levels assumes that loss databases are implemented at the national level. It is also assumed that for loss data sharing, only summary or aggregated statistics may be available.

Data on economic losses should be event based (i.e., data must be related to the specific event). It is preferable to report on direct economic losses excluding indirect losses because direct losses are concrete, comparable, and verifiable. Because there is no accepted methodology for standardized reporting, loss data recording should use national currencies. For loss data sharing, the losses should be converted into a common currency. To determine the overall amount of disaster impacts, economic losses for all affected sectors must be included, avoiding possible gaps or double accounting. For loss data-sharing purposes, only the sum of direct losses over all sectors would be reported. For transparency purposes, this information should come along with a list of the top-level sectors that have been considered in the reporting. Those that are missing should also be reported.

Finally, for the loss recording process to be successful, the practices would need to be strengthened to make the data useful at the national level beyond narrowly defined objectives, for example, for prevention policy and risk assessment.

2.7. CONCLUSIONS

The Sendai Framework and SDGs have become the catalyst for new paradigms of loss data and risk data generation. The anecdotal disaster loss reporting that started thousands of years ago, the more scientific but still hazard-centered loss reporting of hazard-based approaches, and the one in a lifetime event loss recording at coarse spatial unit for compensation or other purposes are to be overcome. The new paradigm identifies losses, exposure, vulnerability, and hazard as variables that can help in generating risk knowledge through loss accounting, forensic analysis, and risk modeling. That risk assessment exercise needs to be repeated regularly because exposure and vulnerability change over time and so does the risk. Risk reduction is an incremental process that is to be monitored.

Following the Sendai Framework recommendation, the collection of damage and losses should be implemented as much as possible at the local level because hazard events are often localized because the energy released occurs on geographically confined spaces. Recording of loss data is also desirable at fine scale because the dissemination of risk information should start at the local level. The empowered community is the one that can react best to future hazardous events. In fact, dissemination of past loss information and risk information in general, a key recommendation of the Sendai Framework, should be a major incentive to perform risk assessments.

Whenever possible, the losses should be quantified relative to the exposure. That practice alone provides standardization of loss accounting over time. Exposure and affected elements do not necessarily need to be included in loss databases but can instead refer to external databases properly constructed to assure interoperability.

The pre-event exposure database will include exposed assets characterized by their use and structural characteristics. That is, the pre-condition for the modeling of future losses provides the advantage of allowing a rapid and precise quantification of the losses if a hazardous event unfolds. Loss databases and exposure and vulnerability should be available in the form of geographically referenced information. In fact, geographical space should be one of the standardizing variables within loss databases.

A loss database should also report on damage and losses to cultural/archeological sites. The loss data on cultural sites, when recorded, are typically not separated from the damage to the built up. That needs to be addressed because cultural sites are a major source of indirect income and thus would account for indirect losses.

The recording of the data in the database is the choice of the country. The granularity of reporting should be at the asset level, and the municipality should be the coarsest level allowed. The decision to record at a coarser level than municipality will still allow national wide risk assessments and accounting, but it will undermine the forensic analysis and local modeling. Also, local recording especially when used against exposed assets is in itself a quality measure. Aggregation of local loss assessments can be combined in different ways to allow for the identification of risk hotspots. That identification is typically not possible from coarse scale reporting that averages out losses in hotspots areas with that in risk free areas.

Following the Sendai Framework recommendation, the databases should be structured to allow for aggregation for the purposes of national and global comparisons. The global comparison remains a major challenge because it implies the global agreement on the loss data indicators as well as those of exposure and vulnerability. Seemingly, simple issues such as hazard loss attribution or the construction of composite indicator to use for global reporting remain a point of debate within countries, supranational organizations, and UN agencies addressing loss data and at the Sendai Framework negotiations. The international working groups that are meeting to discuss the implementation of the Sendai Framework negotiations will eventually propose a set of indicators. Scientific endeavors may want to build on that proposal to address future reporting that may have more strict requirements than those defined by the Sendai Framework. Those scientific endeavors to better report losses in order to refine our understanding of knowledge should be starting now.

REFERENCES

Aitsi-Selmi. A., K. Blanchard, and V. Murray (2015), The 2015 Sustainable Development Goals and the Sendai Framework for Disaster Risk Reduction: a year for policy coherence. http://www.evidenceaid.org/the-2015-sustainable-development-goals-and-the-sendai-framework-for-disaster-risk-reduction-a-year-for-policy-coherence/.

Carrera, L., G. Standardi, F. Bosello, and J. M. Mysiak (2015), Assessing direct and indirect economic impacts of a flood event through the integration of spatial and computable general equilibrium modeling. *Environmental Modelling & Software* (*63*) 109–122.

Centre for Research on the Epidemiology of Disasters (CRED) (2011), Disaster Loss Characterization: Review of Human and Economic Impact Indicator Definitions. Working Paper. Brussels: Centre for Research on the Epidemiology of Disasters, Université catholique de Louvain.

Corbane, C., T. De Groeve, and D. Ehrlich (2015), A European framework for recording and sharing disaster damage and loss data. *The International Archives of the Photogrammetry, Remote Sensing and Spatial Information Sciences*, Volume XL-3/W3, 2015 ISPRS Geospatial Week 2015, *28* Sep – 0.

Cutter, S., and M. Gall (2015), Sendai target at risk. *Nature Climate Change, 5*, 707–709.

De Groeve, T., K. Poljansek, and D. Ehrlich (2013), Recording Disaster Losses: Recommendations for a European approach. Report by the Joint Research Centre of the European Commission, EUR 26111.

De Groeve, T., K. Poljansek, D. Ehrlich, and C. Corbane (2014), Current Status and Best Practices for Disaster Loss Data recording in EU Member States: A comprehensive overview of current practice in the EU Member States. Report by Joint Research Centre of the European Commission, EUR 26879.

EU Expert Working Group on Disaster Damage and Loss Data (2015), Guidance for Recording and Sharing Disaster Damage and Loss Data, Towards the development of operational indicators to translate the Sendai Framework into action, Report by the European Commission Joint Research Centre, EUR 27192.

Gall, M., K. A. Borden, and S. L. Cutter (2009), When Do Losses Count? *Bulletin of the American Meteorological Society, 90* (6), 799–809.

Global Facility for Disaster Reduction and Recovery (2010), Damage, Loss and Needs Assessment. Guidance Notes. Vol. 1–3. The World Bank, Washington.

Guha-Sapir, D., and M. F. Lechat (1986), The Impact of Natural Disasters: A Brief Analysis of Characteristics and Trends. Prehospital Disaster Med. *2*, 221–223.

Guha-Sapir D., and R. Below (2006), Collecting Data on disasters: Easier said than done, Asian Disaster Management News, *12* (2), 9–10.

Guha-Sapir D., and P. Hoyois (2015), Estimating populations affected by disasters: A review of methodological issues and research gaps, UN Note on affected.

INSPIRE Thematic Working Group Natural Risk Zones (2013). D2.8.III.12 INSPIRE Data Specification on Natural Risk Zones–Technical Guidelines, D2.8.III.12 v3.0.

IRDR DATA (2014), Peril Classification and Hazard Glossary. DATA Project Report No. 1. Available from http://www.preventionweb.net/english/professional/publications/v.php?id=36979.

IRDR DATA (2015), Guidelines on Measuring Losses from Disasters. Data Project Report No. 2. Available from http://www.irdrinternational.org/wp-content/uploads/2015/03/DATA-Project-Report-No.-2-WEB-7MB.pdf.

Kaplan, S., and B. J. Garrick (1981), On The Quantitative Definition of Risk. *Risk Analysis, 1* (1), 11–27.

Kirchsteiger, C. (1999), On the use of probabilistic and deterministic methods in risk analysis. *Journal of Loss Prevention in the Process Industries 12* (5), 399–419.

Margottini, C., G. Delmonaco, and F. Ferrara (2011), Impact and losses of natural and Na-Tech disasters in Europe. In Inside Risk: a strategy for sustainable risk mitigation. Ed. S. Menoni and C. Margottini, Springer pp. 93–127.

Meyer, V., N. Becker, V. Markantonis, R. Schwarze, J. C. J. M. van den Bergh, L. M. Bouwer, P. Bubeck, P. Ciavola, E. Genovese, C. Green, S. Hallegatte, H. Kreibich, Q. Lequeux, I. Logar, E. Papyrakis, C. Pfurtscheller, J. Poussin, V. Przyluski, A. H. Thieken, and C. Viavattene (2013), Review article: Assessing the costs of natural hazards–state of the art and knowledge gaps, Nat. Hazards Earth Syst. Sci., *13*, 1351–1373, doi:10.5194/nhess-13-1351-2013.

Mitchell, T., D. Guha-Sapir, J. Hall, E. Lovell, R. Muir-Wood, A. Norris, L. Scott, and P. Wallemacq (2014), Setting, Measuring and Monitoring Targets for Disaster Risk Reduction: Recommendations for post-2015 international policy frameworks. Overseas Development Institute, London, 69 p.

Molinari, D., S. Menoni, G. T. Aronica, F. Ballio, N. Berni, C. Pandolfo, M. Stelluti, and G. Minucci (2014), Ex post damage assessment: an Italian experience. *Nat. Hazards Earth Syst. Sci.*, *14*, 901–916.

Organization for Economic Co-operation and Development (OECD) (2014). Improving the evidence base on the costs: Towards an OECD accounting for Risk management expenditures and losses from disasters, GOV/PGC/HLRF(2014)8. Available from http://www.oecd.org/officialdocuments/public displaydocumentpdf/?cote=GOV/PGC/HLRF(2014)8&docLanguage=En.

Salvati, P., C. Bianchi, M. Rossi, and F. Guitzzetti (2010), Societal landslide and flood risk in Italy. Nat. Hazards Earth Syst. Sci. *10*, 465–483.

United Nations Development Programme (UNDP) (2008), Disaster Loss Database Standards, United Nations Development Program, Working group on disaster data, pp. 50. Available from http://www.gripweb.org/gripweb/sites/default/files/methodologies_tools/Disaster%20database%20standards_black.pdf.

United Nations Development Programme (UNDP) (2013), A comparative review of country-level and regional disaster loss and damage databases. United Nations Development Program, pp. 50. Available from http://www.undp.org/content/dam/undp/library/crisis%20prevention/disaster/asia_pacific/lossanddamagedatabase.pdf.

United Nations Office for Disaster Risk reduction (UNDORR) (2015), Proposed Updated Terminology on Disaster Risk Reduction: A Technical Review. http://www.preventionweb.net/files/45462_backgoundpaperonterminologyaugust20.pdf [accessed 17 May 2017].

United Nations Office for Disaster Risk Reduction (UNISDR) (2015a), Sendai framework for disaster risk reduction 2015–2030. In: UN world conference on disaster risk reduction, 2015 March 14–18, Sendai, Japan. Geneva: United Nations Office for Disaster Risk Reduction. Available at *1654* http://www.unisdr.org/files/43291_sendaiframeworkfordrren.pdf [Accessed 10 February 2016].

United Nations Office for Disaster Risk Reduction (UNISDR) (2015b), Global Assesment Report–2015. Available from http://www.preventionweb.net/english/hyogo/gar/2015/en/home/.

Weichselgartner, J., and P. Pigeon (2015), The role of knowledge in disaster risk reduction. *Int. J. Disaster Risk Science, 6*: 107–116.

Wirtz, A., W. Kron, P. Loew, and M. Steuer (2014), The need for data: natural disasters and the challenges of database management. Nat Hazards, *70*:135–157.

World Bank (2010), Damage loss and needs assessment; Guidance Note No. 2. Global Facility for Disaster Risk Reduction and Recovery, pp. 84 Available from https://openknowledge.worldbank.org/bitstream/handle/10986/19046/880860v20WP0Bo000Damage0Volume20WEB.pdf?sequence=1&isAllowed=y.

Wu, H., and Z. L. Li (2009), Scale issues in Remote Sensing: A Review on analysis, Processing and Modelling. *Sensors 9* (3), 1768–1793.

Part II
Data Storage

3

Overview of Loss Data Storage at Global Scale

Roberto Rudari, Marco Massabò, and Tatiana Bedrina

ABSTRACT

Several databases of loss data are available today to end users and practitioners. The goals supported by such databases are different, and therefore, the level and type of information that can be extracted varies strongly from one another. In addition to that, they are many times limited in reporting of various details of damages, but rarely they record ancillary information about the causes, the context, and the preconditions of the event. All of these pieces of information are produced and stored many times at a national or even at a global scale, but they are not connected to the loss database for additional analysis. Current technology could help in moving toward a more comprehensive concept of loss database that includes information about flood footprints, preconditions, observed and forecast forcing, and pre-existing vulnerability assessments. A possible roadmap that moves the first steps in this direction is proposed.

3.1. INTRODUCTION

Disaster loss data are perceived by all communities contributing to Disaster Risk Reduction (DRR) policies as a cornerstone for an informed process. (See Chapter 1 and Chapter 2 in this book.) The need has been felt at different levels, local, national, regional, and global, and disaster loss databases trying to respond to the respective demands have been developed. Today, with emerging capacity in collecting and managing large quantities of data and with an increased standardization that eases data exchange between tools and actors, it is time to start thinking of how information can be re-organized and made accessible across scales so that the right level of information would be available in a consistent way from the local to the global level. Therefore, this section, though focused on global datasets, will describe the noteworthy initiatives at different scales, namely national, regional, and global that could be

linked together in a unified framework. The data sets will be compared in terms of data collection methodologies and data organization, and differences will be highlighted within the same scale of application and across the different spatial scales. This exercise is compiled with the ultimate goal of studying the feasibility of a standardization effort aimed at enhancing the interoperability between data sets, and ultimately leading to a seamless disaster loss database that could serve the main application fields of such efforts as compensation, loss accounting, and forensic and risk modeling [*De Groeve*, 2013]. (See Chapter 1 and Chapter 2 in this book.)

3.2. EUROPEAN UNION GUIDELINES USED FOR THE CONTEXTUAL ANALYSIS OF THE DATA SETS

The Joint Research Centre European Commission (JRC EU) expert working group on disaster damage and loss data, in consultation and collaboration with European Union (EU) and international institutions,

CIMA Research Foundation, Savona, Italy

Flood Damage Survey and Assessment: New Insights from Research and Practice, Geophysical Monograph 228,
First Edition. Edited by Daniela Molinari, Scira Menoni, and Francesco Ballio.

issued documents on disaster losses recording with the aim of supporting and enhancing risk assessment and risk management as well as exchanging information at the EU level among member states. The Guidance for Recording Disaster Losses[1] [*De Groeve*, 2013] and Guidance for Recording and Sharing Disaster Damage and Loss Data[2] [*EU Disaster Loss Data Working Group*, 2015] provide definitions and recommendations on disaster loss data collection. These efforts were undertaken due to lack of common methods and standards among EU countries in loss data accounting and recording. (See Chapter 2 in this book.)

The Guidance of 2013 offers specifications on loss data, hazard events, conceptual models of loss databases, scope and scale of a database, and requirements for loss data recording at national level of the EU countries. The Guidance of 2015 provides definition of indicators of the disaster damage and loss database and minimum requirements for damage and loss data sharing at European and international levels. The following sections take advantage of the requirements included in the JRC guidelines in order to describe and analyze each database in a standard and comparable manner.

3.3. OVERVIEW OF DATA SETS AT GLOBAL SCALE (EM-DAT, NATCATSERVICE, SIGMA)

The comparative overview on three global data sets, most accepted and used by the global community, Emergency Events Database (EM-DAT) (Centre for Research on the Epidemiology of Disasters [CRED]), NatCatSERVICE (Munich Re), and Sigma (Swiss Reinsurance) is provided in this section. Parameters used by the JRC EU expert working group were taken as main indexes of comparison of databases. Specifically, the databases have been analyzed for their scope and scale (including geo-localization of losses), the disaster nomenclature/classification used, the use of thresholds for events recording, the use of uncertainty evaluation methodology.

3.3.1. Scope and Period Covered

The scope or geographical coverage of examined databases is global. Disasters are recorded on an event entry basis for all of three data sets. The period covered starts from 1900 for EM-DAT, 1980 for NatCatSERVICE, and from 1970 for Sigma to the present.

3.3.2. Methodology for Data Collection and Classifications

These databases don't follow the same methodology of data classification and have their own specificity on data registration. The damage thresholds and attributions of an event to a disaster vary for all three data sets reported in Table 3.1). EM-DAT defines a disaster as having the following criteria: 10 or more fatalities; 100 people affected; declaration of a state of emergency; and a call for international assistance. NatCatSERVICE collects information on a disaster with any property damaged and/or any person affected, injured, or dead without defining specific thresholds. Sigma requires at least one of the following criteria for inclusion in the database: 20 deaths and/or 50 injured and/or 2000 homeless and/or the following insured losses (million $US): 18.3 (marine), 36.7 (aviation), 45.5 (all other losses), and/or more than 70 (total losses).

Similar indicators on losses including damages to housing, property, and economic and human losses are used in the three datasets.

The Integrated Research on Disaster Risk (IRDR) research program peril classification and hazard glossary is used only by the EM-DAT database to refer a disaster to a specific type. Sigma and NatCatSERVICE databases have their own specificity on classification of disasters. EM-DAT and Sigma databases account for natural and man-made disasters. NatCatSERVICE includes exclusively natural catastrophes, excluding drought and man-made (i.e., technical disasters). EM-DAT includes human, economic, and infrastructure disaster impacts.

3.3.3. Information Technology Platform and Data Accessibility

Information Technology (IT) platforms offer search engines and graphical options for data visualization. The country profile[3] of EM-DAT allows users to filter information by the time period, character of disaster (natural or man-made), and country. The advanced search[4] option allows the generation of data sheets based on the overall EM-DAT records. The portal of NatCatSERVICE Connect.Munichre[5] (registration is required), designed exclusively for clients of Munich Re, provides the annual large range of catastrophe portraits with up to 200 attributes. The most important are human loss information; direct economic losses, that is, the tangible monetary impact of a disaster; insured losses; scientific data;

---

[1] *http://publications.jrc.ec.europa.eu/repository/bitstream/111111111/29296/1/lbna26111enn.pdf*

[2] *http://drr.jrc.ec.europa.eu/Portals/0/Loss/JRC_guidelines_loss_data_recording_v10.pdf*

[3] *http://emdat.be/country_profile/index.html*

[4] *http://www.emdat.be/advanced_search/index.html*

[5] *https://register.munichre.com/WEBCRM/LoginSite/login.aspx*

Table 3.1 Global and Regional Disaster Loss Data Repositories.

DLD database	EM-DAT	NatCat SERVICE	SIGMA
IT platform	http://www.emdat.be/	http://www.munichre.com/natcatservice	http://www.sigma-explorer.com/
Coverage scope/scale	Global National/Event based	Global National/Event based	Global National/Event based
Holder institute	CRED, Université Catholique de Louvain in Brussels, Belgium	Munich Reinsurance Company, Munich, Germany	Swiss Reinsurance Company, Zurich, Switzerland
Language	English	English, German, France, Spanish	English, German, France, Spanish
Disaster type	Natural (including epidemics) and man-made disasters and conflicts	Natural disasters (excluding drought and man-made, i.e., technical disasters)	Natural and man-made disasters (excluding drought)
Flood classification	IRDR General river flood, flash flood, storm surge/coastal flood	Flood: River flood Flash flood Storm surge	Flood, flash flood, flood warning, sea level rise, and waves activity
Thresholds and losses	10 or more fatalities; 100 people affected; declaration of a state of emergency; call for international assistance	Entry if any property damage and/or any person affected (injured, dead) Before 1980, only major events	20 deaths, 50 injured, 2,000 homeless (not including affected) Insured losses of 18.3 million (maritime disasters), 36.7 million (aviation disasters), 45.5 million (all other losses) Overall losses more than 70 million in US dollars
Uncertainty	Quarterly cross-checks	Systematic evaluation	Not specified
Time period	From 1900 to the present	From 79 AD to the present Before 1970, only major events	Major losses since 1970 to the present
Data sources	UN agencies, government organizations, and Red Cross and Red Crescent Societies	Priority given to Lloyd's list, Reuters, Reports from clients and branch offices, Insurance press	Newspapers, direct insurance and reinsurance periodicals, specialist publications, and reports from insurers and reinsurers
Data sharing	Free, downloadable CSV format	Partially accessible to public, more services for clients Statistics on significant natural disasters since 1980 in PDF format	Is not accessible Yearly publication of "raw information" listing all disasters for the year available to clients Analysis of data available in graphic form Data not downloadable

socioeconomic information and damaged sectors of economy. The Sigma platform[6] provides access to loss data containing economic losses, insured losses (in USD billions), and the number of victims and homeless.

The access to information varies among databases. EM-DAT offers limited online data access through different search options. All data can be accessed upon request through the website. NatCatSERVICE provides access to data for scientific projects and offers a vast set of analyses on the official website, but full access is mainly provided to insurance clients. The data on the Sigma database are available only for the insurance industry and insurance clients. Annual reports that include aggregated statistics are issued by all three sources.

3.3.4. Data Source and Quality Control

The data sources used for EM-DAT are different from the ones used in the reinsurance databases. In EM-DAT, priority is given to data from United Nations (UN) agencies, government organizations, and Red Cross and Red Crescent Societies. The Munich and Swiss reinsurance companies prefer Lloyd's list, Reuters, reports from clients, and branch offices and insurance press.

EM-DAT has a procedure that includes quarterly crosschecks for data quality assurance and uncertainty assessment, while NatCatSERVICE performs, with the same aim, a systematic evaluation. There is no clear indication of the quality assurance/quality control procedures for Sigma.

3.4. NATIONAL DATA SETS INCLUDING GOOD PRACTICES (SLOVENIA, MOLDOVA, UNITED STATES, COLOMBIA)

Some good practices of loss data accounting at the national level are described in this section. The data sets of Slovenia and Moldova republics, the database of Colombia, developed on the base of DesInventar system, and the Spatial Hazard Events and Losses Database (SHELDUS) in the United States (US). The main characteristics of these repositories are reported in Table 3.2. National data sets are evaluated because they can represent the building stones of a global data set when the method of representing an exposing data is sufficiently standardized, with the clear advantage of possibly offering greater detail in case one country or a specific event needs to be analyzed. Several national thematic databases exist at the national level. Some examples are the Swiss Federal Research Institute WSL database[7] for flood and

landslide events in Switzerland and the FloodCat[8] in Italy, a new national flood database of the Civil Protection Department. The other example is an object-specific database of GFZ German Research Centre for Geosciences, HOWAS21[9]. (See Chapter 5 in this book). This database archives loss data on riverine and pluvial floods related to individual objects (housing, lands, infrastructure, defense structures, and other) in Germany. The collected data is used for two main purposes: validation of flood models and forensic analysis. These are some good examples on data sets that collect detailed event-specific data (for example flooding and landslides) and contribute to improving disaster loss data availability. In this chapter, we will however analyze only three examples in more detail: one in the EU (Slovenia), one in the Balkan area (Moldova), and one in South America (Colombia).

3.4.1. Disaster Loss Database Slovenia

The national database of Slovenia from 2003 accounts for losses of natural hazards and other disasters related to human losses (killed, injured), economic and agriculture sector, and property damages (buildings and infrastructure). The scope of this database is national; its scale is at the asset level. The basic principle for disaster inclusion is direct damage to property or agriculture exceeding 0.03% of planned state budget revenue. Commissions of civil protection at the local, regional, and national levels carry out the damage assessment. The damage assessment is coordinated by the Administration of the Republic of Slovenia for Civil Protection and Disaster Relief (ACPDR) in collaboration with the Association of Municipalities and Towns of Slovenia and the Association of Municipalities of Slovenia.

Damage evaluation methodology[10] includes damage caused by natural disasters and industrial accidents. The damage groups include land, facilities, fixed and current assets (movable property and stocks, agricultural production, multiannual plantations), cultural property, and loss of revenue in a holding.

The web-based application AJDA is a central system operated by ACPDR for calculating, analyzing, and reporting on losses and processing applications of victims caused by natural and industrial hazards. The access

---

[6] http://www.sigma-explorer.com/
[7] http://www.wsl.ch/fe/gebirgshydrologie/HEX/projekte/schadendatenbank/index_EN

[8] http://mydewetratest.cimafoundation.org/#
[9] http://howas21.gfz-potsdam.de/howas21/
[10] Decree on damage evaluation methodology (Official Gazette of the Republic of Slovenia, no. 67/03, government acts register 2002-1911-0067, in effect since 19 July 2003).

Table 3.2 Good Practices of National Loss Data Sets.

DLD database	Slovenia National Database	Moldova National Database	United States SHELDUS	Colombia DesInventar
IT platform	AJDA	GISCUIT	http://hvri.geog.sc.edu/SHELDUS/	http://www.desinventar.net/DesInventar/profiletab.jsp?countrycode=col
Coverage scope/scale	National/Asset level	National/Asset level	National/ Event based	National/Municipal
Holder institute	Administration for Civil Protection and Disaster Relief	Civil protection and emergency situations service	University of South Carolina, United States	Surocidente Seismological Observatory OSSO Corporation, Cali; Valle, Colombia
Language	Slovenian	Romanian, Russian	English	English, Spanish
Disaster type	Natural and industrial accidents	Natural, social-biological emergencies and man-made	Natural and man-made disasters	Natural and man-made disasters
Flood classification	National methodology: Flood	National methodology	NWSPD10-16[1] Derective: Flood Coastal flood Flash flood; Hayden Flood Classification	IRDR and DesInventar methodology
Thresholds and losses	Damage to property or agriculture more than 0.03% of planned state budget revenue	Hydrological emergencies: general floods, spring floods, snow/rain floods and rise of groundwater Cessation of normal living conditions of the population	All events, causing monetary and/or human losses	All events, causing monetary and/or human losses
Uncertainty	Quality assurance/quality control procedures are applied	Quality assurance/quality control procedures are applied	Quality assurance/quality control procedures are applied	Quality Control mechanisms: trends, graphics, maps, statistics
Time period	2003 to the present	1997 to the present	Hard copies for 1960 till 2009 and digital data imports from 2010 to the present	From 1914 till 2012
Data sources	Official sources	Official sources	NCDC[2], National Center on Environmental Information, USGS[3], US Census Bureau, US Bureau of Labor Statistics	Governmental organizations and non-governmental organizations
Data sharing	Internal DB of ACPDR[4] No public access Data not downloadable	Internal DB of CPESS[5] No public access Data not downloadable	Open access to public by subscription Spreadsheet and GIS formats: CSV, FTP, MapInfo (MIF, DXF)	Open access to public Data downloadable in Excel, CSV, XML, SVG, KML formats

[1] National Weather Service Policy Directive
[2] National Climatic Data Center
[3] United States Geological Survey
[4] Administration of the Republic of Slovenia for Civil Protection and Disaster Relief
[5] Civil Protection and Emergency Situation Service

to the database is reserved to authorities, and it is closed to the public. The sources of information on disasters are state civil protections offices, environmental agencies, scientific institutions, auditors, and insurance companies.

3.4.2. National Disaster Database Moldova

The Civil Protection and Emergency Situations Service (CPESS) is a focal point for the data collection of all types of emergencies and their losses. CPESS records all data into a single electronic database and transmits a statistical summary every three months to the National Bureau of Statistics.

The official classification (Government Regulation n. 1076 of 16 November 2010 and the Decree of CPESS n. 139 of 4 September 2012) of emergencies defines man-made, natural, and social-biological emergencies. The floods belong to a group called hydrological emergencies, and they are divided into categories of general floods, spring floods, snow/rain floods, and rise of groundwater. All events causing human losses, breakdown of living conditions, and/or monetary losses are accounted. All of this information is used for the compensation of consequences and disaster recovery. Depending on the area of extension, events are defined in the following different levels: object, local, territorial, national, and transboundary.

CPESS has developed the project "Disaster and Climate Risk Reduction" on the base of cost-effective web mapping platform called GISCUIT[11] to support data recording during emergency situations. The quality of loss data on emergency situations is high, since the data are provided by official assessment commissions controlled by public management authorities. The database is closed to the public, and information is provided to public authorities and private sector after an official request.

3.4.3. SHELDUS, United States

The SHELDUS[12] online database provides information at the national level for the United States (it does not include Puerto Rico, Guam, and other US territories). The information on natural hazards and related human (injured, fatalities), property, crop, and economic losses is registered. The losses are registered in association with a specific event. The accounting of meteorological and hydrological events is based on the National Weather Service Instruction (NWSI) 10-1605,[13] dated 17 November 2005, and they are defined accordingly as coastal flood, flash flood, and general flood (river flooding is included in the flood category).

The SHELDUS database is quality controlled and validated, in addition to a first automatic screening done by the system after the data are entered online. The data sources are the National Climatic Data Center (NCDC), the National Center on Environmental Information (NCEI), United States Geological Survey (USGS), US Census Bureau, US Bureau of Labor Statistics, and others.

The license[14] of terms and conditions provides the right to use the information for scientific, publication, design, research, and other purposes. Aggregated data are available for unlimited downloads after subscription. The raw data are available on a payment basis.

3.4.4. DesInventar, Colombia

The disaster loss database of Colombia is one of the most populated DesInventar-based databases (described in section 3.3; see also Chapter 1 in this book). The database contains 33,817 records on losses of natural and man-made disasters from 1914 till 2012. Regarding flood losses, the number of records is 13,647 within time period from 1984 till 2011. The database is operated by NGOs and hosted by Suroccidente Seismological Observatory OSSO Corporation, Cali, Valle, Colombia. The database contains information from official sources, as well as from unofficial ones.

The database of Colombia[15] uses IRDR Peril Classification and the Hazard Glossary.[16] The losses data includes standard DesInventar indexes such as human losses, physical losses, and multi-sectoral economic losses, which are supplemented by additional national specific losses. The data are provided in open access for the public and are available for analysis at the official DesInventar website.[17]

3.4.5. Conclusions

Disaster losses databases at the national level described above represent the building stone of global database. In particular, in the global data set, the following main characteristics should be retained from national data sets:

---

[11] http://giscuit.com/mm
[12] http://hvri.geog.sc.edu/SHELDUS/

[13] http://www.ncdc.noaa.gov/stormevents/pd01016005curr.pdf
[14] http://hvri.geog.sc.edu/SHELDUS/docs/END_USER_LICENSE_AGREEMENT.pdf
[15] http://www.desinventar.net/DesInventar/profiletab.jsp?countrycode=col
[16] http://www.desinventar.net/definitions.html
[17] http://www.desinventar.net/DesInventar/profiletab.jsp

1. Loss data should be accounted for at the asset level. Slovenia and Moldova provide accurate and detailed information at the asset level.

2. Do not include the threshold limit for recording disasters data. Moldova, United States, and Colombia databases account for all disaster events and do not apply threshold limits. All disasters are recorded and remain in the archives. This includes disasters that caused small losses, not only those that caused large losses.

3. Graphical and mapping tools for dynamic analysis of data are used. National DesInventar databases or the Sigma Explorer[18] platform provide the ability for data visualization and processing that facilitate the usability of data by operators.

3.5. NATIONAL DATA SETS IN A REGIONAL CONTEXT AND GLOBAL CONTEXT (EUROPEAN UNION EFFORT, THE COMMONWEALTH OF INDEPENDENT STATES EFFORT, DESINVENTAR DATABASE)

A step forward in the direction indicated in the present contribution is represented by regional initiatives that set standards of data exchange for inter-comparison purposes. These initiatives are the starting point for a larger framework that can allow data sharing on a global level to be a standard practice. Some good practices in this sense are described in this section, such as, the EU member states guidelines and the Commonwealth of Independent States (CIS) practice provides an example of a framework for accounting disaster data and related losses at the regional level. Similarly, the DesInventar methodology provides a standard applied under the guidance of the UN in many developing counties and works in a global context (details are reported in Table 3.3).

3.5.1. European Union Effort

The EU effort is led by JRC, and the main core is developed in consultation with a working group composed both by member states representatives and an expert in Disaster Loss Data (DLD) collection and use. Such effort produced three reference documents: "Recording Disaster Losses: Recommendations for a European approach" [De Groeve et al., 2013]; "Current Status and Best Practices for Disaster Loss Data recording in EU Member States" [De Groeve et al., 2014], and "Guidance for Recording and Sharing Disaster Damage and Loss Data" [EU DLD WG, 2015]. These documents trace a

path to the guidelines for the minimum standards required to share data at the EU level and, although not compulsory, represent a great guidance for the member states to achieve a minimum level of standardization in order to set a European Panorama and most importantly to optimally tune the EU instruments related to disasters, including the "Solidarity Fund".[19]

These guidelines are badly needed. According to the sixth technical workshop report on a EU approach for recording damage and loss data [JRC, 2015], there are differences in national disaster damage and loss database and methodologies for data recording by EU member states. The current status of the implementation of JRC guidance on sharing loss data demonstrates the following:

• EU national databases recorded events at the level of territorial units of minimum resolution according to Nomenclature of Territorial Units for Statistics (NUTS3).

• Flood is the only hazard covered by all countries.

• Because of the absence of defined indicators and thresholds, in some EU countries, the effects of disasters are not registered.

• Some gaps and missing data due to the variety of national actors in charge of data loss recording are still present.

The main actions of the current activity of the JRC working group follow: update the comparative analysis about EU national database and Sendai targets; continue sharing damage and loss data on the JRC minimum requirements; and start to implement the platform for flood event and loss data recording data that is compliant to Infrastructure for Spatial Information in the European Community (INSPIRE) through pilot projects. (For a deeper discussion of JRC activities, see Chapter 2 in this book.)

3.5.2. Commonwealth of Independent States Countries Effort

The Interstate Council for Emergency Situations of natural and man-made disasters established in 1993 is a part of the Executive Committee[20] of CIS. The Council collects information on disasters and losses two times at year, coming from all Interstate Council members (Azerbaijan, Armenia, Belarus, Kazakhstan, Kyrgyzstan, Moldova, Russian Federation, Tajikistan,

---

[18] http://www.sigma-explorer.com/

[19] http://eur-lex.europa.eu/legal-content/EN/TXT/?uri=URISERV%3Ag24217
[20] http://www.cis.minsk.by/

Table 3.3 National Databases in a Regional and Global Context EU Databases, DesInventar and CIS Countries Database.

DLD database	EU databases	DesInventar database	CIS[1] countries database
IT platform	Internal	http://www.desinventar.net/index_www.html	http://www.cis.minsk.by/
Coverage scope/scale	EU member states NATS3 level	Global Country/NUTS2 and NUTS3	CIS countries
Holder institute	Governmental and Civil Protection authorities	UNISDR - Global Assessment Report Team, Geneve, Switzerland	National/Country DB holder: Executive Committee of CIS, Interstate Council for Emergency Situations Performer: MoES, Minsk, Belarus
Language	Local	English, Multilingual	Russian
Disaster type	Natural and man-made disasters	Natural and man-made disasters	CIS classifier of natural and man-made disasters (Resolution n. 16 August 2002)
Flood classification	Classifications vary	IRDR	Natural and man-made disasters
	Flood covered by all EU member states	Flood: Coastal flood Flash flood Ice jam flood Riverine flood	Flood, rain/snow floods, flash floods, low waters, rise of groundwater
Thresholds and losses	Various	All disasters caused casualties or damages	Official CIS Classifier thresholds
Uncertainty	Various	Quality control mechanisms: trends, graphics, maps, statistics	Quality assurance/quality control on significant events by CIS member states
Time period	Various	1994 to the present	2004 to the present
			Half-yearly reports
Data sources	Civil Protection authorities	NGOs, governmental institutions, civil protection agencies	Civil protection authorities, ministries, assessment commissions
Data sharing	No public access	Open access for public	Online public access to annual reports
		Excel, CSV, XML, SVG, KML	Data not directly downloadable

[1]Commonwealth of Independent States

Turkmenistan, Uzbekistan, and Ukraine), and distributes it among all partners.

The official CIS Classifier[21] of emergency situations of both a natural and man-made nature of CIS countries is adopted by the Interstate Council for Emergency Situations of the natural and man-made CIS by Resolution n. 16 of 15 August 2002.[22] The half-year reports contain summarized data on human losses, economical losses, and multi-sectoral damages for each subtype of emergency situation. Human losses include fatalities, directly affected people (injured, ill), and the total number of evacuated persons. The economical losses are calculated in thousands of US dollars. The losses in housing sector and industrial sectors are divided in three groups: destroyed, damaged, and saved. The data on disasters are provided by state authorities of emergency situations of CIS countries.

The Executive Committee publishes[23] annual summaries on natural and man-made disasters from 2010 till 2014 available for public. The more detailed reports may be provided through an official request to the Executive Committee[24].

3.5.3. DesInventar

DesInventar[25] is a Disaster Loss Accounting System started in 1994 by LA RED (Network of Social Studies in the Prevention of Disasters in Latin America). The current development is supported by the United Nations Development Programme (UNDP) and the United Nations Office for Disaster Risk Reduction (UNISDR), which sponsored similar systems in the Caribbean, Asia, Africa, and Europe. The database is growing, and currently 82 countries (Africa, Americas, Asia, Europe, Carribean, Oceania) are publishing the disaster loss data through the DesInventar system.

The DesInventar methodology is based on IRDR standardized classification and definitions. The description of an event defines cause[26] and effects[27] occurred. The losses registered include human losses, physical damage (houses, infrastructure, affected sectors), and

economic loss (in local currency and US dollars). One of the prerequisites of methodology is that the information must be spatially disaggregated in order to show the effects of disasters at the level of administrative unit of NUTS2 or NUTS3 resolution.

The DesInventar[28] software product supports many international standards as well as Open Geospatial Consortium OGC, XML, Global IDEntifier number generator system (GLIDE), and Google API. The interface can be customized and adapted to national needs. The open access to data is provided through Analysis module. The data are downloadable in most common formats, such as CSV, Excel, XML, and MS Access, that ensure compatibility with other databases and that use the same standards. (A deeper description of initiatives linked to the implementation of DesInventar is supplied in Chapter 1 in this book.)

3.5.4. Comparison of Different Frameworks

Table 3.4 provides comparison of requirements on loss data collection of natural and man-made disasters of different frameworks: EU JRC, Sendai, DesInventar, CIS countries, and the European Union Floods Directive (EUFD)/2007 (only floods). The main indicators (human losses, damages in housing, and direct loss of all economic sectors) are generally the same for all revised methodologies, except Sendai. It requires the most detailed description of affected people. The registration of number of damages in the housing sector is required at the national level (Document 3, Proposed Guideline of National Disaster Loss Database, 2015). The accounting of direct economic loss to sectors, including housing, is recommended at the global level (Document 4, Indicator System for the Sendai Framework, 2015). The geographic and temporal information of each NUTS2/NUTS3 or Unit of Management (UoM) are required by EU JRC, EUFD 2007/60/EC, DesInventar, and CIS countries data sets (at the national level). The direct economic loss is accounted for per all sectors in the EU framework. The EU JRC framework is the most detailed in registration of direct economic losses in monetary value by different sectors and includes the ownership of suffered losses. An interesting point for the CIS methodology is that it requires recording of the trace data on undamaged properties and territory, besides the destroyed and damaged ones. For a more detailed comparison, please refer to Table 3.2.

[21] http://www.e-cis.info/foto/pages/24928.pdf

[22] http://cis.minsk.by/reestr/ru/printPreview/text?id=2165&serverUrl=http://cis.minsk.by/reestr/ru

[23] http://e-cis.info/index.php?id=891

[24] http://e-cis.info/page.php?id=20054

[25] http://www.desinventar.net/index_www.html

[26] http://www.desinventar.net/definitions.html

[27] http://www.desinventar.net/effects.html

[28] http://www.desinventar.net/index_www.html

Table 3.4 Comparative Table on Disaster Losses Recording for EU Minimum Requirements, Sendai Requirements, EUFD/2007, Desinventar, CIS Regional Level.

EU JRC < Yearly reports/data for specific event >	SENDAI Framework
IRDR Classification <N aturalHazardClassification >	IRDR Classification
Geographic information of each NUTS2/NUTS3 or UoM	-
Temporal information *< validFrom >* *< validTo >*	-

HOUSING

Houses destroyed < total number >	-
Houses damaged < total number >	-
Education centers < total number >	Number of educational facilities damaged and destroyed by disasters (hazard events)
Health facilities < total number >	Number of health facilities damaged and destroyed by disasters (hazard events)

HUMAN LOSS

Deaths < number of persons >	Deaths < per 100,000 >
Directly affected < number of persons >	Affected < per 100,000 >
Missing < number of persons >	Missing < per 100,000 >
-	Injured or ill < per 100,000 >
-	Evacuated < per 100,000 >
-	Relocated people < per 100,000 >
-	People living in damaged houses < per 100,000 >
-	People living in destroyed houses < per 100,000 >
-	People who received relief or compensation < per 100,000 >

ECONOMIC LOSS

Direct loss for all sectors < total in monetary value >	Direct economic loss in relation to global GDP
Agriculture *< total in monetary value >*	Agricultural loss due to extensive disasters (hazard events)
Industrial *< total in monetary value >*	Direct economic loss due to industrial facilities damaged and destroyed by extensive disasters (hazard events)

EUFD2007/60/EC < 6-year reports >	DesInventar	**CIS Regional level** < Half-yearly reports/data by type of event >
Guidance for Reporting under the Floods Directive (2007/60/EC)	IRDR Classification	CIS classification of emergency situations of natural and man-made character
Geographic information of each UoM	Geographic information of each NUTS2/NUTS3	-
Temporal information < validFrom > < validTo >	Temporal information < validFrom > < validTo >	-
Consequences to Property: Homes	Houses destroyed < number per event > Houses damaged < number per event >	Houses/constructions destroyed < total number > Houses/constructions damaged < total number >; Houses/ constructions saved due to relief operations < total number >
Adverse consequences to the community, such as education	Education centers affected < number per event >	-
Health and social work facilities (such as hospitals)	Hospitals affected < number per event >	-
Deaths	Deaths < number of persons >	Deaths total/children < number of persons >
Human health immediate or consequential impacts	Directly affected < number of persons >	Directly affected: injured, ill total/ children < number of persons >
	Missing < number of persons >	-
-	Injured < number of persons >	-
-	Evacuated < number of persons >	Evacuated < number of persons >
-	Relocated < number of persons >	-
-	-	-
-		
-	-	People saved due to relief operations
-	Direct loss for all sectors < total in monetary value local/US dollars >	Direct loss for all sectors < total in thousands of US dollars >
-	-	-
-	-	-

(Continued)

Table 3.4 (Continued)

EU JRC < Yearly reports/data for specific event >	SENDAI Framework
Commerce < *total in monetary value* >	Direct economic loss due to commercial facilities damaged and destroyed by extensive disasters (hazard events)
Tourism < *total in monetary value* >	-
Housing < *total in monetary value* >	Direct economic loss due to housing units damaged by disasters (hazard events) Direct economic loss due to housing units destroyed by disasters (hazard events)
Education < total in monetary value >	Direct economic loss due to disaster damage to critical infrastructure among them health and educational facilities
Health < total in monetary value > Electrical < total in monetary value > *Water supply* < *total in monetary value* > *Transport* < *total in monetary value* > **Owner:** Individuals, businesses, government, NGOs **Status of ownership** (who bears the loss?): Individuals, business, government, NGOs, insurance companies < *in monetary value* >	

Disaster Damage to Critical Infrastructure

-	Number of times basic services have been disrupted due to disasters (hazard events)
-	-
-	Number of disaster damage to critical infrastructure
- -	Number of lengths of road damaged and destroyed by disasters (hazard events)
-	-
-	-
-	-

[1]http://ec.europa.eu/environment/emas/pdf/general/nacecodes_en.pdf

EUFD2007/60/EC <6-year reports >	DesInventar	CIS Regional level < Half-yearly reports/data by type of event >
-	-	-
-	-	-
	-	-
-	-	-
-	-	-
-	-	-
-	-	-
-	-	-
Adverse consequences to sectors of economic activity (NACE codes[1])	-	-
Rural Land Use: agricultural activity (livestock, arable, and horticulture), forestry, mineral extraction, and fishing	Agriculture damages in crops < Hectares > Livestock < Number >	Agricultural losses: Crops, fodder, forestry, livestock, fowl Destroyed Agricultural losses Damaged Agricultural losses Saved due to relief operations
-	Energy affected	-
-	Water supply affected	-
-	Transport affected	Techniques Destroyed < number >
	Damages in roads < Mts >	Techniques Damaged < number > Techniques Saved due to relief operations < number >
Environment: Water body Status, Protected Areas, Pollution Sources	-	Territory Damaged < Hectares >
Cultural Heritage: Cultural Assets, Landscape	-	Material values saved due to relief operations
Storage and communication	-	-

3.6. THE USE OF GLOBAL DATA SETS: A CHANGE IN PARADIGM

From a first look at the global data set presented, it is clear that the driving force behind them is loss accounting to support policies at the global or regional level, either for private companies (e.g., reinsurers) or international founding institutions (e.g., investments banks). With the information that is currently stored, little can be done for forensic analysis and risk modeling. Causal links are rarely stored, and quantitative information about forcing is often not organized in a manner to be used for the event analysis. In many cases, the geographic information is also vague due to the nature of the sources used to feed such databases. On the other hand, to bend such databases so that they can meet those requirements would be unpractical and hard to sustain in the long run. Many complementary sources of information exist that could serve this purpose once combined with DLDs. In this section, some of the most relevant initiatives and data sets

Table 3.5 Complement Data Sets on Global and Regional Level with Risk and Assessment-related Disaster Information.

DLD database	Copernicus EMS	Relief Web	GDACS
IT platform	http://emergency.copernicus.eu/mapping/#zoom =2&lat=13.5027&lon =−30.915&layers=0B000000T	http://reliefweb.int/disasters	http://www.gdacs.org/alerts/default.aspx?profile=archive
Coverage scope/scale	Global Country/Region	Global Event based	Global Event based
Holder institute	Joint Research Center, Brussels, Belgium	UN OCHA[1]	United Nations guidance EU JRC
Disasters type	Natural disasters, man-made emergency situations, and humanitarian crises	Natural and man-made disasters	Multi-hazard
Flood classification	Flood Flash flood	IRDR classification: Flood Flash flood	Internal
Thresholds and losses	All emergency events	Alert, ongoing, past disasters	Impact assessment service
Uncertainty	Validation protocol developed JRC and users feedback	Not specified	Not specified
Time period	2012 to the present	1981 to the present	From 2002 earthquakes, 2011 cyclones, and 2006 floods to the present
Data sources	Civil Protection Authorities and Humanitarian Aid Agencies	UN Offices, Red Cross And Red Crescent Societies, government organizations, news agencies, national services	ReliefWeb, UNOSAT, EMM, JRC, NASA, DFO
Data sharing	Data not downloadable Mapping products in JPG format	Data downloadable through API	Data downloadable in KML format and through RSS and API

[1]UN Office for the Coordination of Humanitarian Affairs
[2]The National Association of Radio Distress-Signalling and Infocommunications
[3]Asian Disaster Reduction Centre

are presented and discussed as possibly complementing a classical loss database at the regional or global scale.

3.6.1. Databases for Forensics

In this section, guides and databases important for forensics purposes are reported. The data sets produced in the contexts described in this section relate to pre- and post-disaster information and necessary details for analysis of events, such as maps, land use, land cover, cadastre, and other important information for assessment and analysis of metadata. As an example, both the EUFD 2007/60/EC and the Post-Disaster Needs Assessment (PDNA) Methodology provide instruction on disaster assessment and registration procedures. The Copernicus Emergency Services, however, produces a valuable source of ancillary data (e.g., delineation or grading maps) that could be easily linked and included in DLD's for forensic purposes. The reliable information is collected from authorized sources. The general characteristics of data sets are reported in Table 3.5.

RSOE EDIS	Dartmouth Flood Observatory	GLIDE	EEA WISE
http://hisz.rsoe.hu/alertmap/index2.php	http://floodobservatory.colorado.edu/Archives/index.html	http://glidenumber.net/glide/public/search/search.jsp?	http://www.eea.europa.eu/data-and-maps/data/european-past-floods
Global	Global	Global	EU member states
Event based	NUTS2/NUTS3	Event based	
RSOE[2] Budapest, Hungary in cooperation with General-Directorate of National Disaster Management and Crisis Management Centre of the Ministry of Foreign Affairs	University of Colorado, Boulder, CO, US	ADRC[3], Kobe, Japan	DG Environment, Joint Research Centre, and Eurostat and the European Environment Agency
Natural and man-made disasters	Large flood events	Natural and man-made disasters	Floods
Internal: Flood Flash flood Flood warning Sea level rise and waves activity	Internal, only large flood events	IRDR Flood Flash Flood Wave Surge	Floods Directive 2007/60/EC
All events that may cause disaster, emergency or/and human losses	Significant damage to structures or agriculture	All the events that may cause disaster or emergency; human losses, physical losses, economic losses	All flood events with social (number of fatalities), property (including homes), environmental and economic impacts
Uncertainty procedures carried out by data sources organizations	Not specified	Does not provide the statistical accuracy of data	Information assessed and processed by the ETC-ICM and the EEA
2004 to the present	1985 to the present	1936 to the present	1980 to the present
Official information, screening of internet press publications	Data obtained by NASA, JAXA, ESA, and other space agencies, news and governmental sources	CRED, UNSDR, UNDP, JAXA, WMO, Pacific Disaster Center and other, news Statistics, Charts, and Tabular Reports	Reports of EU member states for the EU Flood Directive (FD), relevant national authorities and global databases on natural hazards
Data downloadable thrpugh RSS, API. Access also via Android APP	Data downloadable in HTML tables, Excel, XML, GIS MapInfo, SHP		Data downloadable in MS Access and CSV tabular formats

3.6.1.1. European Union Floods Directive Requirements.

The EUFD 2007/60/EC[29] of the European Parliament and of the council on the assessment and management of flood risks has been in force since 26 November 2007. The directive prescribes EU member states provide an assessment of inland waters as well as all coastal waters across the whole territory of the EU that are at risk of flooding. The information obtained in the framework of the directive should be provided for public access.

The definitions and characteristics of floods are determined by EUFD and provided in Guidance for Reporting under the Floods Directive (Document No. 0[30]). The list comprises Source of Flooding, Mechanism of Flooding, and Characteristics of Flooding. The list of types of consequences of flooding is also defined and includes Human Health (Social), Environment, Cultural Heritage, and Economic.

3.6.1.2. Post-Disaster Needs Assessments.

The PDNA[31] tool represents a coordinated approach of UN agencies and programs, the World Bank, donors, and non-governmental organizations in assistance to governmental authorities during post-disaster damages, losses, and recovery needs. Among principle PDNA activities is the collection of pre-disaster and post-disaster data and information and assessment of the disaster's effects and impacts. The assessment effects regard the community and principal economic sectors, such as housing, agriculture, commerce, manufacturing, infrastructure, transport, tourism, telecommunications, water supply, culture, education, environment, and others. The comprehensive assessment combines qualitative and quantitative information checked by ad hoc assessment teams. *PDNA Guidelines* (2013) developed by the EU in collaboration with the UN Development Group and the World Bank provide a common platform for partnership and coordinated action in post-disaster assessment and recovery planning.

3.6.1.3. Copernicus Emergency Management Service.

Copernicus EMS[32] of European Earth observation program has been operational since April 2012. Copernicus services are designed for disaster risk reduction purposes and preparedness activities during natural disasters, man-made emergency situations, and humanitarian crises. The maps are accompanied by detailed legends and comments on the purpose, core users, and consequences of an event (crisis information, infrastructure, transportation, settlements, and others).

The information derived from satellite remote sensing is completed by available in situ or open data sources. The produced maps are provided for free downloading in GeoTIFF, GeoPDF, GeoJPEG, and vector (Shapefile and KML) formats. (For deeper discussion of the Copernicus EMS see Chapter 14 in this book.)

3.6.2. Complementing Information for Disaster Loss Databases

The data sets of information related to disasters at a European and global level are described in this section. They can be used to complement or to better organize disaster loss databases with important ancillary information. The general characteristics of these data sets are provided in Table 3.5.

3.6.2.1. GLIDE.

The Global IDEntifier[33] or GLIDE number generator system is intended to facilitate linkages among records in various disaster databases. The GLIDE methodology is based on CRED criteria and definitions. For identification of disasters, the Asian Disaster Reduction Center (ADRC) proposed applying a globally common unique identification code to each new disaster event. The GLIDE number is assigned automatically to a disaster event by the Automatic GLIDE Generator system. This alpha-numeric codification consists of two letters, identifying the disaster type; year of the disaster; a six-digit sequential disaster number; and International Organization for Standardization (ISO) code for country, where an event has taken place. Among losses in the GLIDE archives, human losses, physical damages, and economic losses (US dollars) are registered. The GLIDE[34] search engine provides simple tools for data querying. The statistics, charts, and tabular reports are available for users. Because the information comes from different sources (official organizations, partners,[35] and news publications) event descriptions differ by content, details, and quality of information. Usually a general description of an event is provided. The need of a unique event identifier to be attached to the records of a disaster loss database is a key feature to enable event-wise analysis, and therefore, it is considered to be a fundamental parameter for the database. In this case, the use of a global identifier is even more relevant.

[29] *http://www.envir.ee/sites/default/files/flooddirective.pdf*

[30] *http://icm.eionet.europa.eu/schemas/dir200760ec/resources/Floods%20Reporting%20guidance%20final.pdf*

[31] *http://www.recoveryplatform.org/pdna/*

[32] *http://emergency.copernicus.eu/mapping/sites/default/files/files/CopernicusEMS-Service_Overview_Brochure.pdf*

[33] *http://glidenumber.net/glide/public/about.jsp*

[34] *http://glidenumber.net/glide/public/search/search.jsp?*

[35] *http://glidenumber.net/glide/public/institutions.jsp*

3.6.2.2. European Environment Agency Water Information System for Europe. European Environment Agency Water Information System for Europe (EEA WISE) is an official viewer of the EUFD 2007/60/EC[36]. The central access point Water Data Centre[37] provides water-related data via the European past floods data set[38].

The general aim of the EEA WISE is to ensure the consistency of the flow of information with EU water legislation such as the Water Framework Directive 2000/60/EC, the Drinking Water Directive 98/83/EC, the Bathing Water Directive 2006/7/EC, and INSPIRE Directive. EEA WISE information could be easily linked to the databases providing geospatial information to the disaster loss database. The information can be downloaded by users in MS Access and CSV tabular formats.

3.6.2.3. Global Disaster Alert and Coordination System. The Global Disaster Alert and Coordination System[39] (GDACS) is a multi-hazard alert service under UN guidance. The worldwide alert framework provides information exchange and coordination in the first phase after major sudden-onset disasters. The multi-hazard disaster impact assessment service is managed by EU JRC. JRC developed a data sharing platform[40] that captures data on disaster events published by different organizations. The narratives contained in the GDACS database could be valuable information to be included in a national DLD. The data provided for the public is available in KML format, as well as through RSS and API feeds.

3.6.2.4. Radio Distress-Signalling and Infocommunications Emergency and Disaster Information Service. Similarly to GDACS, the National Association of Radio Distress-Signalling and Infocommunications (RSOE) operates Emergency and Disaster Information Service (EDIS)[41] and provides information from all events worldwide that have caused a disaster or an emergency.

The collected data are archived and sent to officials (government, national disaster management, World Health Organization [WHO], European Centre for Disease Prevention and Control [ECDC][42] and other partners) for validation. The RSOE EDIS service products are an alert map (real-time data) and a database on past and ongoing emergencies[43] available in open access through RSS, API, and Android APP services.

3.6.2.5. Relief Web. Relief Web[44] digital service provides descriptions of ongoing and past disasters accompanied by reports, analysis, appeals, situation snapshots, thematic maps, data assessment, and financial data. The information from more than 4,000 global sources, related to emergency management and relief, is collected by Relief Web archive. The loss data contains mainly human losses and physical losses for an event. Locations of ongoing disasters are placed on an interactive global map[45] together with disaster information. A dedicated API distributes data to the public.

3.6.2.6. Dartmouth Flood Observatory. Dartmouth Flood Observatory (DFO) operates the Global Active Archive[46] of Large Flood Events, which contains a data on significant floods (from 1985 to the present). Its methodology[47] defines main cause, severity class, geographical extension, magnitude, and other. To each registration a serial number is assigned, which is a successive digital number. The losses include human losses (dead, displaced) and economical losses (US dollars). The damages on infrastructure and agricultural sectors, including hectares of crops or arable land flooded, or total amount of land flooded are provided in comments. Each flood event record with affected area has GIS files[48] in Mapinfo and Shapefile interchange formats available for download. The information is provided also in HTML, Excel, and XML formats.

3.6.2.7. Remarks on Complementary Information for Disaster Loss Databases. The initiatives presented in this section are driven by diverse purposes but contain some common features that are the ideal complement to enrich and add value to a DLD conceived in a classical way. The most relevant items that should be retained follow:

1. The possibility of describing an event via a unique event identifier (e.g., improved and more flood-focused GLIDE-like number) so that it would be possible to link

---

[36]http://www.eea.europa.eu/themes/water/interactive/floods-directive-viewer

[37]http://water.europa.eu/data-and-themes

[38]http://www.eea.europa.eu/data-and-maps/data/european-past-floods

[39]http://portal.gdacs.org/about

[40]http://www.gdacs.org/alerts/default.aspx?profile=archive

[41]http://hisz.rsoe.hu/alertmap/index2.php#

[42]ecdc.europa.eu/

[43]http://hisz.rsoe.hu/alertmap/search/index.php

[44]http://reliefweb.int/about

[45]http://reliefweb.int/disasters

[46]http://floodobservatory.colorado.edu/Archives/index.html

[47]http://floodobservatory.colorado.edu/Archives/ArchiveNotes.html

[48]http://floodobservatory.colorado.edu/Archives/MasterListrev.htm

information gathered in different databases. This important point has been taken firmly on board by the Global Flood Partnership[49] (GFP) [*De Groeve et al.,* 2015] that involved World Meteorological Organization (WMO) and other interested institutions. At present the Global Flood Observatory (GFO) uses both the GLIDE number (if created) and the DFO serial number, which is a continuous numbering in the DFO database. The other databases described in this chapter practice the use of internal serial numbering, and not coded information about the event such as a GLIDE number.

2. The delineation of the flood extent and water depth, mainly with the support of different EO data sources, if possible in a multi-temporal fashion so that information about the event dynamics can be inferred and data can be more effectively linked to detailed studies by means of physically based models [*Pulvirenti et al.,* 2014].

3. A textual description of the event development where information about the triggering factors and boundary conditions can be derived by experts for a forensic reconstruction of single events.

4. Information about reference scenarios (characterized in terms of frequency) so that it would be possible to benchmark the different events with respect to such reference scenarios.

5. The DLD database in formats compatible with a standard GIS software could improve visualization and spatial analysis performed through databases. For example, SHELDUS and DFO databases support Shapefile and MapInfo GIS formats. The DesInventar database provides an option of using overlay on a thematic map with Google or VirtualEarth map. Thus, information obtained from spatial analysis in DesInventar could be visualized on geospatial mapping platforms. These convenient tools facilitate the process of data visualization and allow perform spatial analysis through a DLD database.

3.7. CONCLUSIONS: TOWARD A COMPREHENSIVE GLOBAL DATA SET

The analysis presented in this section highlights clearly how valuable information is currently collected and made available by many institutions and initiatives. It is also clear that a tremendous added value would be created by combining these different sources of information across scales on one side and connecting complementary information on the other.

Two main points can be identified in a possible roadmap leading to what we can call a "comprehensive global dataset."

First, there is the need for even stronger initiatives of standardization to enhance the interoperability of the different data sets both at the same scale and in the information aggregation at different scales. In this sense, the JRC guidelines for EU level reporting including minimum standards, the CIS country methodology, or the DesInventar initiative all represent decisive steps forward in this direction. Inter-comparison and further alignment of such initiatives in order to reach an international standard could allow in future a more effective data information exchange and interpretation.

Second, there is the need to link, if not integrate, more and more information related to how the event was triggered, how it developed and eventually impacted the territory to create the recorded damage or loss so that an effective forensic analysis or a comprehensive risk modeling can be really supported.

In this sense, the Global Flood Record initiative within the Global Flood Partnership has the aim to support this idea within the scientific and the practitioner community, to test its feasibility with pilot initiatives, and eventually to create a suitable environment for its development in practice.

Together, with damage or loss data, the information on saved lives, assets, and properties during rescue operations would be relevant. The CIS states databases, as was mentioned in section 3.2, are unique databases recording this type of information. The data on saved lives and assets would be useful for interpretation and evaluation of rescuers' actions during mitigation of disaster effects, which is important for forensic analysis.

ACKNOWLEDGMENTS

The authors thank reviewers for their insightful comments during the process of preparing this chapter. The authors also acknowledge UNISDR and PPRD East2 project's contributions to the publication.

REFERENCES

De Groeve, T., K. Poljansek, and D. Ehrlich (2013), Recording Disaster Losses: Recommendations for a European approach, Luxembourg: Publications Office of the European Union, ISSN 1831-9424 (online) doi:10.2788/98653.

De Groeve, T., K. Poljansek and D. Ehrlich, C. Corbane (2014), Current Status and Best Practices for Disaster Loss Data Recording in EU Member States: a comprehensive overview of current practice in EU Member States, Luxembourg: Publications Office of the European Union, ISSN 1831-9424 doi: 10.2788/18330.

De Groeve, T., J. Thielen-del Pozo, R. Brakenridge, R. Adler, L. Alfieri, D. Kull, F. Lindsay, O. Imperiali, F. Pappenberger, R. Rudari, P. Salamon, N. Villars, and K. Wyjad (2015) Joining Forces in a Global Flood Partnership.

---

[49] *http://portal.gdacs.org/Global-Flood-Partnership*

Bull. Amer. Meteor. Soc., *96*, ES97–ES100. doi: http://dx.doi.org/10.1175/BAMS-D-14-00147.1.

EU Expert Working Group on Disaster Damage and Loss Data (2015), Guidance for recording and sharing disaster damage and loss data–Towards the development of operational indicators to translate the Sendai Framework into action, Luxembourg: Publications Office of the European Union, ISSN 1831-9424, doi: 10.2788/186107.

Joint Research Centre of the European Commission (2015), Implementation of Guidance on sharing loss data among EU Countries and coherence with the Sendai framework for disaster risk reduction, Sixth technical workshop on an EU approach for recording damage and loss data, Ispra, Italy.

Pulvirenti, L., N. Pierdicca, G. Boni, M. Fiorini, and R. Rudari (2014), Flood Damage Assessment through multi-temporal COSMO-SkyMed data and Hydrodynamic Models: the Albania 2010 Case Study. *IEEE Journal of Selected Topics in Applied Earth Observations and Remote Sensing*, Vol. 7m No. 7, pp. 2848–2855, DOI: 10.1109/JSTARS.2014.2328012.

Pottering, H.-G, and M. Lobo Antunes (2007), Directive 2007/60/of the European Parliament and of the Council of 23 October 2007 on the assessment and management of flood risks, Official Journal of the European Union, L 288/27–34.

Post-Disaster Needs Assessment Guidelines, Volume A (2013), European Union, World Bank, UN Development Group.

Resolution of the Council of Ministers of Foreign Affairs of the Commonwealth of Independent States on the activities of the Interstate Council for Emergency Situations of the Natural and Man-made (2007), 25 April, Astana.

Serje, J. (2015), Proposed Guideline of national disaster loss database, Document 3 of Sendai Framework for DRR, UNISDR Global Assessment Report team.

UNISDR (2015), Considerations on Developing a System of Indicators Based on the Sendai Framework for Disaster Risk Reduction 2015-2030: A proposal for monitoring progress, Draft to support the process for discussing on indicators, follow up and review process, Geneva.

4

Direct and Insured Flood Damage in the United States

Melanie Gall

ABSTRACT

A wealth of hydrological data, historic as well as real-time, are available in the United States. The focus of the information is largely on observational data (e.g., water levels and flows, water quality, meteorological conditions, etc.) and flood risk communication (e.g., real-time streamflow visualizations, 100-year floodplain maps). In contrast, there is a paucity of information on societal and economic impacts of floods. The National Flood Insurance Program, which is a federal flood insurance program, provides some understanding regarding insured flood losses but insurance penetration is low even in high-risk areas. This results in significant uninsured damage particularly in high risk areas along the Atlantic and Gulf coasts and in watersheds with the potential for catastrophic flood impacts such as the Mississippi River and its tributaries.

An alternative or supplementary metric to insured flood losses are direct flood losses as compiled by the National Centers for Environmental Information. Hidden to most users of direct flood loss data though is the fact that these figures are largely educated guessing. Aside from tornado damage, there is no systematic surveillance of post-event damage by the National Weather Service or other relevant agencies such as the Federal Emergency Management Agency or the U.S. Geological Survey. Thus, loss accounting in the United States is fraught with uncertainties. Several issues coalesce and contribute to high uncertainties in flood loss data: reliance on third party information, tangential importance to the mission of federal agencies, lack of automated estimation workflows, and gaps in the observational network. To facilitate current processes and improve the accuracy of loss estimates, best estimation practices need to be developed, shared, and trained to aid non-NWS sources to improve their estimation accuracy. This best practice should also include the communication of estimation uncertainty/credibility.

4.1. INTRODUCTION

Floods persistently occur in the United States (US). Along with hurricanes and severe weather, floods are the leading cause of socioeconomic impacts from natural hazards [*Hazards and Vulnerability Research Institute (HVRI)*, 2015]. Floods have killed 3,345 people, caused about $285 billion in direct damage and $51 billion in federally insured losses since 1978 [*Federal Emergency Management Agency (FEMA)*, 2015a; *National Weather*

Service (NWS), 2015a]. The most devastating impacts from floods originate from hurricanes and their associated storm surge (e.g., Hurricane Katrina in 2005, Superstorm Sandy in 2012) as well as from riverine flooding mostly in the large watersheds of the Mississippi, Ohio, and Tennessee rivers (e.g., 1913, 1993, 2011, etc.).

Much of the US flood experience and risk management policies can be explained in the context of select historic floods, especially in these watersheds. From the first flood control laws (Swamp Land Acts of 1849 and 1859), the establishment of the US House Committee on Flood Control in 1916, and the passing of the initial Flood Control Act in 1917, to the creation of the Federal Flood Insurance

Department of Geography, University of South Carolina, Columbia, South Carolina, USA

Flood Damage Survey and Assessment: New Insights from Research and Practice, Geophysical Monograph 228, First Edition. Edited by Daniela Molinari, Scira Menoni, and Francesco Ballio.

Program in 1978 and the 2000 Disaster Hazard Mitigation Act, the involvement of the US federal government in flood risk management is extensive [*Arnold*, 1988]. This involvement expresses itself also in the amount of tax dollars spent on post-flood recovery and flood control structures. Since 1936, the US Army Corps of Engineers (USACE) has emplaced more than 8,500 miles of levees/dikes and around 400 reservoirs. As of 2009, the USACE estimates the cumulative costs of these construction and maintenance projects at more than $120 billion and claims that "for every dollar spent, approximately six dollars in potential damage have been saved" [*USACE*, 2009]. In addition, FEMA has spent more than $2.8 billion on flood risk mapping since 2003, of which less than a third was covered by funds raised through the National Flood Insurance Program (NFIP) [*Government Accounting Office (GAO)*, 2010, 2015a]. As a result of catastrophic flood events related to Hurricane Katrina in 2005 and Hurricane Sandy in 2012, the NFIP itself owes the US taxpayer about $23 billion [*GAO*, 2015b]. Given the insurance program's inability to cover large-scale events, legislators attempted to reform the program in 2012 and 2014 with limited success in achieving actuarial soundness. Now, the NFIP has evolved from a risk management measure to a fiscal liability for the US government, second only to

Social Security. As of October 2015, the NFIP holds more than 5 million policies for a value of more than $1.2 trillion in assets [*FEMA*, 2015b, 2015c] with half of these assets located in coastal floodplains [*National Oceanic and Atmospheric Administration*, 2011].

The financial and legislative resources and infrastructure devoted to flood risk management do not extend to the monitoring of the impacts of floods on society. Although the rising trend of flood losses is clear, the exact amount and degree of increase vary by data source (Figure 4.1). While knowledge on flood losses is better compared to, for instance, landslides or drought, loss estimation is still crude and incomplete [*Gall et al.*, 2009; *Smith and Katz*, 2013]. Systematic data collection of direct flood losses goes back to 1960 but is fraught with uncertainties. With flood insurance being a federal program, there is some knowledge on insured losses. However, data on uninsured or indirect losses are entirely absent or only available for a few case studies [*Hallegatte*, 2008; *Smith and Katz*, 2013]. Overall, the United States have a very patchy and somewhat limited understanding of the societal impacts of floods.

The following sections of this chapter discuss in more detail the sources of loss data, data uncertainties, and what is known about flood losses. The chapter concludes

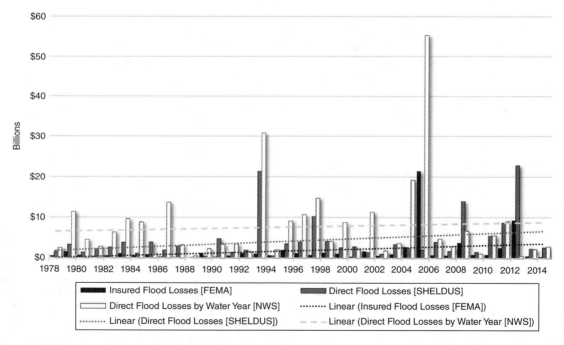

Figure 4.1 Direct and insured flood losses from three different data sources: (1) insured flood losses/claims paid through the National Flood Insurance Program as reported by FEMA, (2) direct flood losses as reported by the National Weather Service (NWS), and (3) direct flood losses as reported by the Spatial Hazard Events and Losses Database for the United States (SHELDUS). Please note that NWS losses are reported by water year (October through September) and not by calendar year as done by FEMA and SHELDUS. In addition, SHELDUS distinguishes between tropical and non-tropical flooding. Here, only non-tropical flooding is reported. As a result, storm surge damage is excluded, which explains the low loss estimates for 2005 (Hurricane Katrina). All trend lines show an upward trend with direct losses according to the NWS being highest, followed by direct losses as reported by SHELDUS and then insured losses.

with a discussion of future data needs and the implications of a limited understanding of flood losses for flood risk reduction activities.

4.2. FLOOD LOSS PATTERNS IN THE UNITED STATES

Gilbert F. White famously wrote in his dissertation, "floods are 'acts of God' but flood losses are largely acts of man" [*White*, 1945]. Loss statistics confirm this statement. Most of the flood losses in the United States occur in states with the potential for catastrophic riverine or coastal flooding and with high numbers of people and assets in flood zones.

In the past, states like Texas and Florida led in NFIP's loss statistics though Hurricane Katrina in 2005 and Superstorm Sandy in 2012 changed this. The State of Louisiana currently leads the nation in flood insurance payouts with nearly $17 billion [*FEMA*, 2015b]. These losses were to a large extent precipitated by Hurricane Katrina though other storms such as Hurricane Rita and Hurricane Gustav added to the tally. Second in payouts is Texas with $6.1 billion in insurance payments. Here, too, the majority of losses were triggered by storm surge (e.g., Hurricane Ike in 2011). Third and fourth in the list are the states of New Jersey with $5.7 billion in paid claims along with New York, which received payouts of $5.2 billion. The coastal damage from Superstorm Sandy's storm surge in 2012 catapulted New Jersey and New York into leading positions in the list of flood claim recipients surpassing the State of Florida. In fact, Florida used to rank in the top three due to multiple catastrophic hurricanes in the 1990s through the mid-2000s. In 2004, Florida was crisscrossed by four hurricanes (one category 2 hurricane, two category 3 hurricanes, and one category 4 hurricane) [*Blake et al.*, 2011]. Still, the damage from Superstorm Sandy was catastrophic enough to surpass the flood insurance payouts received by Floridians up to that point.

It is tropical flooding, i.e., storm surge from hurricanes, not riverine or flash flood events that are literally driving the NFIP into ruin. In the United States, wind damage from hurricanes is covered by separate wind policies. Storm surge damage and all other flood damage are covered under the NFIP. Despite NFIP payouts being capped, the sheer number of homes affected by storm surge is extensive. Coastal homes are frequently a total loss after being inundated and structurally damaged by storm surge. Although riverine and flash floods are certainly capable of sweeping a home off its foundation, the area of high flow velocity and structurally damaging impacts is much smaller than the impact area of storm surge. In addition, coastlines are built-up compared to more rural settings resulting in high exposure and higher damage figures. It is therefore not surprising that Superstorm Sandy, which affected one of the most densely (and wealthy) areas in the country ranks as the third costliest year in US History after Hurricane Katrina (2005) and Hurricane Andrew (1992), which was a category 5 hurricane that made landfall south of Miami.

Insured losses tell only part of the flood loss picture. Figure 4.2 visualizes direct non-tropical flood damage, i.e., property and crop damage caused by riverine, flash floods, coastal flooding, etc. Although the majority of losses in the top-ranking NFIP states are clustered along the coast, the impact from riverine and flash flooding is more dispersed throughout the country, largely along major watersheds and/or urban centers. Interestingly, Figure 4.2 features Ocean and Monmouth counties in New Jersey as having the highest non-tropical flood losses with more than $10 billion each. This is due to the fact that the catastrophic impacts from Superstorm Sandy are categorized as coastal flooding and not storm surge according to SHELDUS's underlying data source, the NWS *Storm Data* database since Sandy was a post-tropical cyclone at landfall [*Blake et al.*, 2013].

Counties with the third and fourth highest direct non-tropical flood damage are Linn County in Iowa and Grand Forks in North Dakota. Cedar Rapids, county seat of Linn County, is the second most populous city in Iowa and has experienced numerous floods along the Cedar River, at a tributary of the Mississippi River, mostly recently in 2013 and 2008. Grand Forks, North Dakota experienced catastrophic flooding in 1997 caused by the Red River, which drains into Lake Winnipeg.

When adjusting direct loss estimates not only for inflation but also for population, i.e., the number of people living in a county at the time of the event (resulting in per capita as the unit of measurement), the impacts of non-tropical floods become even more pronounced (Figure 4.3). Tunica County, Mississippi experienced more than $103,098 in per capita flood losses since 1978 compared to Ocean, New Jersey with only $18,268 in direct flood losses per person. Tunica County, with a population of 10,778 (2010 Census), experienced catastrophic flooding in 2011 when the Mississippi River reached historic streamflow levels that triggered the concurrent operation of the Birds Point–New Madrid, Morganza, and Bonnet Carré floodways to avoid downstream flooding of Baton Rouge and New Orleans in Louisiana [*NWS*, 2011]. Other counties with significant amounts of flood losses per capita are clustered in the state of Missouri due to their proximity to the Missouri River and Mississippi River.

4.3. SOURCES OF FLOOD INFORMATION IN THE UNITED STATES

In contrast to the paucity of loss information, there is a wealth of hydrological data available in the United States. The focus of the information is largely on observational data (e.g., water levels and flows, water quality, meteorological conditions, etc.) and flood risk communication (e.g., real-time streamflow visualizations, 100-year floodplain maps). Operational data originate from

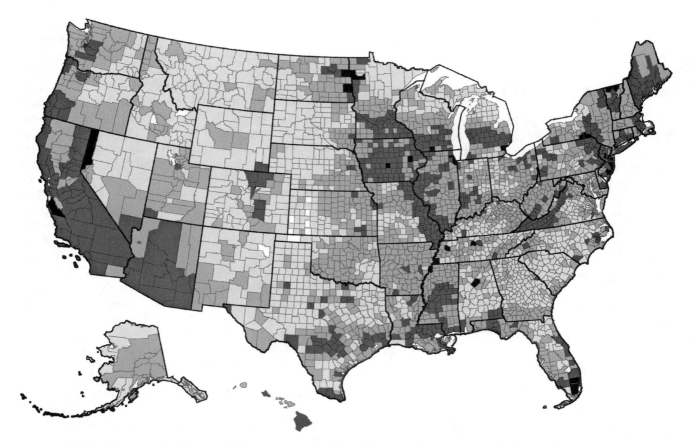

Figure 4.2 Spatial distribution of direct, non-tropical flood losses at the county level. The four loss categories range from light ($551 to $5 million), medium ($5 million to $50 million), and dark gray ($50 million to $500 million) to black ($500 million to $10 billion). Direct losses include property and crop losses. Losses are adjusted to 2014 dollars. Period of record is from 1960 to 2014. Data source: SHELDUS Version 14.

the U.S. Geological Survey (USGS), which operates the National Streamflow Information Program. The program consists of a nationwide stream gauge network of more than 7,500 stations (Figure 4.4) and is funded collaboratively. About half of the program funds come from federal agencies such as the USGS, USACE, and the Bureau of Reclamation. State and local entities provide the other half of program funding [*USGS*, 2010]. Streamflow information is critical in regard to flood forecasting, design of bridges and culverts, water resource management, reservoir operations, and more. Depending on the implemented technology, a stream gauge records streamflow in 15- to 60-minute intervals and transmits the records every one to four hours to a USGS office using mostly satellite transmission [*USGS*, 2015]. Historical streamflow data are particularly important for the calculation of flood recurrence intervals as part of the 100-year floodplain delineation. In addition to its network of stationary stream gauges, the USGS also operates rapid deployment gauges, self-contained, temporary stations, which are generally deployed on

structures such as a bridge. These mobile stations are extremely valuable along ungauged streams and/or during flood emergencies.

Real-time streamflow data feed into many decision support and visualization tools such as the USGS Flood Inundation Mapper and most importantly into NWS river forecasts and warnings. The NWS issues three types of flood alerts, advisories, watches, and warnings, depending on the probability and severity of the flood threat. Through its local forecast offices, the NWS collects hazardous weather information including floods and relays the information to the National Centers for Environmental Information (NCEI), which was formerly known as the National Climatic Data Center.

The NCEI is responsible for the archival of past weather information (wind speed, hail size, etc.) and maintains the so-called *Storm Data* database that consistently documents the date (day, month, year), location (state, county, and latitude/longitude), magnitude, and type of event. It is the nation's only original source of direct loss information from climatological, meteorologi-

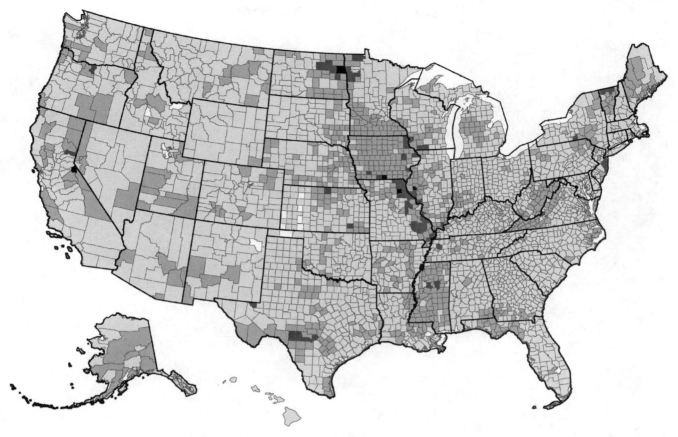

Figure 4.3 Spatial distribution of direct, non-tropical flood losses per capita at the county level. The four loss categories range from light (up to $1,000 per capita), medium ($1,000 to $10,000 per capita), and dark gray ($10,000 to $50,000 per capita) to black ($50,000 to $100,000 per capita). Losses are adjusted to 2014 dollars, and the county population counts at the time of the flood event. Period of record is from 1960 to 2014. Data source: SHELDUS Version 14.

cal, and hydrological hazards. The database distinguishes among 48 different types of natural hazards including seven different flood types: coastal flood, flash flood, flood, lakeshore flood, seiche, storm surge/tide, and tsunami. Riverine, glacial lake outburst, and ice-jam floods are all subsumed under the general "floods" category [*NWS*, 2007]. The database is available online and can be accessed at https://www.ncdc.noaa.gov/stormevents/. The period of record varies depending on hazard type. Flood loss reporting starts in 1960 although online records are only available from 1996 onward. Data prior to 1996 are accessible through either the hard copy reports of *Storm Data* or via a bulk data download (comma-separated file format) containing all hazards and events for the entire period of record (1955 through present). *Storm Data* reports four different types of direct losses: number of fatalities, injuries, dollar amounts for property, and/or crop losses.

A second source of original flood loss information, though focused exclusively on insured flood losses, is FEMA's Federal Insurance and Mitigation Administration (FIMA), which manages the NFIP. Floods are the only natural hazard in the United States insured by the federal government. All other hazards are either covered by the private insurance sector through homeowners' policies (e.g., hail, tornado, wildfire) or supplementary policies and/or riders (e.g., wind/hurricane, earthquake), or are uninsurable (e.g., landslides). The NFIP insures mostly residential homes though commercial policies are available. The amount of coverage the NFIP provides is capped at $250,000 for residential ($100,000 for personal property) and $500,000 for commercial policies ($500,000 for personal property). Flood insurance is only available to homeowners of communities that participate voluntarily in the program. The purchase of flood insurance is furthermore voluntary for anyone outside the Special Flood Hazard Zone, i.e., the 100-year floodplain. Any homeowner inside the 100-year floodplain who holds a federally back mortgage through lenders such as the Federal National Mortgage Association (e.g., Fannie Mae) or the

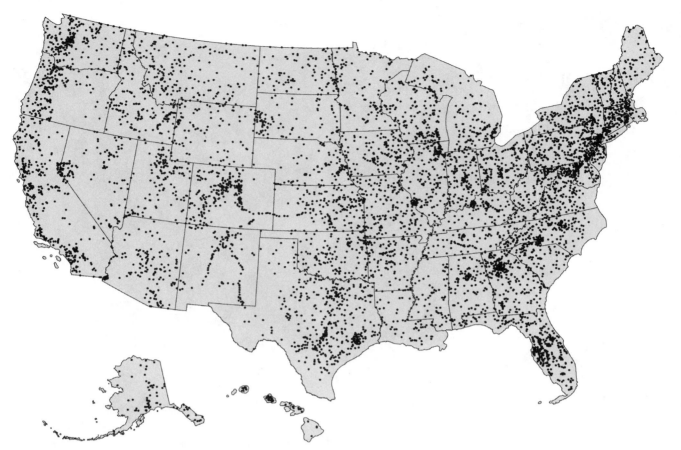

Figure 4.4 Map of the U.S. National Streamflow Information Program and its about 7,500 stream gauges.

Federal Home Loan Mortgage Corporation (e.g., Freddie Mac), however, is obliged to purchase flood insurance. Insurance premiums vary significantly between flood risk zones. Broadly speaking, the NFIP distinguishes three flood zones in the 100-year floodplain: a) coastal zones subject to storm surge and wave height of 3 feet or more (V-zones), b) coastal zones subject to storm surge and wave height between 1.5 and 3 feet (coastal A-zones), and c) flood zones subject to rising waters and waves less than 1.5 feet in height (A-zones). Additionally, the NFIP delineates the 500-year floodplain where residents have a 0.2% annual chance of flooding. Any areas with a lower annual flood chance are designated as X-zones.

The NFIP offers insurance statistics online [*FEMA*, 2015d]. Publicly accessible data focus exclusively on claim and policy information by state, calendar year, fiscal year (October through September), and major flood events. The period of record starts in 1978. For privacy protection reasons, the information is highly aggregated by year, flood event, or location (local county or community) and does not compare to the level of temporal or spatial accuracy provided by NCEI's *Storm Data* and other data sources.

A third source of primary flood loss data is the commercial flood insurance market and industry-affiliated research entities such as the Insurance Information Institute, Property Claim Services, or catastrophic risk modelers. However, in the commercial arena of flood information, data on flood insurance claims and payouts are confidential business information. As a result, loss data, that is, primary, modeled, or value-added data, are only shared with paying members, clients, or researchers.

In addition to these three sources of primary data on flood losses, there are providers of modeled or value-added flood loss data. An originally free data provider of value-added primary flood losses, the web-based SHELDUS maintained by the Hazards and Vulnerability Research Institute at the University of South Carolina, now charges a fee for access to flood loss data. Although SHELDUS relies heavily on data generated by NCEI, there are significant differences to *Storm Data* such as the period of record (1960 to present), geo-referencing of losses, and a systematic way of handling events involving multiple hazard types (e.g., floods and thunderstorms). The differences to NCEI and uncertainties inherent in flood loss data are discussed in more detail in the next section.

Whenever "observational" loss estimates are unavailable, loss modeling is a possible solution to fill in the gaps. Modeling flood losses is not as challenging in the United States as in many other countries, thanks to the availability of FEMA's Hazards U.S. Multi-Hazard (HAZUS) software. The software is free and easily integrates with the Environmental Systems Research Institute's (Esri) ArcGIS allowing users familiar with geographic information systems (GIS) to model flood losses. HAZUS entails a generic building stock, demographic data, and damage functions that the user can either retain or modify to achieve loss models with less uncertainty. Digital elevation data, crucial to the functionality of HAZUS, are also freely available (USGS's 3D Elevation Program) and seamlessly integrate into HAZUS. Modeling flood losses is a valuable alternative for case study analyses. Practically though, it is not feasible to model flood losses for all flooding events that occur nationwide given the computational resource intensiveness of HAZUS and the inability of the software to discern losses at scales below the census block. As a result, HAZUS is largely used for planning purposes or in case study settings and not for rapid loss assessments of historic flood events. It is important to note that HAZUS may be well suited for county-level loss estimation but not for more accurate loss estimation or flood route modeling since it trades ease of use and ready-made data sets for accuracy and user-provided information [Gall et al., 2007; Banks et al., 2015].

Although there are several sources of flood losses, the quality and reliability of the information is challenging. Existing data are incomplete, both spatially and temporally, and contain multiple layers of uncertainty introduced during data collection as well as processing [Gall et al., 2009, 2011]. This becomes obvious when comparing data sets from different providers who essentially utilize the same underlying data. Examples of such discrepancies are highlighted in the following section using NCEI's Storm Data and SHELDUS as examples.

4.4. UNCERTAINTIES IN US FLOOD LOSS ACCOUNTING

There are numerous limitations to the flood loss statistics provided in the previous sections. Some were noted such as the effect of the differential treatment of tropical and non-tropical flood losses. Others have been outlined more systematically elsewhere (see Chapter 2). The following discussion will therefore focus on select problems unique to the US context and its approach to flood loss data collection and processing.

At first glance, direct and insured flood loss information in the United States appear to be reliable and collected comprehensively due to its provision by trusted government agencies such as NCEI and FEMA. It is true that insured flood loss data are reliable and accurate because the data are distinctly tied to insurance payouts for damage to buildings and content. However, the reliability and accuracy of direct flood losses is uncertain since loss figures are largely estimates and not observational assessments.

Gall et al. [2009] identified six sources of uncertainty in loss estimates from natural hazards. These uncertainties are introduced by over-reporting or underreporting of certain hazard types (hazard bias), changes in estimation methodologies over time (temporal bias), inclusion or exclusion of losses based on thresholds (threshold bias), inclusion or exclusion of loss types (accounting bias), over-reporting or underreporting of losses depending on "ruralness" and remoteness (geography bias), and choices in data processing (systemic bias). In the case of flood losses in the NCEI, the hazard, threshold, and systematic bias are of particular significance because these biases are at the root of diverging loss figures between NCEI and SHELDUS, although they use the same underlying data set.

Hidden to most users of direct flood loss data is the fact that loss accounting is largely educated guessing. Aside from tornado damage, there is no systematic surveillance of post-event damage by the NWS. The NWS generates flood loss data by either estimating the damage following standard procedures [NWS, 2007] or by utilizing third-party estimates. Due to resource constraints, these estimates are not verified. It is upon the Storm Data preparer to use best judgment regarding the credibility of external estimates. These external estimates originate from sources such as first responders (law enforcement, fire, and rescue personnel), emergency managers, amateur radio users, broadcast media, newspapers, and other federal and local agencies (e.g., Department of Highways, Park Service, and Forest Service), as well as the public. In addition, the reporting of losses is not mandatory and depends on the availability of "credible" estimates, unlike the physical, temporal, and spatial parameters of a flood event, which are available for every instance in Storm Data. Consequently, not all events in Storm Data come with loss information. And not all events where losses are reported provide data across all four loss categories. Missing or no loss information is commonplace.

In Storm Data, loss estimates of floods are more readily available and have a higher degree of reporting than any other hazard type in the database. This is not the result of better sources or more easily identifiable damage but is the product of procedural guidance. In fact, the "U.S. Army Corps of Engineers requires the NWS to provide monetary damage amounts (property and/or crop) resulting from any flood event" [NWS, 2007]. Such a mandate does not exist for other weather phenomena. Although a Storm Data preparer may elect to provide no information in the absence of a credible estimate for any non-flood event, he/she must enter a monetary damage amount for floods. As a result, flood events are more frequently reported with loss estimates than any other hazard in *Storm Data*. This USACE directive skews the share of flood losses within the database and artificially raises the relative significance of floods versus hurricanes, severe weather, and so forth.

This distortion of loss estimates is amplified by the use of categorical loss values rather than exact dollar amounts for events prior to 1995, when logarithmic loss categories were in use. For example, flood damage worth $100,000 would be entered as a categorical value of '5' (i.e., loss between $50,000 and $500,000) instead of the exact dollar amount. After 1995, categorical values were abandoned and exact estimates provided. These categorical estimates are a key factor in differences between flood loss figures derived from *Storm Data* vs. SHELDUS (Figure 4.1). The NWS *Storm Data* shown in Figure 4.1 utilize the mid-point for pre-1995 categorical values (e.g., $250,000) whereas SHELDUS reports the lower bound of the loss category (e.g., $50,000). SHELDUS utilizes the lower bound of loss categories to reduce epistemic uncertainty and provide conservative estimates where actual losses are likely to be higher but not lower. The outcome of these two procedural choices leads to higher flood loss estimates by *Storm Data* although the underlying data are identical; it is simply a product of data processing.

As mentioned previously, there are semantic differences between these data sets in terms of how losses are classified. The NWS reports flood-related losses of around $55 billion for 2005 (Hurricane Katrina) whereas SHELDUS estimates stay well below $5 billion (Figure 4.1). How is this possible when the underlying data source is identical? The reason lies in the classification of storm surge losses. SHELDUS attributes storm surge losses to hurricanes and not to floods. It hereby explicitly separates tropical from non-tropical flooding and avoids any double counting in hurricane loss reporting. In 2012 on the other hand, storm surge from Superstorm Sandy was deemed "coastal flooding" by the NWS, and therefore, the numbers are shown as flood losses in SHELDUS. The losses reported by the NWS only appear lower than the losses from SHELDUS due to the reporting by water years. Superstorm Sandy happened at the end of October and as such, its losses are part of water year 2014.

Other aspects of data processing that lead to different outcomes relate to issue of multi-hazard and/or multi-county events. Depending on the magnitude of the flood, impacts may be felt across several communities, counties, and states. It is therefore often difficult to estimate exact flood losses by location (e.g., county, forecast zone). Particularly for events prior to 1995, *Storm Data* esti-

mated lump sums rather than providing individual estimates for each county. It also reports these lump sums as many times as there are counties involved, essentially resulting in a double counting when querying the database on a county-by-county level. SHELDUS, on the other hand, distributes lump estimates evenly across the affected counties to avoid any double counting and to generate county-specific loss profiles.

The issue of multi-hazard events is related. Floods are generally speaking a secondary hazard caused by prolonged or excessive rainfall, snow, or ice melt, hurricanes, high tides, dam failures, etc. Because NCEI focuses on weather phenomena rather than losses, its event reports list all primary and secondary hazards. This means that instead of just reporting the flood-related damage, all related hazards are identified along with a lump estimate making it difficult for the end user to connect damage to a particular hazard. SHELDUS, again, evenly distributes the lump estimate to all hazard types involved. This essentially reduces the amount of damage assigned to flooding although the damage was most likely triggered by flooding.

Even though both NCEI and SHELDUS ingest the "same" loss information, the outputs they produce differ significantly. To overcome these discrepancies and improve the flood loss accounting in general, significant changes and improvements should be implemented.

4.5. FUTURE DATA NEEDS

The best flood loss data have complete coverage, both temporally and spatially, and are easily accessible and of high quality. The issue of data quality was touched on in the preceding section. The following section focuses on data availability. Several issues coalesce and contribute to paucity of flood loss data: reliance on third party information, tangential importance to the mission of federal agencies, lack of automated estimation workflows, and gaps in the observational network.

4.5.1. Reduce Reliance on Third Party Information

At present, damage estimates from external sources represents the "best" available information. In a reanalysis of loss estimates from 1926 through 2003, *Downton and Pielke* [2005] found that "NWS flood damage estimates do not present an accurate accounting of actual costs, nor do they include all of the losses that might be attributable to flooding." The researchers concluded that estimates for catastrophic floods tend to be less flawed but inaccuracies arise from a) underestimation and b) omission of moderate and smaller floods. Since the NWS continues to rely on external sources since Pielke's reevaluation, it is fair to assume that these issues still persist. In fact, the procedural guidance of NCEI's *Storm Data* is unequivocal: "Because of time and resource constraints, information from these [outside] sources may be unverified by the NWS" [*NWS*, 2007]. Clearly, the NWS is capable and well positioned to generate estimates through field observations since the organization consistently conducts post-event damage surveys for all tornados. Unfortunately, the agency has neither the resources nor the mission to engage in such activities, or even to verify external guesstimates.

To facilitate the current process and improve the accuracy of loss estimates from external sources, best estimation practices need to be developed, shared, and trained to aid non-NWS sources to improve their estimation accuracy. This best practice should also include the communication of estimation uncertainty/credibility. Data end users do not have the experience or background to judge which sources produce more trustworthy estimates than others, nor are they often aware of the original source. By communicating uncertainties, end users have the chance to screen records and omit records from their analysis if the uncertainty is too high.

4.5.2. Elevate Importance of Loss Accounting

There is no single federal agency with loss accounting as part of their core responsibilities. This is surprising given that numerous agencies consider loss reduction part of their mission. Even for NCEI, archiving loss information is only peripheral when its key concern is the archival of climate and weather data. For example, the mission of the NWS is to "provide weather, water, and climate data, forecasts and warnings for the protection of life and property and enhancement of the national economy" [*NWS*, 2015a]. FEMA declares "to support our citizens and first responders to ensure that as a nation we work together to build, sustain, and improve our capability to prepare for, protect against, respond to, recover from and mitigate all hazards" [*FEMA*, 2014]. Though until the revision of its mission statement in 2011 [*Serino*, 2011], FEMA aimed to "reduce the loss of life and property and protect the Nation from all hazards..." [*FEMA*, 2008]. And the USGS states to "serves the Nation by providing reliable scientific information to describe the Earth; (and) minimize loss of life and property from natural disasters..." [*USGS*, 2014].

Arguably all three agencies are concerned with the reduction of losses from floods and other natural hazards, though none is explicitly reliable for assessing losses beyond what the NWS is already doing. Clearly, estimating losses through field surveys or verifying third-party estimates requires resources: time, people, and money, which the NWS does not have. Without a commitment of extra resources, no agency will or can improve the quality of loss estimates.

4.5.3. Automated Estimation Workflows

If more resources cannot be committed for field observations and/or loss verification, then alternative strategies must be developed. Such alternatives are modeling or statistical estimation of flood losses as well as the use of volunteered information.

FEMA's HAZUS software offers a starting point for modeling the impact from medium and large floods. Modeling losses using HAZUS could be achieved by making it a distributed task involving various stakeholders (e.g., regional NWS forecast offices, universities, state emergency management agencies, etc.). However, HAZUS is certainly not suited to model losses for small-scale events due to its data and methodological constraints. Given the availability of a fairly long period of record for flood losses (more than 50 years), estimating losses using statistical approaches (e.g., Monte Carlo simulations) may be a more promising route [*Smith and Matthews*, 2015]. Statistical estimation is certainly a less resource-intensive method compared to HAZUS and one that could be implemented at the operational level. This would allow for rapid, near real-time loss estimation. As observational data become available, statistical estimates could then be replaced in the database of record.

4.5.4. Expand Observational Network

As mentioned previously, loss estimates originate either from the NWS (e.g., post-tornado surveys) or represent third party data. Reports from the general public via social media outlets, however, are not yet considered credible information. In the 21st century, where real-time data are abundant and people act as sensors [*Goodchild and Glennon*, 2010], data volunteered by the public could and should be incorporated. By expanding the network of storm spotters or similar, the NWS could develop a network of verified (and trained) volunteers using a systematic template [*Mileti*, 1999] to report loss information. Studies show that social media information is mostly a reliable predictor though there are avenues of abuse and misinformation [*Gil de Zúñiga et al.*, 2012]. A plethora of third party estimates would essentially corroborate and distill the correct estimate.

Flood loss estimation also suffers from the limited network of stream gauges across the US where not every creek, tributary, or river is gauged. If there is no observation of a flood via a stream gauge or similar, and no reported loss, then no flood occurred. This status quo leads to an underreporting and underestimation of floods, especially small floods and floods in less populated areas. Many floods occur along ungauged streams, and non-catastrophic losses may simply not make headlines, etc. Unfortunately, the size and extent of the USGS stream gauge network has remained stagnant over the past decades [*USGS*, 2010]. This is particularly problematic for flooding in small watersheds and along creeks where the system responds very fast and stream gauges are critical for timely issuance of flash flood warnings.

4.6. CONCLUSION

The lack of a comprehensive loss baseline for floods is a blind spot in America's collective understanding of the burden of flooding. It leaves legislators and planners in the dark regarding the effectiveness of flood risk management. Quality loss data are management tools, outcome measures and tangible metrics that can help evaluate the nation's performance on flood mitigation and flood risk reduction. An adoption of performance metrics, which incorporate loss data (see Sendai Framework), would be a first step toward more accountability and transparency at government levels. As Burby states, "the local government paradox is that while their citizens bear the brunt of human suffering and financial loss in disasters, local officials pay insufficient attention to policies to limit vulnerability" [*Burby*, 2006].

Unfortunately, there is no accountability in regard to flood risk management policies in the US, which explains the current status quo of flood loss data. Only a few states, like California, hold local communities and the state financially liable for levee failures. For the rest of the country, floods remain "acts of man," and the federal government is largely exempt from any liabilities from flood-related claims under the Flood Control Act of 1928 and the Federal Torts Claim Act [*Brougher*, 2011]. This means that citizens will continue to carry the burden of flood losses though some homeowners (and few renters), particularly those in high risk zones, will be able to pass that burden onto the federal government by means of flood insurance. America goes to great lengths to administer a flood insurance program and to construct and maintain costly flood control projects but neglects flood loss accounting.

ACKNOWLEDGMENTS AND DATA

The author is affiliated with the Spatial Hazard Events and Losses Database (SHELDUS). Stream gauge data used in this chapter originate from the USGS. Flood loss data were retrieved from the Spatial Hazard Events and Losses Database for the US (http://sheldus.org), the National Weather Service's Hydrologic Information Center (http://www.nws.noaa.gov/hic/), and FEMA's Policy and Claims Statistics for the National Flood Insurance Program (https://www.fema.gov/policy-claim-statistics-flood-insurance). All loss figures are adjusted for inflation using 2014 as the base year. Per capita losses from SHELDUS are adjusted by using a county's population of the year when the flood occurred, not 2014 population.

REFERENCES

Arnold, J. L. (1988), The Evolution of the 1936 Flood Control Act, *Environ. Hist. Rev.*, *15*(1), 134, doi:10.2307/3984682.

Banks, J. C., J. V. Camp, and M. D. Abkowitz (2015), Scale and Resolution Considerations in the Application of HAZUS-MH 2.1 to Flood Risk Assessments, *Nat. Hazards Rev.*, *16*(3), 04014025, doi:10.1061/(ASCE)NH.1527-6996.0000160.

Blake, E. S., C. W. Landsea, and E. J. Gibney (2011), *The deadliest, costliest, and most intense United States Tropical Cyclones from 1851 to 2010 (and other frequently requested hurricane facts)*, National Hurricane Center (NHC), Miami, FL.

Blake, E. S., T. B. Kimberlain, R. J. Berg, J. P. Cangialosi, and J. L. I. Beven (2013), *Tropical Cyclone Report Hurricane Sandy*, Miami, FL.

Brougher, C. (2011), *Flood damage related to Army Corps of Engineers projects: selected legal issues*, Washington, DC.

Burby, R. J. (2006), Hurricane Katrina and the Paradoxes of Government Disaster Policy: Bringing About Wise Governmental Decisions for Hazardous Areas, edited by W. L. Waugh, *Ann. Am. Acad. Pol. Soc. Sci.*, *604*(1), 171–191, doi:10.1177/0002716205284676.

Downton, M. W. and R. A. J. Pielke (2005), How accurate are disaster loss data? The case of U.S. flood damage, *Nat. Hazards*, *35*(2), 211–228.

Federal Emergency Management Agency (2008), *FEMA Strategic Plan, Fiscal Years 2008–2013*, Washington, DC.

Federal Emergency Management Agency (2014), FEMA's Mission Statement, FEMA.gov, Available from: https://www.fema.gov/media-library/assets/videos/80684 (Accessed 22 December 2015).

Federal Emergency Management Agency (2015a), Loss Statistics Country-Wide as of 10/31/2015, Available from: http://bsa.nfipstat.fema.gov/reports/1040.htm.

Federal Emergency Management Agency (2015b), Insurance in Force as of October 31, 2015, Available from: https://www.fema.gov/insurance-force-month (Accessed 21 December 2015).

Federal Emergency Management Agency (2015c), Policies in Force by Month as of October 31, 2015, Available from: https://www.fema.gov/policies-force-month (Accessed 21 December 2015).

Federal Emergency Management Agency (2015d), Policy and claim statistics for flood insurance.

Gall, M., B. J. Boruff, and S. L. Cutter (2007), Assessing flood hazard zones in the absence of digital floodplain maps: Comparison of alternative approaches, *Nat. Hazards Rev.*, *8*(1), 1–12, doi:10.1061/(ASCE)1527-6988(2007)8:1(1).

Gall, M., K. A. Borden, and S. L. Cutter (2009), When do losses count? Six fallacies of natural hazards loss data, *Bull. Am. Meteorol. Soc.*, *90*(6), 799–809, doi:10.1175/2008BAMS2721.1.

Gall, M., K. A. Borden, C. T. Emrich, and S. L. Cutter (2011), The unsustainable trend of natural hazards losses in the United States, *Sustainability*, *3*(11), 2157–2181, doi:10.3390/su3112157.

Government Accountability Office (2010), *FEMA Flood Maps: Some Standards and Processes in Place to Promote Map Accuracy and Outreach, but Opportunities Exist to Address Implementation Challenges (GAO-11-17)*, Washington, DC.

Government Accountability Office (2015a), *Status of FEMA's implementation of the Biggert-Waters Act, as Amended (GAO-15-178)*, Washington, DC.

Government Accountability Office (2015b), *High-Risk Series: An Update (GAO-15-290)*, Washington, DC.

Gil de Zúñiga, H., N. Jung, and S. Valenzuela (2012), Social Media Use for News and Individuals' Social Capital, Civic Engagement and Political Participation, *J. Comput. Commun.*, *17*(3), 319–336, doi:10.1111/j.1083-6101.2012.01574.x.

Goodchild, M. F. and J. A. Glennon (2010), Crowdsourcing geographic information for disaster response: a research frontier, *Int. J. Digit. Earth*, *3*(3), 231–241, doi:10.1080/17538941003759255.

Hazards and Vulnerability Research Institute (HVRI) (2015), 1960–2014 U.S. Hazard Losses. Hazards and Vulnerability Research Institute. Available from http://hvri.geog.sc.edu/SHELDUS/docs/Summary_1960_2014.pdf.

Hallegatte, S. (2008), An adaptive regional input-output model and its application to the assessment of the economic cost of Katrina., *Risk Anal.*, *28*(3), 779–799, doi:10.1111/j.1539-6924.2008.01046.x.

Mileti, D. S. (1999), *Disasters by Design: a reassessment of natural hazards in the United States*, Joseph Henry Press, Washington, DC.

National Oceanic and Atmospheric Administration (2011), NOAA's State of the Coast: Federally-insured assets along the coast, *2011* (September 30). Available from: http://stateofthecoast.noaa.gov/insurance/welcome.html.

National Weather Service (2007), *Storm Data Preparation*, Department of Commerce, National Oceanic & Atmospheric Administration, National Weather Service (NWS), Silver Spring, MD.

National Weather Service (2011), *Spring 2011 Middle & Lower Mississippi River Valley Floods*, Silver Spring, MD.

National Weather Service (2015a), About NOAA's National Weather Service,

National Weather Service (2015b), Hydrologic Information Center–Flood Loss Data, Available from: http://www.nws.noaa.gov/hic/.

Serino, R. (2011), Written testimony of the FEMA for a Senate Homeland Security and Government Affairs' Ad Hoc Subcommittee on Disaster Recovery and Intergovernmental Affairs hearing titled "Accountability at FEMA: Is Quality Job #1?", Homeland Security, Available from: https://www.dhs.gov/news/2011/10/20/written-testimony-fema-senate-homeland-security-and-government-affairs-ad-hoc (Accessed 22 December 2015).

Smith, A. B. and R. W. Katz (2013), US billion-dollar weather and climate disasters: data sources, trends, accuracy and biases, *Nat. Hazards*, *67*(2), 387–410, doi:10.1007/s11069-013-0566-5.

Smith, A. B. and J. L. Matthews (2015), Quantifying uncertainty and variable sensitivity within the US billion-dollar weather and climate disaster cost estimates, *Nat. Hazards*, *77*(3), 1829–1851, doi:10.1007/s11069-015-1678-x.

US Army Corps of Engineers (2009), *Value to the Nation: Flood Risk Management*, Alexandria, VA.

US Geological Survey (2010), USGS National Streamflow Information Program–2010 Update, Available from: http://water.usgs.gov/nsip/status.html (Accessed 22 December 2015).

US Geological Survey (2014), About USGS, Available from: http://www.usgs.gov/aboutusgs/(Accessed 22 December 2015).

US Geological Survey (2015), USGS Current Water Data for the Nation, Available from: http://waterdata.usgs.gov/nwis/rt (Accessed 22 December 2015).

White, G. F. (1945), *Human adjustments to floods: a geographical approach to the flood problem in the United States*, University of Chicago, Chicago.

5

HOWAS21, the German Flood Damage Database

Heidi Kreibich[1], Annegret Thieken[2], Sören-Nils Haubrock[3], and Kai Schröter[1]

ABSTRACT

HOWAS21, the flood damage database for Germany (http://howas21.gfz-potsdam.de), was established at the German Research Centre for Geosciences GFZ in 2007 and contains object-specific flood damage data resulting from pluvial and fluvial floods in Germany. It is focused on direct tangible damage. Affected objects are classified as private households, commerce and industry, traffic areas and roads, watercourses, and hydraulic structures. The data sets contain various information about flood impact, exposure, and vulnerability. The minimum requirement is that they contain information about affected economic sectors, direct loss in monetary terms, water depth, flood event, spatial location, and the method of data acquisition. The main purpose to record flood damage data in HOWAS21 is to derive loss estimation models and support forensic analysis.

5.1. INTRODUCTION

Systematically collected and comparable natural hazard damage data are an essential component for ex-post event analysis and disaster response as well as for risk assessment and management [*Hübl et al.*, 2002]. Four main application areas for damage data were identified in *De Groeve et al.* [2014]: 1) Loss compensation: Many loss databases in Europe are based on a collection of claims used in compensation mechanisms. For example, the European Union Solidarity Fund (EUSF) requires loss data collected within a given timeframe to substantiate claims; 2) Loss accounting aims at documenting trends and consequently evaluating disaster risk reduction policies; 3) Forensic analysis aims to improve disaster

management from lessons learnt via identifying loss drivers by measuring the relative contribution of exposure, vulnerability, coping capacity, mitigation, and response to the resulting damage; and 4) Disaster risk modeling aims to estimate losses of future disasters. Detailed, object-specific loss data are required for developing loss models and for calibrating and validating models. The information needs for the four application areas are overlapping, even if the forensic and modeling applications require information at higher detail.

Studies that collect and analyze object-specific flood loss data are commonly undertaken with the primary goal to derive loss models or causal relationships. Several surveys tried to overcome the lack of detailed damage data [e.g., *Ramirez et al.*, 1988; *Joy*, 1993; *Gissing and Blong*, 2004; *Thieken et al.*, 2005; *Zhai et al.*, 2005; *Kreibich et al.*, 2007]. Detailed information about flood losses, event and object characteristics, warning variables, social factors, etc., of flood-affected residential and commercial properties was gathered. Relatively early, water depth and the characteristic of the affected object, e.g., use

[1] Section Hydrology, GFZ German Research Centre for Geosciences, Potsdam, Germany

[2] Institute of Earth and Environmental Science, University of Potsdam, Potsdam, Germany

[3] Beyond Concepts GmbH, Osnabrück, Germany

Flood Damage Survey and Assessment: New Insights from Research and Practice, Geophysical Monograph 228,
First Edition. Edited by Daniela Molinari, Scira Menoni, and Francesco Ballio.

of the building or affected economic sector, were identified as the most important loss-determining variables [*Grigg and Helweg*, 1975]. Thus, depth-damage curves were internationally accepted as the standard approach for urban flood loss assessments [*Smith*, 1994]. However, when deriving depth-damage curves, the data, even from larger databases, show large scatter, since only a part of the data variance is explained [*Blong*, 2004; *Merz et al.*, 2004]. Flood damage depends on many factors, e.g., flow velocity, duration of inundation, sediment concentration, contamination of floodwater, availability and information content of flood warning, and the quality of external response in a flood situation [*Smith*, 1994; *Wind et al.*, 1999; *Penning-Rowsell and Green*, 2000; *Kreibich et al.*, 2005, 2009; *Thieken et al.*, 2005; *Totschnig and Fuchs*, 2013). *Thieken et al.* [2005] and *Merz et al.* [2013] investigated single and joint effects of impact (i.e., flood characteristics) and resistance factors (e.g., characteristics of the building at risk and like type or structure) on flood loss ratios of private households.

However, there are still general shortcomings in empirical damage assessment: many studies of flood damage only include detailed assessments of a relatively small number of objects [*Blong*, 2004; *Mazzorana et al.*, 2014]. Since additional information about the affected object, flood warning, precautionary measures, etc., is often missing in damage data collections, their effects cannot be analyzed and considered in loss modeling.

HOWAS21, the flood damage database for Germany (http://howas21.gfz-potsdam.de), aims to overcome these shortcomings. It was established at the German Research Centre for Geosciences GFZ in 2007 and is developed to contain object-specific flood damage data resulting from pluvial and fluvial floods. The main purpose to record losses in HOWAS21 is to derive loss estimation models and to support forensic analysis.

5.2. OTHER FLOOD DAMAGE DATABASES

Most natural hazard damage databases that include flood damage are event-specific databases, that is, they contain aggregated national damage costs. Such documentations of disastrous events provide a basis for ex-post analysis of the event [*Hübl et al.*, 2002] and for loss accounting, e.g., trend analyses investigating whether damage due to natural hazards increases over time [*Bouwer*, 2011]. The most prominent examples of event-specific damage databases are the global databases NatCatSERVICE from Munich Re (www.munichre.com; *Kron et al.*, 2012) and the Emergency Events Database (EM-DAT) International Disaster Database (www.emdat.be; *Choryński et al.*, 2012). The NatCatSERVICE database contains overall and insured loss figures and fatalities of natural catastrophes. The natural events are classified in geophysical (e.g., earthquake), meteorological (storm), hydrological (flood,

mass movement wet), and climatological events (extreme temperature, drought, wildfire). The EM-DAT International Disaster Database contains worldwide data on the occurrence and impact of natural disasters (floods, droughts, storms, mass movements, etc.) and technological disasters and complex emergencies from 1900 to the present. The database is free and fully searchable through its website, also allowing users to download available data. An example of a national event-specific database is the Swiss flood and landslide damage database hosted at the Swiss Federal Research Institute WSL [*Hilker et al.*, 2009]. Information on flood and mass movement damage has been systematically collected in the database since 1972. The estimated direct financial damage as well as fatalities and injured people are documented using press articles as the main source of information. Several other event-specific databases exist, and overviews are provided by *Tschoegl et al.* [2006] and *Gall et al.* [2009]. Several initiatives to improve the availability and usefulness of damage data, including an EU initiative to harmonize and standardize loss databases, are currently ongoing. (See Chapter 2 in this book).

For forensic analyses and for the development and validation of loss models, predominantly object-specific damage data are needed. Only these can provide insights into the damaging processes via investigating causal relations between hazard, exposure, and vulnerability characteristics, and the amount of loss [e.g., *Downton et al.*, 2005; *Jonkman*, 2005]. However, object-specific databases are rare. Probably the best-known example for a synthetically generated national database of flood damage is the one of the Flood Hazard Research Centre (FHRC) from Middlesex University, United Kingdom (UK). It contains synthetic damage data collected via what-if-analyses, i.e., expert estimations about which damage is expected in case of a certain flood situation, e.g., "Which damage would you expect if the water depth was 2 m above the building floor?" The developed absolute flood damage curves for the UK are published in the Multi Coloured Manual [*Penning-Rowsell et al.*, 2005 and updated for 2010), as well as in its predecessors [*Penning-Rowsell and Chatterton*, 1977; *Parker et al.*, 1987]. The Austrian Federal Railway (ÖBB) holds an object-specific flood damage database for railway infrastructure in Austria [*Moran et al.*, 2010]. The database collects information on the damaged object, on the damage (several classes of structural damage and financial damage), on the flood hazard characteristics, and possible mitigation measures for railway infrastructure in Austria. In Italy, a new national database is under development that shall be integrated into the Italian Civil Protection system [*Molinari et al.*, 2014]. The main objective is the development of specific depth–damage curves for Italian contexts. So far, new procedures for data collection and storage have been developed and were applied at the local level for the residential and commercial sectors [*Molinari et al.*, 2014].

5.3. HOWAS21 DATABASE CONCEPT AND STRUCTURE

HOWAS21 is a relational web-based database for object-specific flood damage data focused on direct tangible damage to assets resulting from riverine or pluvial floods in Germany. Its objective is the compilation and homogenization of flood damage data (including hazard, exposure, and vulnerability characteristics) for the development and validation of flood loss models [e.g., *Kreibich et al.*, 2010] and forensic analysis-like quantification of loss mitigation potentials of precautionary measures [e.g., *Kreibich et al.*, 2005; *Hudson et al.*, 2014]. There is no funding or mandate for HOWAS21 to collect damage data for enlarging its content and keeping it up-to-date. Therefore, HOWAS21 relies on voluntary data contributions from surveys and data acquisition campaigns, which happen to be undertaken particularly after flood events in Germany. This lack of funding for continuous, structured data collection and database maintenance is a disadvantage, which leads to data gaps and difficulties in keeping the contained damage data up-to-date.

The GFZ German Research Centre for Geosciences in Potsdam, Germany carries out administration of the database. GFZ is responsible for compiling, reviewing, and maintaining consistency of data, assigning access rights, and verifying user requests.

The utilization of HOWAS21 follows a community-based concept. Three user groups, organized on different levels of data usage, have access to the database in varying degrees:

1. World: The interested public can search in the database and access a range of general information and evaluations. The user interface has an option to search in the database according to selected criteria. These include a structured query, filtered by catchment area, regions (provinces and municipalities), periods (event year), sectors, collection methods, campaign, and a combination of these criteria.

2. Registered user group I: This group includes all institutions that provided a certain amount of data with appropriate quality to HOWAS21. They have full access to the entire database.

3. Registered user group II: Users from academia and non-commercial projects, who did not provide data, can apply for a restricted project-specific use. In return, they need to provide feedback on project results, particularly results based on HOWAS21 data, and in case the user collects flood damage data later on, these have to be included in HOWAS21.

Registered users (group I and group II) need to agree to the terms of usage that define the scope of use, reporting requirements, and prohibition to disseminate data. Currently 54 users from science, insurance, authorities, and engineering consultancy are registered to HOWAS21. Among these, 12 organizations have provided data to HOWAS21 (user group I).

Unfortunately, most organizations are only interested in extracting data; hardly any organizations are willing to contribute flood damage data. The community-based use concept is not working well. However, a significant amount of new flood damage data from recent data acquisition campaigns [*Kienzler et al.*, 2015; *DKKV*, 2015] are envisaged to be included into HOWAS21 within the next two years.

For HOWAS21 as well as for proposing a standardized flood damage data collection, a catalogue of relevant items was derived via a multi-step online expert survey using the Delphi-approach. The survey aimed at identifying a catalogue of items that are needed for damage and risk analyses in different damage sectors and that should thus be collected whenever flood damage is recorded. To cover the requirements of different professional fields, 55 experts from governmental agencies, (re-)insurance companies, and science and (engineering) consultancy were included in the panel [*Elmer et al.*, 2010]. In the Delphi survey, the same questionnaire was filled out three times with the respondents receiving feedback on the earlier responses. The idea being that respondents can learn from the views of others and that as such a consensus is reached.

The structure of HOWAS21 is based on the outcome of this Delphi survey. Additionally, a manual was developed that outlines the theoretical framework for flood damage assessment and suggests criteria for damage documentation. For the latter, the core attributes for each sector were supplemented by evaluation methodologies (e.g., measurement units, check lists). In addition, suggestions for metadata, general event documentation, and aggregated damage reports are presented in the manual [*Thieken et al.*, 2010].

In HOWAS21, the damage cases related to individual objects are classified into six damage sectors:

1. private households, particularly residential buildings and contents

2. commercial and industrial sector, including public municipal infrastructure (administration, social issues, education, etc.) as well as agricultural buildings

3. agricultural and forested land

4. public thoroughfare, including roads and transport infrastructure

5. watercourses, including flood defense structures

6. urban open spaces

All attributes of the damage cases are grouped into three main tables per sector as shown exemplarily in Table 5.1. Damage cases include information about the flood characteristics at the location of the affected object, object characteristics, and damage information as well as

Table 5.1 Exemplary Overview of the Main Damage Information Tables for Private Households.

Flood characteristics at the location of the affected object[1]	Object characteristics and damage information[2]	Damage mitigation[3]
Start, end, duration of inundation at object	Location of the building	Knowledge about hazard maps
Name of river causing the inundation	Building type and characteristics (number of stories, age, quality, net dwelling area, intrusion paths and intake sill, building material, use of the cellar, etc.)	Precautionary measures
Maximum water depth	Value of the building, building loss, loss ratio	Early warning, lead time
Maximum flow velocity	Contents value, contents loss, loss ratio	Emergency measures
Contamination, flotsam		Effectiveness of measures
Local return period		
Hazard peculiarities, description of hazard at object		

[1]Same table is used for all sectors.
[2]Specific table per sector.
[3]Table for private households and commercial/industrial sector only.

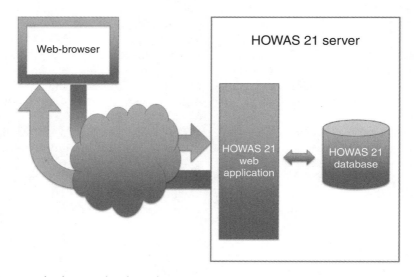

Figure 5.1 HOWAS21 database and web application.

information about damage mitigation. For each case, additional meta-information is stored including information about the flood event and the data acquisition campaign. The latter contains information about the survey type and method, the date/period of the survey, sample description, details about the methods used to measure and process the individual attributes as well as a documentation of underlying assumptions.

Minimum content requirements for damage cases to be incorporated into HOWAS21 have been defined as follows: information about the economic sector of the affected object, monetary loss, inundation depth, the flood event, and spatial location of the affected object at least on the level of zip codes or municipalities. HOWAS21

data are in an anonymous format respecting personal rights according to data privacy regulations.

5.4. TECHNICAL DESIGN AND IMPLEMENTATION

HOWAS21 consists of two major software components: the database and the web application (Figure 5.1). The database contains and manages access to all records that have been collected in multiple research campaigns and initiatives. Data are stored in different tables, and their relationships are modeled and allow for a well-structured, consistent representation of flood damage data and corresponding information. The HOWAS21 web application complements the database by providing a user-friendly data access interface on the Internet. With the help of the web application, HOWAS21 data can be visualized, analyzed, and downloaded easily using a standard web browser.

The database has been designed to cater to the complexity and heterogeneity of flood damage data in multiple sectors. Flood event data are stored and related to object-specific damage. Corresponding metadata is comprehensive and allows for putting the damage information into context. Selected attributes can be used to filter and analyze the data. The HOWAS21 database has been implemented using the relational database management software PostgreSQL, which runs as a database service at the German Research Center for Geosciences, GFZ.

Although the database contains and manages all of the "scientific assets" of HOWAS21, the web application portion makes them available to the user community (Figure 5.2). The user interface is provided by a website that is accessible directly at http://howas21.gfz-potsdam. de, or following the digital object identifier (DOI) http:// dx.doi.org/10.1594/GFZ.SDDB.HOWAS21. It provides comprehensive documentation and information with respect to the available data and scientific background. Users are provided query functionality in the database on selectable criteria, such as catchment areas, damage sectors, or data collection campaigns.

Users can register with HOWAS21 via the web portal. If access is granted and a "contract of using data" is signed, login credentials are provided to the new user. Registered users can log in and access additional functionality to analyze and download data. Finally, contact details and a feedback form allow users to ask questions and provide input to the HOWAS21 team in order to continually improve the quality of the services provided.

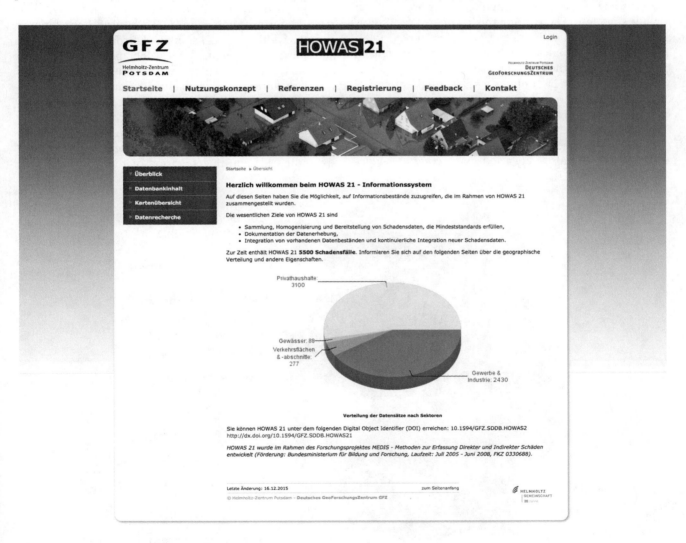

Figure 5.2 HOWAS21 web application.

5.5. DATA SOURCES: SURVEYS AND DATA ACQUISITION CAMPAIGNS

HOWAS21 is designed for empirical as well as synthetic flood damage data (data collected by experts via what-if analyses). However, so far only empirical damage data are contained. HOWAS21 contains about 6000 object-specific flood damage data sets from flood events between 1978 and 2011 in Germany (Figure 5.3). A large part of the data origins from the "old" HOWAS-database that was gathered and maintained by the German Working Group on water issues of the Federal States and the Federal Government (LAWA) from 1978 to 1994 [*Buck and Merkel*, 1999; *Merz et al.*, 2004]. Loss data of HOWAS were collected via on-site expert surveys by damage surveyors of the insurance companies, which were responsible for insurance compensation. After flood events, some representative municipalities were chosen for investigation. There, a complete survey of all damaged buildings was undertaken [*Buck and Merkel*, 1999]. The damage estimates are considered to be reliable because they were the basis for financial compensation.

Further, HOWAS21 contains a substantial amount of damage data from computer-aided telephone interviews with private households and companies who suffered flood damage in 2002, 2005, and 2006 [e.g., *Kienzler et al.*, 2015]. Further loss data from private households and companies also collected via computer-aided

telephone interviews after flood events in 2010, 2011, and 2013 will be added within the next two years. To collect loss data via computer-aided telephone interviews in the aftermath of a flood event, lists of all affected streets are compiled possibly with the help of flood masks derived from radar satellite data, or publicly available information like official reports and press releases [e.g., *Kreibich et al.*, 2007; *Thieken et al.*, 2007; *Kreibich et al.*, 2011]. Commonly, institutes for social or marketing research are commissioned to undertake interviews with the individual that has the best knowledge of the occurred flood damage within a given private household or company (see Chapter 7 in this book).

For traffic infrastructure and water management sectors, only few loss datasets are contained in HOWAS21, and all are collected via on-site expert inspection. Loss data for traffic infrastructure is limited to 246 inundated sections of the road infrastructure in the City of Dresden affected during the flood in 2002. Physical characteristics (length, width, sidewalks, etc.), the road classifications, and some other features were documented. The flood damage was recorded in two ways. First, the absolute loss was derived from files from city administration that contain the repair and reconstruction costs of the reconstruction projects. Second, experts from the city administration were asked to rate the physical damage magnitude witnessed immediately after the flood on a six-point scale and the condition of the road before the

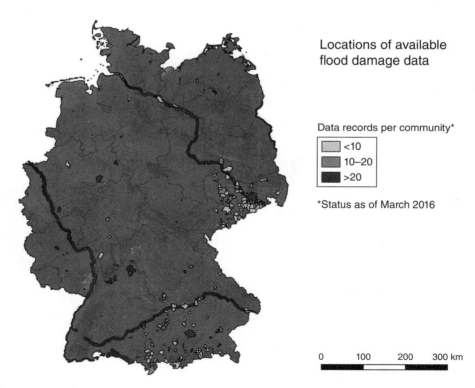

Figure 5.3 Locations of available flood damage data in HOWAS21.

flood on a five-point scale [*Kreibich et al.*, 2009]. A similar procedure was applied for the collection of the 75 damage cases along watercourses and hydraulic structures in the City of Dresden affected during the 2002 flood.

5.6. DATA QUALITY CONCEPT

Loss data analyses, e.g., by *Merz et al.* [2004], *Downton and Pielke* [2005], and *Downton et al.* [2005] revealed a range of problems, such as limited accessibility, inconsistencies, lack of details, or data errors. In order to determine the general quality of loss data for damage analysis, concepts for assessing and visualizing data quality, which can be traced back to *Wang and Strong* [1996] as well as *Funtowicz and Ravetz* [1990], were implemented in HOWAS21.

In general, data quality (DQ) is a multidimensional concept, in which accuracy, reliability, timeliness, relevance, and completeness are often mentioned criteria [see *Wand and Wang*, 1996]. Although there is no agreed definition of DQ, DQ from a consumer's point of view can best be defined as "fitness for use" [*Strong et al.*, 1997], which implies that DQ is relative [*Tayi and Ballou*, 1998]. To better assess DQ, *Wang and Strong* [1996] performed a multi-stage survey among data consumers and established a conceptual hierarchical framework of DQ, in which four categories are distinguished: intrinsic DQ, contextual DQ, representational DQ, and accessibility DQ. These were defined and further broken down to sub-dimensions by *Wang and Strong* [1996].

Since efforts to improve data quality tend to focus narrowly on accuracy [*Wang and Strong*, 1996], the framework can be used to improve the data quality in a whole field with a broader scope. In the case of property-level loss data, particular improvements concerning the accessibility of data as well as their interpretability, consistency, up-to-dateness, completeness, and associated relevancy for derivation of loss models are necessary.

Following the idea of conceptualizing DQ as "fitness for use," the derivation of depth-damage curves from damage data at the asset/property scale stored in HOWAS21 was taken as use case. Depth-damage curves are an internationally accepted standard approach for estimating flood losses in a given sector (e.g., at residential buildings) in dependence of the water level [*Smith*, 1994]. In HOWAS21, DQ is assessed and calculated in real-time after the user has selected a data subset and a depth-damage curve was fitted to the data. The following DQ dimensions are considered in HOWAS21 (Table 5.2). The intrinsic DQ and the reputation of the data source as well as the data accuracy (free-of-error) are evaluated. With regard to contextual DQ, the timeliness of the selected subset is assessed as well as the sample size of the data used to derive a depth-damage curve. Furthermore,

the relevance of the selected subset is evaluated by scoring the goodness of fit between the selected data and the fitted (linear, polynomial, or square-root) regression as well as the homogeneity of the selected data subset. Dimensions concerning the accessibility and representational DQ were neglected.

For the concrete assessment of DQ dimensions of a given data set, the Numeral Unit Spread Assessment Pedigree (NUSAP)-approach proposed by *Funtowicz and Ravetz* [1990] was adapted. NUSAP enables a qualitative goodness or uncertainty assessment of data and models. It has been successfully applied in different fields such as environmental modeling [*Funtowicz and Ravetz*, 1990], observational uncertainty analysis [*Ellis et al.*, 2000a, 2000b], emission monitoring data [*van der Sluijs and Risbey*, 2001], and flood event documentations [*Uhlemann et al.*, 2014]. In HOWAS21, this approach was adapted by defining rules for scores from 0 to 4 for each DQ dimension. For a selected data subset, a score is determined on the basis of the rules outlined in Table 5.1. For example, the timeliness of a data subset is assumed to be very good if the data are younger than 2 years, good if they are 2 to 5 years old, fair if they are 5 to 10 years old, and poor if they are older than 10 or even 15 years (compare Table 5.2). Such an assessment is done for each DQ dimension listed in Table 5.2. In order to visualize the quality of the data subset at hand, a kite diagram is created as proposed by *van der Sluijs and Risbey* [2001]. An example is shown in Figure 5.4. For this data subset, the sample size is evaluated as very well (score of 4), the timeliness of the data as fair (score of 2), while all other dimensions are assessed to be of poor quality (score of 1 or 0).

It is obvious that some meta-information on the data source, such as the methods used for data collection, etc., have to be available. This information is required from data providers and stored in HOWAS21 as meta-information for each data collection campaign. For some data fields (e.g., water level, asset value, loss), information on the type of measurement is stored for each database entry (item).

By these approaches, users of the database HOWAS21 are made aware of the quality of selected data sets without restricting data queries that might be considered to be useless for some tasks. Instead, by providing such quality information, users are motivated to select more appropriate samples for their task at hand.

5.7. EXEMPLARY DATA ANALYSES AND USE

The main objective of HOWAS21 is the compilation and homogenization of flood damage data (including hazard, exposure, and vulnerability characteristics) for the development and validation of flood loss models, e.g., depth-damage curves. In the following discussion, examples of

Table 5.2 Rules for the Assessment of Data Quality for Data Subsets Selected in HOWAS21.

DQ Dimension	Reputation	Accuracy	Timeliness	Sample size	Goodness of fit	Homogeneity of the data subset
Relevant data field	Campaign: type/method of data collection	Measurement of water level and financial loss	Event date	Count of valid data	R^2 of fitted regression	Region, flood type, campaign, building type, age of buildings (construction period)
4	100% on-site expert surveys (by building surveyors)	100% measured water levels and proofs of payments of repair works and replacement costs OR consistent loss valuation with a price and service listing	<2 years	>500 valid data	$R^2 > 0.8$	Data from the same* region (Bundesland), similar flood types, types of data collection, building types, and construction periods
3	Mixture of on-site expert surveys and other proofs of payments (to determine financial losses)	Measured data (water level and loss) with a low to moderate percentage (<50%) of calculated or estimated data	>2–5 years	251–500 valid data	>0.6–0.8	Data with similar flood types, types of data collection, building types and construction periods
2	Written polls and/or combination of unit values with on-site surveys and/or proofs of payments	Measured data (water level and loss) with a high percentage (>50%) of calculated or estimated data	>5–10 years	101–250 valid data	>0.4–0.6	Data with similar flood types, building types and construction periods
1	Different types of data collection including web-based or telephone polls	High percentage (>50%) of average or unit data values	>10–15 years	50–100 valid data	>0.2–0.4	Data with similar building types
0	Different types of data collection with rough estimates	High percentage (>50%) of rough estimates (e.g., loss estimate immediately after the event)	>15 yrs	<50 valid data	<= 0.2	Mixture of data

S
C
O
R
E

* same or similar: <10% other types and <25% "no data"

depth-damage curves; that is, square-root, linear, and polynomial curves for residential buildings are presented (Figure 5.5). Two different samples of empirical damage data were used for curve development: 1) Residential building damage cases irrespective of catchment and year of flood occurrence (n = 1171); and 2) Only residential building damage cases affected in 2005 in the Danube catchment (n = 124). The absolute building loss of all cases was transferred to the year 2005 by means of the construction price index [*Statistisches Bundesamt*, 2010]. Suitability of the depth-damage curves for residential loss assessments in the Danube catchment was evaluated as follows: Validation of the depth-damage curves, i.e., calculation of error statistics, is based on 40 damage cases from the 2005 flood in the Danube catchment, which were not used for curve development (Table 5.3).

Results of the error statistics show that the curves derived on data from the 2005 flood in the Danube catchment perform better than those curves derived on data from different regions and flood events (Table 5.3). This confirms previous findings, which revealed that the transfer of loss models in time and space is critical, and leads to increased uncertainty [*Thieken et al.*, 2008; *Schröter et al.*, 2014]. However, all depth-damage curves show high errors leading to significant uncertainties in flood loss modeling, which has been shown before [e.g., by *Merz et al.*, 2004 and *de Moel and Aerts*, 2011]. Quantifying and communicating these uncertainties is crucial for flood risk assessments and consequently for improved decisions on flood risk management.

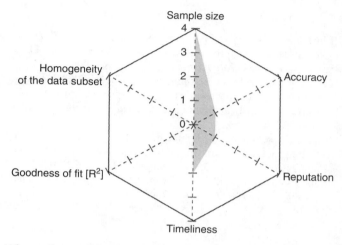

Figure 5.4 Kite diagram for a data subset of HOWAS21 showing data quality dimensions assessed on the basis of the rules listed in Table 5.2.

5.8. CONCLUSIONS

HOWAS21, the German web-based flood damage database, is designed for object-specific flood damage data focused on direct tangible damage to assets resulting from riverine or pluvial floods in Germany. Its objective

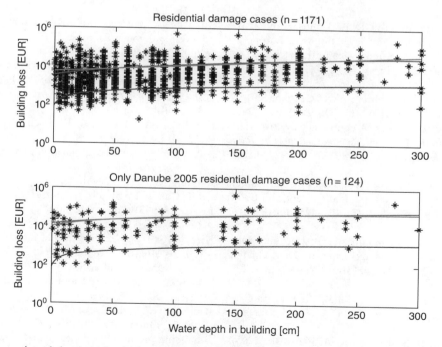

Figure 5.5 Examples of absolute depth-damage curves for residential building loss, i.e., square-root (blue), linear (red), and polynomial (green) curves calculated on the basis of two different samples of empirical damage data (top and bottom). *See electronic version for color representation.*

Table 5.3 Comparison of Depth-Damage Curve Performance Evaluated on Basis of 40 Damage Cases from the 2005 Flood in the Danube Catchment (MBE: Mean Bias Error, MAE: Mean Absolute Error, RMSE: Root Mean Square Error).

Empirical basis	Residential building damage cases irrespective of catchment and year of flood occurrence (n = 1171)			Only residential building damage cases affected in 2005 in the Danube catchment (n = 124)		
Curve type	square-root	linear	polynomial	square-root	linear	polynomial
MBE	−17780	−11189	−10921	−17780	1038	915
MAE	17850	15877	15921	17850	18822	18959
RMSE	33594	30536	30448	33595	28739	28974

is the compilation and homogenization of flood damage data including hazard, exposure, and vulnerability information for the development and validation of flood loss models and forensic analysis. It is easy to use and maintain due to its web application. One of its strengths is its data quality concept and management feature as well as the strict minimum requirements for data entries.

REFERENCES

Blong, R. (2004), Residential building damage and natural perils: Australian examples and issues, *Building Research and Information*, 32, 379–390.

Bouwer, L. M. (2011), Have disaster losses increased due to anthropogenic climate change?, *Bull. Am. Met. Soc.*, 92(1), 39–46.

Buck W. and U. Merkel (1999), Auswertung der HOWAS Schadendatenbank (Analysis of the HOWAS data base), Institut für Wasserwirtschaft und Kulturtechnik der Universität Karlsruhe, HY98/15.

Choryński, A., I. Pińskwar, W. Kron, R. Brakenridge, and Z. W. Kundzewicz (2012), Catalogue of large floods in Europe in the 20th century, In: Z. W. Kundzewicz (2012), *Changes in Flood Risk in Europe*, CRC Press, DOI: 10.1201/b12348-5, 27–54.

De Groeve, T., K. Poljansek, D. Ehlrich, and C. Corbane (2014), Current Status and Best Practices for Disaster Loss Data recording in EU Member States: A comprehensive overview of current practice in the EU Member States, *Report by Joint Research Centre of the European Commission*, JRC92290, doi: 10.2788/18330.

De Moel, H. and J. C. J. H. Aerts (2011), Effect of uncertainty in land use, damage models and inundation depth on flood damage estimates, *Nat Hazards*, 58(1), 407–425, doi:10.1007/s11069-010-9675-6.

DKKV (2015), Das Hochwasser im Juni 2013: Bewährungsprobe für das Hochwasserrisikomanagement in Deutschland, *DKKV-Schriftenreihe Nr. 53*, Bonn, ISBN 978-3-933181-62-6, 207 S.

Downton, M. W. and R. A. Pielke Jr. (2005), How accurate are Disaster Loss Data? The Case of U.S. Flood Damage, *Natural Hazards*, 35, 211–228.

Downton, M. W., J. Z. B. Miller, and R. A. Pielke Jr. (2005), Reanalysis of U.S. National Weather Service Flood Loss Database, *Natural Hazards Review*, 6, 13–22.

Ellis, E. C., R. G. Li, L. Z. Yang, and X. Cheng (2000b), Changes in Village-scale nitrogen storage in China's Tai Lake region, *Ecological Applications*, 10 (4), 1074–1096.

Ellis, E.C., R. G. Li, L. Z. Yang, and X. Cheng (2000a), Long-term change in village-scale ecosystems in China using landscape and statistical methods, *Ecological Applications*, 10 (4), 1057–1073.

Elmer, F., I. Seifert, H. Kreibich, A. H. Thieken (2010), A Delphi method expert survey to derive standards for flood damage data collection, *Risk analysis*, 30, 1, 107–124.

Funtowicz, S. O. and J. R. Ravetz (1990), *Uncertainty and Quality in Science for Policy*, Dordrecht, Kluwer.

Gall, M., K. A. Borden, and S. L. Cutter (2009), When do losses count? Six fallacies of natural hazards loss data, *Bull. Am. Met. Soc.*, 90(6), 799–809.

Gissing, A. and R. Blong (2004), Accounting for Variability in Commercial Flood Damage Estimation, *Aust. Geogr.*, 35(2), 209–222.

Grigg, N. S. and O. J. Helweg (1975), State-of-the-art of estimating flood damage in urban areas, *Water Resour. Bull.*, 11(2), 379–390.

Hilker, N., A. Badoux, and C. Hegg (2009), The Swiss flood and landslide damage database 1972–2007, *Nat. Hazards Earth Syst. Sci.*, 9, 913–925.

Hübl, J., H. Kienholz, and A. Loipersberger (Eds.) (2002), *DOMODIS – Documentation of Mountain Disasters – State of Discussion in the European Mountain Areas*, Interpraevent, Klagenfurt.

Hudson, P., W. Botzen, H. Kreibich, P. Bubeck, and J. Aerts (2014), Evaluating the effectiveness of flood damage mitigation measures by the application of propensity score matching, *Nat. Hazards Earth Syst. Sci*, 14, 1731–1747.

Jonkman, S. N. (2005), Global Perspectives on Loss of Human Life Caused by Floods, *Natural Hazards*, 34, 151–175.

Joy, C. S. (1993), The cost of flood damage in Nyngan, *Climatic Change*, 25, 335–351.

Kienzler, S., I. Pech, H. Kreibich, M. Müller, and A. H. Thieken (2015), After the extreme flood in 2002: changes in preparedness, response and recovery of flood-affected residents in Germany between 2005 and 2011, *Nat. Hazards Earth Syst. Sci*, 15, 505–526.

Kreibich, H., A. H. Thieken, T. Petrow, M. Müller, and B. Merz (2005), Flood loss reduction of private households due to building precautionary measures–Lessons Learned from the Elbe flood in August 2002, *Nat. Hazards Earth Syst. Sci*, 5, 1, 117–126.

Kreibich, H., M. Müller, A. H. Thieken, and B. Merz (2007), Flood precaution of companies and their ability to cope with the flood in August 2002 in Saxony, Germany, *Water Resources Research*, *43*.

Kreibich, H., K. Piroth, I. Seifert, H. Maiwald, U. Kunert, J. Schwarz, B. Merz, and A. H. Thieken (2009), Is flow velocity a significant parameter in flood damage modelling?, *Nat. Hazards Earth Syst. Sci*, *9*, 5, 1679–1692.

Kreibich, H., I. Seifert, B. Merz, and A. H. Thieken (2010), Development of FLEMOcs–A new model for the estimation of flood losses in companies, *Hydrological Sciences Journal*, *55*, 8, 1302–1314

Kreibich, H., I. Seifert, A. H. Thieken, E. Lindquist, K. Wagner, B. Merz (2011), Recent changes in flood preparedness of private households and businesses in Germany, *Regional Environmental Change*, *11*, 1, 59–71.

Kron W., M. Steuer, P. Löw, and A. Wirtz (2012), How to deal properly with a natural catastrophe database–analysis of flood losses, *Nat. Hazards Earth Syst. Sci.*, *12*, 535–550.

Mazzorana, B., S. Simoni, C. Scherer, B. Gems, S. Fuchs, and M. Keiler (2014), A physical approach on flood risk vulnerability of buildings, *Hydrol. Earth Syst. Sci.*, *18*, 3817–3836.

Merz, B., H. Kreibich, and U. Lall (2013), Multi-variate flood damage assessment: a tree-based data-mining approach, *Nat. Hazards Earth Syst. Sci.*, *13*, 1, 53–64.

Merz, B., H. Kreibich, A. Thieken, and R. Schmidtke (2004), Estimation uncertainty of direct monetary flood damage to buildings, *Nat. Hazards Earth Syst. Sci.*, *4* (1), 153–163.

Molinari, D., S. Menoni, G. T. Aronica, F. Ballio, N. Berni, C. Pandolfo, M. Stelluti, and G. Minucci (2014), Ex post damage assessment: an Italian experience, *Nat. Hazards Earth Syst. Sci.*, *14*, 901–916.

Moran, A., A. H. Thieken, A. Schöbel, and C. Rachoy (2010), Documentation of flood damage on railway infrastructure, in: Data and mobility. Transforming information into intelligent traffic and transportation, edited by Düh, J., Hufnagl, H., Juritsch, E., Pfliegl, R., Schimany, H., and Schönegger, H., Springer.

Parker, D., C. Green, and C. S. Thompson (1987), *Urban flood protection benefits: A project appraisal guide*, Gower Technical Press, Aldershot.

Penning-Rowsell, E. C. and J. B. Chatterton (1977), *The benefits of flood alleviation: A manual of assessment techniques*, Gower Technical Press, Aldershot.

Penning-Rowsell, E. C. and C. Green (2000), New Insights into the appraisal of flood-alleviation benefits: (1) Flood damage and flood loss information, *J. Chart. Inst. Water E.*, *14*, 347–353.

Penning-Rowsell, E. C., C. L. Johnson, S. M. Tunstall, S. M. Tapsell, J. Morris, J. Chatterton, and C. Green (2005), *The Benefits of Flood and Coastal Risk Management: A Manual of Assessment Techniques*, Flood Hazard Research Centre, Middlesex University Press, Middlesex, UK.

Ramirez, J., W. L. Adamowicz, K. W. Easter, and T. Graham-Tomasi (1988), Ex post analysis of flood control: benefit-cost analysis and the value of information, *Water Resour. Res.*, *24*, 1397–1405.

Schröter, K., H. Kreibich, K. Vogel, C. Riggelsen, F. Scherbaum, and B. Merz (2014), How useful are complex flood damage models? *Water Resources Research*, *50*, 4, 3378–3395.

Smith, D.I. (1994), Flood damage estimation–A review of urban stage-damage curves and loss functions, *Water SA*, *20*(3), 231–238.

Statistisches Bundesamt [Federal Statistical Agency] (2010) *Preise–Preisindizes für die Bauwirtschaft, Februar 2010*, Statistisches Bundesamt, Fachserie 17 Reihe 4, Wiesbaden.

Strong, D.M., Y. W. Lee, and R. Y. Wang (1997), Data Quality in Context, *Communications of the ACM*, *40*(5), 103–110.

Tayi, G. K. and D. P. Ballou (1998), Examing data quality, *Communications of the ACM*, *41*(2), 54–57.

Thieken, A. H., M. Müller, H. Kreibich, and B. Merz (2005), Flood damage and influencing factors: New insights from the August 2002 flood in Germany, *Water Resources Research*, *41*, 12, W12430.

Thieken, A. H., H. Kreibich, M. Müller, and B. Merz (2007), Coping with floods: preparedness, response and recovery of flood-affected residents in Germany in 2002, *Hydrological Sciences Journal*, *52*, 5, 1016–1037.

Thieken, A. H., A. Olschewski, H. Kreibich, S. Kobsch, and B. Merz. (2008), Development and Evaluation of FLEMOps a New Flood Loss Estimation MOdel for the Private Sector, I:315–24. WIT Press. doi:10.2495/FRIAR080301.

Thieken, A. H., I. Seifert, and B. Merz (2010), *Hochwasserschäden: Erfassung, Abschätzung und Vermeidung*, Oekom.

Totschnig, R. and S. Fuchs (2013), Mountain torrents: Quantifying vulnerability and assessing uncertainties, *Engineering Geology*, *155*, 31–44.

Tschoegl, L., R. Below, and D. Guha-Sapir (2006), An analytical review of selected data sets on natural disasters and impacts. *Report for the UNDP/CRED Workshop on Improving Compilation of Reliable Data on Disaster Occurrence and Impact*, Bangkok, Thailand.

Uhlemann, S., A. H. Thieken, and B. Merz (2014), A quality assessment framework for natural hazard event documentation: application to trans-basin flood reports in Germany, *Nat. Hazards Earth Syst. Sci.*, *14*, 189–208.

van der Sluijs, J. and J. Risbey (2001), *Uncertainty assessment of annual VOC emissions from paint in the Netherlands*.

Wand, Y. and R. Wang (1996), Anchoring Data Quality Dimensions in Ontological Foundations, *Communications of the ACM*, *39* (11), 86–95.

Wang, R. and D. Strong (1996), Beyond Accuracy: What Data Quality Means to Data Consumers, *Journal of Management Information Systems*, *12*(4), 5–34.

Wind, H. G., T. M. Nierop, C. J. de Blois, and J. L. de Kok (1999), Analysis of flood damages from the 1993 and 1995 Meuse floods, *Water Resources Research*, *35*, 3459–3465.

Zhai, G. F., T. Fukuzono, and S. Ikeda (2005), Modeling flood damage: Case of Tokai flood 2000, *Journal of the American Water Resources Association*, *41*, 77–92.

Part III
Data Collection

6

Best Practice of Data Collection at the Local Scale: The RISPOSTA Procedure

Nicola Berni[1], Daniela Molinari[2], Francesco Ballio[2], Guido Minucci[3], and Carolina Arias Munoz[2]

ABSTRACT

This chapter describes the experience gained in the Umbria region (Central Italy) in developing RISPOSTA (Reliable InStruments for POST event damage Assessment), a new procedure for the collection of damage data at the local scale, after flood events.

First, the objectives of the procedure are discussed (i.e., desirable outcomes) as well as when and where the procedure can be applied (i.e., the juridical and the physical context). Then, the logical structure of RISPOSTA is described along its four logical axes: 1) timeline of activities, 2) actors, i.e., who does what, 3) activities included in the procedure, and 4) the sector-based approach.

The third part of the chapter supplies a detailed description of activities included in the procedures and of tools developed for their implementation: 1) the collection of data regarding flooded areas, damage to residential buildings, and damage to industrial/commercial premises, for which a field survey is required, 2) data gathering from (and sharing with) responsible stakeholders for the other exposed sectors, and 3) data coordination. The chapter ends with a critical analysis of strengths and limits of the current procedure as well as of desirable improvements in order to increase the comprehensiveness of resulting damage scenarios. Results of stress tests are presented in order to discuss the feasibility and robustness of the procedure.

6.1. INTRODUCTION: WHY AND WHERE TO APPLY RISPOSTA

This chapter describes a new procedure called RISPOSTA (Reliable InStruments for POST event damage Assessment) for the collection of damage data in the aftermath of floods.

The procedure defines information to be collected, how, when, and by whom, in order to provide consistent and reliable data to its users. The latter are identified as all the authorities/people dealing with the consequences of a flood in the emergency and the recovery phases. In this period, having an extensive and comprehensive picture of the flood event, of the different types of damage occurred, and of their main causes, is crucial to guide recovery toward the reduction of the vulnerability of the affected area and then of the potential losses in case of a future event. Collecting required data to feed such a "complete event scenario" [*Menoni*, 2001] is the main objective of the procedure (see Chapter 11).

The time development of the procedure follows the evolution of the emergency and recovery phases, from the double perspective of the processes and of the administrative needs.

[1] Umbria Region Civil Protection Authority, Foligno (PG), Italy

[2] Department of Civil and Environmental Engineering, Politecnico di Milano, Milan, Italy

[3] Department of Architecture and Urban Studies, Politecnico di Milano, Milan, Italy

Flood Damage Survey and Assessment: New Insights from Research and Practice, Geophysical Monograph 228,
First Edition. Edited by Daniela Molinari, Scira Menoni, and Francesco Ballio.

In the emergency and post-emergency phase, between 1 and 20 days after the flood, key information required by users include the main features of the physical event, the affected areas, the state of essential services, available resources to cope with the event, people requiring assistance, and so on.

In the recovery phase, between 1 and 6 months after the flood, information needs include a comprehensive scenario of damage occurred, of existing priorities of intervention, and of used and still required resources. A reliable loss accounting is crucial at this stage as a base to define properly the best recovery strategies, including those related to public and private compensation of damages.

On a longer term (say, between 6 and 12 months after the event), the previous scenario must be completed including indirect damages that typically manifest several months after the flood (such as, loss of income because of industrial/service disruption, environmental damage, and psychological damage). Such an integrated scenario allows analysis and improvement, if required, to existing or planned risk mitigation strategies, by means of a forensic analysis of the event as well as improved estimation of expected damages.

Defining complete event scenarios implies collecting data from a variety of stakeholders (i.e., local authorities, utilities companies, private citizens who collect/own data for the affected items within their competency), at different moments after the event. On the other hand, information must be available to be used for different purposes (i.e., from emergency management to loss accounting to disaster forensic or risk modeling) and by different actors. A key role is coherently played in the procedure by a "data coordinator" being in charge of collecting and organizing data (from all the different sources, at different times); without such a role, it would be impossible to acquire consistent and reliable information to be used to feed multi-purposes scenarios. For the Italian institutional context, the coordinator role is naturally appointed to Regional Civil Protection Authorities (RCPA in the remainder of this chapter). Such bodies play a key role when a flood affects more than one municipality or in the case of trans-regional floods (that is in the case of events "b" and "c" according to the national law on Civil Protection 225/92), acting as the reference authority for all actors involved in the emergency and recovery phases.

RISPOSTA was developed by taking inspiration from other existing methodologies for data collection, such as the PDNA methodology [*Global Facility for Disaster Risk Reduction (GFDRR)*, 2013], developed with the leading role of the United Nations, the methodology developed by Australian authorities [*Emergency Management Australia (EMA)*, 2002], and the guidelines supplied by the World Meteorological Organization (WMO) specifically for post-flood losses assessment [*Associated Programme of Flood Management (APFM)*, 2013]. In detail, such methodologies have been adapted and extended to the context under investigation, in order to maintain their main distinctive features but also to include successful practices and procedures for data collection currently implemented in Italy; with respect to this, for example, the positive experience gained in Italy in surveying damage to buildings after earthquakes was considered in the development of the procedure. On the other hand, new practices/procedures were introduced whenever required.

Moreover, in order to assure the feasibility of the procedure, RISPOSTA was defined in close collaboration with the Civil Protection Authority of the Umbria region (Central Italy). Several stakeholders (such as, local authorities, utilities companies, and trade associations) were also involved, at different levels, during the development of the procedure. It is worth noting that RISPOSTA was designed to be applied in the Italian physical context, that is, in the case of small-scale or scattered events, with a limited number of affected items in every flooded area. In fact, as it will be better explained in section 6.2, RISPOSTA adopts a mix of direct surveys on the field and cooperative procedures aimed at gathering data from a variety of stakeholders.

In the next sections, RISPOSTA is described in detail as far as both its main distinctive features and operational tools are concerned. The description refers specifically to the implementation of the procedure in the Umbria region.

6.2. THE LOGICAL STRUCTURE OF RISPOSTA: THE FOUR AXES

Activities identified in the procedure are organized according to four main logical axes: time, actors, actions, and exposed sector. In the next sub-sections, the four axes are described in detail.

6.2.1. Time of Action

Time is the first logical axis, which delineates when activities included in the procedure must be performed. Time of activities was defined by considering the following:
• the evolution of the flood event and of consequent damaging processes (i.e., direct damage can be surveyed soon after the flood but indirect damage can manifest several days/months after the event and may have unfavorable repercussions that can last years)
• the need for information of different users during the emergency and the recovery phase (see previous section)
• the administrative procedures for recovery management and damage compensation in effect at the national

and the European level (for example, direct damage must be surveyed within 90 days after the event so that the RCPA can perform the loss accounting required to declare the "State of Emergency").

6.2.2. Actors: Who Does What?

The second logical axis classifies activities according to actors, that is, the people and entities who collect and use the data. With respect to this, the procedure includes as much as possible practices and procedures already in place in Italy.

The analysis of the Italian juridical and institutional context allowed the identification of a variety of actors to be included in the procedure. In addition to the RCPA at a regional level, they were identified as all the municipal, provincial, and regional authorities responsible for the different exposed items (e.g., roads, civil works, public buildings), utilities companies, trade associations, and private citizens (including business owners). All of them collect and use (damage) data to deal with the emergency and recovery phases.

In general, a distinction can be made between the RCPA and the other stakeholders. The latter collect and use damage data related only to items within their competency (e.g., provincial authorities for roads, utilities companies for essential service infrastructures, and citizens for private residences/premises). On the contrary, the RCPA is interested in having a more comprehensive overview of flood impacts, being on the frontline of emergency management and acting as the coordinator of data during the emergency and the recovery phase.

Another distinction can be made between public and private actors. Public stakeholders (i.e., municipal, provincial, and regional authorities) are legally enforced to collect and communicate their data to the RCPA, in order to ease both the emergency and the recovery phases. Accordingly, they are compulsorily involved in the procedure. On the other hand, private actors are neither obliged to collect data nor communicate them to the RCPA (with the exception of key data for the emergency management such as the disruption of essential services, environmental contamination, etc.). Their involvement in the procedure is then on a voluntary base. As a matter of fact, utilities companies usually collect damage data in order to deal with the emergency and the return to normalcy as quickly as possible, thus minimizing indirect damage. Private citizens collect data only when public funds are available for damage compensation. In this case, they must communicate such information to the RCPA in order to access available funds. Municipal authorities and trade associations act as a link between private citizens and the RCPA, respectively, for damage to residences and commercial/industrial premises. Indeed, they collect, homogenize, and synthesize all the requests from citizens and send related information to the RCPA.

6.2.3. Actions: Activities to be Performed

The third logical axis classifies three groups of activities according to the type of actions implemented in the procedure. Whenever "centralized" actors already collect damage data within their competency (this is the case, for example, of utilities companies and municipal authorities), protocols must be negotiated with data owners in order to acquire and organize available knowledge. Such action is referred to in the procedure as "data gathering." The main ideas at the base of data gathering are to 1) avoid data duplication, 2) minimize data collection efforts by integrating existing successful practices, and 3) guarantee data sharing and integration. Protocols define which data should be shared, at which time, and with which format. In some cases, negotiation may include the request of acquiring integrative information with respect to what has already done by the data owner.

By contrast, when data collection is currently performed by "scattered" actors and/or on a voluntary base (like in the case of private citizens), a *field survey* is necessary to guarantee the availability and consistency of data of interest. Accordingly, specific procedures and methods were developed within RISPOSTA.

The third action included in the procedure is related to *data coordination*, including all elaborations required for the definition of complete event scenarios (see Chapter 11).

Data coordination is performed in the Umbria region by means of an ad hoc developed information system, which is described in detail in Chapter 15.

6.2.4. Sectors: Homogeneous Groups of Exposed Items

The last axis distinguishes among homogeneous groups of exposed items. Indeed, damage data collection (and analysis) is performed per "exposed sectors" in the procedure. Sectors include (a) residential buildings, (b) industrial and commercial premises (i.e., businesses), (c) farms, (d) infrastructures, (e) public items, (f) emergency costs, (g) people, (h) environmental and cultural heritage, and (i) the physical scenario. The physical scenario sector does not pertain to exposed items; it includes all information regarding the flood characteristics (first of all, identification of flooded areas).

Sectors were chosen by balancing reporting and collection needs. The first ones refer to the need of creating complete event scenarios that are comprehensive and clear at the same time. Describing flood impacts per "exposed sectors" allows then, on the one hand, rationalizing the system under investigation and, on the other, highlighting interconnections among its different parts. To this aim, sectors reflect the classifications adopted by existing methodologies (e.g., WMO and Post-Disaster Needs Assessment [PDNA]) as well as what is recommended by the floods directive in terms of principal targets of flood risk assessments/management plans.

Collection needs refer, instead, to minimize efforts required by data collection by integrating existing practices and procedures when possible. Coherently, sectors were identified according to the possibility of adopting homogeneous actions for data collection within each sector. In line with the actions identified in the previous sub-section, two situations can be identified:

• sectors (from c to h) with a centralized data collection process, for which data gathering from external stakeholders is sufficient
• sectors (a, b, i) with a scattered data collection process, for which a field survey of damage data is introduced as a complement to data gathering

6.3. STATE OF IMPLEMENTATION OF THE RISPOSTA PROCEDURE

The structure of the RISPOSTA procedure is portrayed in Table 6.1, with respect to its logical axes. At its present level of implementation, RISPOSTA considers only one actor, the RCPA. Further stakeholders are indeed involved in the process, but no mandatory activities are still defined for them as proper actors of the procedure; on the other hand, RCPA coordinates with them every time a flood occurs, in order to get data of interest. Negotiations are in place, however, in order to optimize and automatize data gathering by means of an active involvement of all identified actors in the procedure.

In spite of the fact that RCPA is the only formal actor, a further actor appears in the procedure: in the Umbria region, the Civil Protection Authority is supported by an Expertise Centre (i.e., Politecnico di Milano) for all the activities that can be performed remotely, thus reducing efforts required by the authority to perform the procedure.

In the next sections, single activities in Table 6.1 are described in detail, by analyzing the different actions included in the procedure.

6.4. FLOODED AREAS, RESIDENTIAL BUILDINGS, AND INDUSTRIAL/COMMERCIAL PREMISES (DIRECT SURVEY CENTERED)

As explained before, when data collection is scattered among different stakeholders or when existing practices do not supply enough information to define a comprehensive scenario of flood impacts, data gathering must be integrated with the direct field survey of data of interest. This occurs for the characterization of the flooded area and for the investigation of damage to residential buildings and industrial/commercial premises.

Survey tools and methods were developed and tested jointly with the RCPA and surveyors (i.e., public administration officials and technical volunteers asked by the RCPA to help with data collection). Procedures were refined through several training sessions and field applications after flood events (see Box 6.1 and Box 6.2): feasibility and effectiveness of the procedure and quality of the collected data were evaluated and addressed in joint meetings.

6.4.1. The Procedure for the Survey of Flooded Areas

Table 6.2 summarizes the activities embedded in the procedure, times of action, and responsible actors as far as the survey of the flooded area is concerned. Most of activities are performed by the RCPA with the support of trained technicians for field survey. The Expertise Center is responsible of organizing and coordinating the survey.

Relevance of these activities for damage assessment is twofold. First, it is important to know the extension of the flooded area as a basis for the collection of flood damage data. The usual practice is to derive such information from satellite or aerial images acquired soon after the event. However, such tools may not be applicable and may furnish unreliable results for events with relatively short time scales (e.g., flash floods) because areas wetted at the time when the images are acquired may be a significant underestimation of the maximum flood extension. In such cases, typical of mountain areas, a field survey is required.

Second, from the perspective of risk modeling, the extent of the flooded area is not enough to explain observed damage, whereby other hazard variables are required like water depth, water velocity, sediment and contaminants loads, etc. Quantification of such variables is usually derived by numerical modeling whose accuracy strongly influences the reliability of damage estimation. Information allowing for model calibration and validation (typically, observed water depths and velocities in the flooded area) can be recorded on the field during the survey of the flooded area, with only minor additional efforts for the survey teams.

Activities included in the procedure consist in detail of the following [*Ballio et al.*, 2015]:

1. Acquisition of pre-existing knowledge on the hazard by extracting information from existing databases (typically hazard assessment and information on past events provided by competent authorities), monitoring network, and hazard models. The objective is to integrate available information on flood-prone areas to be used as the basis for field surveys.

2. Acquisition of data on the physical event. Monitoring data, notifications from experts in the field, and satellite and aerial images collected soon after the event by national authorities and fire brigades are acquired, when available.

Table 6.1 Main Activities Included in RISPOSTA, According to the Logical Axes of the Procedure (Time, Actors, Actions, and Exposed Sectors).

ACTIVITIES	ACTOR		ACTION			SECTOR								
	RCPA	Expert. Centre	Survey	Gathering	Coordination	Residences	Businesses	Farms	Infrastructures	Public items	Emergency	People	Environ./cultural	Physic. scenario
Acquisition of pre-existing knowledge on the hazard	X			X										X
Acquisition of pre-existing knowledge on exposure and vulnerability	X			X		X	X	X	X	X	X	X	X	
Set-up and management of the IS		X			X									
Data sharing		X			X									
Event														
Acquisition of data on the physical event	X			X										X
Data sharing		X			X									
2–3 days														
Survey of the flooded area/water elevation	X		X											X
Organization and coordination of the survey (flooded areas)		X	X											X
Data analysis (field survey)		X			X									
Data validation (physical event)		X			X									
Inputting data (physical event)		X			X									
Data sharing		X			X									
20 days														
Survey of damage to residences and businesses	X		X			X	X							
Organization and coordination of the survey (residences and businesses)		X	X			X	X							
Acquisition of damage data from the Regional Emergency Room (SOUR)		X		X				X	X	X	X	X	X	
Data analysis (field survey)		X			X									
Data validation (residences/businesses/SOUR)		X			X									
Inputting data (residences/businesses/SOUR)		X			X									
Data sharing		X			X									
90 days														
Survey of damage to residences (optional)	X		X			X								
Organization and coordination of the survey (residences)		X	X			X								
Acquisition of damage data from the responsible stakeholders (optional)		X		X				X	X	X	X	X	X	
Acquisition of monetary damage data		X		X		X		X	X	X	X	X	X	
Data validation (all sectors except businesses/monetary damage data)		X			X									
Inputting data (all sectors except businesses/monetary damage data)		X			X									
Data analysis (field survey, complete event scenario)		X			X									
Data sharing		X			X									
6 months														
Survey of damage to businesses	X		X			X								
Organization and coordination of the survey (businesses)		X	X			X								
Acquisition of damage data from the responsible stakeholders		X		X				X	X	X	X	X	X	
Acquisition of monetary damage data		X		X				X	X	X	X	X	X	
Data validation (all sectors except residences/monetary damage data)		X			X									
Inputting data (all sectors except residences/monetary damage data)		X			X									
Data analysis (field survey, complete event scenario)		X			X									
Data sharing		X			X									
12 months														

Box 6.1 Stress test of the procedure for the survey of flooded areas.

The procedure for the survey of the flooded areas was tested for the first time after the flood that hit the Umbria region in November 2014. No satellite or aerial images were available. Therefore, hazard assessment maps available in flood and landslide management plans (i.e., the so-called Piani di Assetto Idrogeologico [PAI]) were used to define regions to be surveyed. Two teams of experts in topography were equipped with a paper map of the identified area and common measurement tools (e.g., global positioning system [GPS] localizer, measuring tape, compass, and camera). The teams spent four hours surveying an area of about 1.5 km² in the town of Foligno.

The experience was successful because it allowed having a definition of the flooded areas a few days after the occurrence of the flood. In the absence of this procedure, for two previous flood events in the same region, information on flooded areas became available only several months after the flood.

Teams experienced some problems in 1) interpreting the hydraulic phenomena so that minor incoherence was observed by comparing the surveyed flooded areas with the topography of the region, and 2) measuring water elevation because sometimes it was taken in correspondence of high slopes with no reference to a reliable point on the DTM. The time spent for completing the survey was also a matter of concern. Lessons learnt indicated the necessity of specific training of the surveyors, as well as the need for simple mobile applications supporting the survey. Such applications should enable the drawing of both the flooded area perimeter and surveyed points on a geo-referenced digital map, taking and storing measurements, and integrating data by means of geo-referenced photographs.

Box 6.2 Stress tests of the procedure for the survey of damage.

The forms for residential buildings and industrial/commercial premises were tested after the floods that hit the Umbria region in November 2012 and November 2013. In total, more than 50 surveyors were involved including researchers, civil protection personnel, and trained qualified technicians.

The surveyors were organized in teams of two people. Each team was supplied with a map of the area to be surveyed, where buildings to be assessed were identified. Each team spent 30 to 60 minutes surveying a building, depending on the use of the building (e.g., big industrial premises may require a survey of more than 60 minutes), on the building type (e.g., multi-apartment houses require more time than a single house) and on the willingness of private owners to cooperate.

The surveyors were equipped with common measurement tools (GPS localizer, measuring tape, laser distance meter, etc.). They felt familiar with both instruments and forms, because they were trained in their use by means of civil protection drills.

After the November 2013 floods, a test was also conducted to verify the reliability of collected information. To this aim, two different teams were asked to survey independently the same set of buildings. Generally, a very good coherence/homogeneity of acquired information was observed as data collected by the two teams coincided in about 90% of the cases.

Nevertheless, after a first validation, about 23% of completed forms resulted in a status of incomplete and/or not usable because important information on hazard, exposure, vulnerability, or damage was lacking. This can be explained, on the one hand, by singular features of surveyed buildings that were difficult to assess, as they were not included in the forms. Difficulty included unusual shapes, materials, etc. Indeed, despite the presence of an open field in the forms to take note of these singularities, surveyors tended to fill in only pre-defined fields. On the other hand, experience has shown that, during the field survey, it is difficult to follow the order of the questions on the forms. Thus, surveyors have to jump from one question to another, as well as to different pages of the forms. This may imply that certain information is not recorded as surveyors miss filling in some fields.

The Information and Communications Technology (ICT) application for data collection, described in Chapter 15, was conceived to give more flexibility in the compilation of the forms, to guarantee the recording of all key information, as well as to speed up both the data collection and the input of data into the system database.

Table 6.2 Scheme of the RISPOSTA Procedure for Data Collection on the Physical Event: Activities to be Performed, Times of Actions, and Responsible Actors (Extract from Table 6.1).

ACTIVITIES	ACTOR		ACTION			SECTOR
	RCPA	Expert. Centre	Survey	Gathering	Coordination	Physic. scenario
Acquisition of pre-existing knowledge on the hazard	X			X		X
Event						
Acquisition of data on the physical event	X			X		X
2–3 days						
Survey of the flooded area/water elevation	X		X			X
Organization and coordination of the survey		X	X			X
20 days						

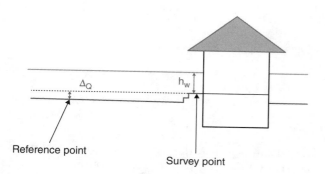

Figure 6.1 Typical survey context for water elevation measurement.

3. Survey of the flooded area. On the basis of information from points (1) and (2), the Expertise Centre identifies regions to be surveyed on a map. Then, the regions are divided in sub-regions and assigned to teams of specifically trained surveyors. Survey simply consists of looking for watermarks and analyzing the topography of the region under investigation, in order to identify the perimeter of the flooded area on the map. Information reported by citizens can also be useful for defining such borders.

4. Survey of water elevation. On the bases of the flooded area perimeter and of data coming from the monitoring network, the Expertise Centre performs a qualitative hydraulic analysis of the event in order to identify possible points for water elevation measurement. These typically are located in correspondence of singularities: slope changes, trees, rocks, buildings, artifacts, etc. Points are then identified on maps, related to different flooded regions, which are distributed among the survey teams. The survey is conducted by measuring water depth at each point and to assure that the measure can be easily referred to the available digital terrain model (DTM). Figure 6.1 is emblematic of a typical survey context. Water depth (h_w) can be measured by considering the watermark on the house, but the DTM is affected by

high uncertainty at building locations just because of the presence of buildings. The measure must then be referred to a "reference point" where the DTM is more accurate (i.e., plain areas with no artifacts) by measuring the elevation difference (Δ_Q).

6.4.2. The Procedure for the Survey of Damage: General Features

Activities to be performed with respect to the field survey of damage to residential and industrial/commercial premise are summarized in Table 6.3. As discussed in section 6.2, the timing of action is defined by considering the needs of data users, also with respect to the Italian regulations on damage compensation.

Activities to be performed consist in detail of the following [*Ballio et al., 2015*]:

1. Acquisition of pre-existing knowledge on the exposure and vulnerability of buildings/premises by extracting information from existing databases (e.g., cadastral or risk maps, Aree Vulnerate Italiane (AVI) catalogue made by Italian National Research Council). Such information (like surface area, number of floors, age, level of maintenance, type of activity, number of employees, etc.) can be uploaded into the forms implemented in the damage survey.

2. Survey of damage to buildings/premises. The survey is carried out using the procedure described in *Molinari et al.* [2013]. Put briefly, on the basis of the flooded area, items and premises to be surveyed are identified on a map by the Expertise Centre and subdivided among teams of surveyors. The survey is performed in the field by means of ad-hoc forms (see sections 6.4.3 and 6.4.4) a few months after the event. However, a second survey can be carried out some months later, with the aim of collecting information on longer-term damage (typically indirect damage) not defined at the time of the first survey. In the case of residential buildings, such a survey is optional

Table 6.3 Scheme of the RISPOSTA Procedure for Data Collection on Damage to Residential Buildings and Industrial/Commercial Premises: Activities to be Performed, Times of Actions, and Responsible Actors (Extract from Table 6.1).

ACTIVITIES	ACTOR		ACTION			SECTOR	
	RCPA	Expert. Centre	Survey	Gathering	Coordination	Residences	Businesses
Acquisition of pre-existing knowledge on exposure and vulnerability	X			X		X	X
Event							
Survey of damage	X		X			X	X
Organization and coordination of the survey		X	X			X	X
90 days							
Survey of damage to residences (optional)	X		X			X	
Organization and coordination of the survey (residences)		X	X			X	
Acquisition of monetary damage data		X		X		X	
6 months							
Survey of damage to businesses	X		X				X
Organization and coordination of the survey (businesses)		X	X				X
Acquisition of monetary damage data		X		X			X
12 months							

because it is possible that indirect damage (such as time spent outside the house or loss of rental income) does not manifest or is already known at the time of the first survey. In the case of industrial/commercial premises, the second survey is required because indirect damage (like loss of income because of activity disruption, loss of orders, unemployment, damage due to humidity) can be assessed only after some months after the event. On the occasion of the second survey, data considered not satisfactory or missing after the first collection can be further acquired. Forms are pre-compiled according to existing knowledge on the exposure and vulnerability of the buildings/premises, as well as with data from previous surveys.

3. Acquisition of monetary damage data. During the survey, damage is assessed in physical units (e.g., number of damaged doors, square meters of damaged floor). In a second step, the monetary damage value is assessed on the basis of the compensation requests submitted by owners, according to regional and/or national authority funds available at different times after the occurrence of the event.

6.4.3. The Procedure for the Survey of Damage: Forms for Residential Buildings

The forms for the survey of residential buildings were designed to collect information related to loss accounting, damage compensation, and the evaluation of the effectiveness of ongoing/planned risk mitigation strategies. This means that, through collected data, a forensic analysis of the flood can be carried out; on the other hand, data can be used to feed/validate damage models. Information typically includes hazard, exposure, and vulnerability factors, direct physical damage (like damage to doors, walls, technical equipment like the electrical and plumbing systems, furniture, etc.), and mitigation actions taken during the warning period and before the event. Furthermore, some information on indirect damage is collected like the number of days spent outside the building, and the time and costs needed for cleanup [*Molinari et al.*, 2013].

The detail of information acquired is included in Table 6.4.

The forms require the collection of two kinds of data: "objective" data, such as water depth, number of floors, etc., and "subjective" data, such as level of maintenance, the valuation of which may change according to the subjective view of the collector. In order to avoid inconsistencies, criteria are provided for the evaluation of subjective data. For example, the "level of maintenance" is classified according to the definition adopted by the Real Estate and Property Price Database.

The forms were designed to be used in different situations that may derive from the specific features of the houses to be surveyed (which may be apartment buildings, small detached houses, semi-detached houses, etc.). In order to cover the largest possible number of different situations, they are organized into colored sheets corresponding to the building as a whole, the common areas

Table 6.4 Information Collected by Means of the forms for Damage to Residential Buildings [from *Molinari et al., 2013*].

Section	Description	Aspects
Form A: General information		
1. General information	Includes aspects to identify building locations and to describe under what condition the survey was carried out	• geographic coordinates • land registry coordinates • address • who carried out the survey • detail of the interviewed person
2. Building features	Includes aspects to characterize building exposure/vulnerability	• building typology (i.e., detached house, apartment building, semi-detached house, etc.) • period of construction • building structure (e.g., concrete, masonry, wood, steel) • surface • number of floors (and presence of cellar) • building elevation
3. Description of flood event	Includes aspects that are important for characterizing stress on the building	• Permanence of water into building • water depth outside the building • presence of sediments/ contaminants
4. Description of the damage	Includes aspects that are important for identifying affected parts of the buildings and forms to be compiled	• affected parts (i.e., number of housing units, common areas, number of attached buildings, structural damage) • forms to be compiled (i.e., A, B, C, D)
FORM B: Damage to housing unit. N.B. This form must be filled in for every unit in the building.		
1. General information	Includes aspects: • for identifying the property • for describing affected floors • for describing residents	• owner • damaged floors • number of residents, children, elderly people, disabled people
2. Damage to affected floor X N.B. This section needs to be filled in for every affected floor in the unit.	Includes: • further aspects that are required to fully characterize the exposure/vulnerability/ location of the Xth floor as well as the stress on it • all aspects that are required to characterize the direct damage to the floor • aspects relating to indirect damage • aspects relating to mitigation actions	• surface • level of maintenance • technological systems • use (e.g., residential, commercial, storage, etc.) • maximum water depth inside the building at floor Xth • damage to coating/plaster, windows and doors, floor, technological systems, contents • damage due to high velocity • loss of usability • clean-up cost • mitigation actions: type of action, time of action, motivation
FORM C: Damage to common areas		
1. General information	Includes aspects for describing affected floors	• damaged floors
2. Damage to affected floor X N.B. This section needs to be filled in for every affected floor in the common areas.	Includes the same aspects as form B- section 6/2	

(Continued)

Table 6.4 (Continued)

Section	Description	Aspects
FORM D: Damage to attached building. N.B. This form must be filled in for every attached building.		
1. General information	Includes aspects: • for identifying the building locations • for identifying the property	• geographical coordinates • land registry coordinates • owner
2. Building features	Includes aspects for characterizing building exposure/vulnerability.	• period of construction • building structure (e.g., concrete, masonry, wood, steel) • surface • number of floors (and presence of cellar) • building elevation
3. Description of flood event	Includes aspects that are important for characterizing stress on the building	• permanence of water into building • water depth outside the building • presence of sediments/contaminants
4. Description of the damage	Includes aspects that are important for identifying damaged floors	• affected floors
5. Damage to affected floor X N.B. This section need to be filled in for every affected floor in the building	Includes the same aspects as form B- section 6.2	

(entrance, stairs), and individual dwellings. They may be completely or partly filled in, depending on the specific characteristics of the building to be surveyed [*Molinari et al.,* 2013].

Although damage models usually supply and consider damage at a whole building level, compensation is made for each housing unit (including those in the same building). The forms were therefore designed to collect data at both the whole building and sub-building level, satisfying both damage modeling and loss compensation needs.

6.4.4. The Procedure for the Survey of Damage: Forms for Industrial/Commercial Premises

As for residential buildings, the forms for industrial/commercial premises were designed in order to meet information needs for all users. In particular, besides damage to the building structure and its main functional parts (as for residences) damage to the production/commercial unit is also recorded. Information registered in the forms includes damage to machinery and production plants, equipment, raw materials and finished products, and stock. Data to be collected include also actions performed to mitigate the damage, recovery costs, and indirect damage, including lost working days, lost clients, and the consequences for labor [*Molinari et al.,* 2013].

The indicators that are specifically relevant for industries have been drawn, on the one hand, from previous reports available on floods affecting industrial areas in France [see *Ledoux,* 2000]. On the other hand, current practices and procedures for loss accounting and damage

compensation were considered; i.e., information required from business owners by the RCPA, after the 2012 flood, is included in the forms.

The forms are organized into colored sheets corresponding to the difference types of damage that may occur in the premise (such us damage to building structure, damage to machinery, indirect damage, etc.). They may be completely or partly filled in, depending on the occurrence of certain types of damage, guaranteeing flexibility in the implementation of the procedure. Also in this case, evaluation criteria are provided for subjective data.

The detail of information acquired by means of the forms is included in Table 6.5.

6.5. OTHER SECTORS (DATA GATHERING CENTERED)

As previously discussed, for a number of sectors (namely, farms, infrastructures, public items, people, environmental and cultural heritage, and emergency costs) no field survey is necessary because data can be obtained from responsible stakeholders who collect them.

The first step in the definition of the procedure was the identification of data owners/collectors, for every affected sector. Then activities to be performed were identified as summarized in Box 6.3. Table 6.6 summarizes identified activities; timing of action is defined by considering both the nature of data and Italian regulations on damage compensation. The main actor is the Expertise Centre because data gathering can be mostly performed remotely.

Table 6.5 Information Collected by Means of the forms for Damage to Residential Buildings.

Section	Description	Aspects
Form A: General information		
1. General information	Includes aspects to identify building locations and to describe under what condition the survey was carried out	• geographic coordinates • land registry coordinates • address • who carried out the survey • detail of the interviewed person
2. Premises features	Includes aspects to characterize premise vulnerability	• type of activity (commercial/industrial) • commercial/industrial sector • number of employees • seasonal criticalities (yes/no) • special plant (yes/no)
3. Building features	Includes aspects to characterize building exposure/vulnerability	• building typology (i.e., single building/single warehouse/multiple warehouse, building portion, warehouse portion) • property/rent • period of construction • building structure (e.g., concrete, masonry, wood, steel, prefab) • external areas • level of maintenance • surface • number of floors (and presence of cellar) • building elevation
4. Description of flood event	Includes aspects that are important for characterizing stress on the building	• permanence of water into building • water depth outside the building • presence of sediments/contaminants
5. Description of the damage	Includes aspects that are important for identifying affected parts of the buildings, damage to employees and forms to be compiled	• affected parts (i.e., damage to building structure and plants, damage to machinery, production plants, equipment and furniture, damage to store and archives, recovery and mitigation costs, damage to mobile goods, indirect damage: usability, activity disruption) • damage to employees • forms to be compiled (i.e., A, B, C, D, E)
FORM B: Damage to building structure and plants		
1. Direct damage to building structure	Includes aspects for identifying main damage to the building structure	• damaged floors • structural damage • damage to external coating/plaster • surface
2. Direct damage to affected floor X N.B. This section needs to be filled in for every affected floor in the building.	Includes: • further aspects that are required to fully characterize the vulnerability of the floor as well as the stress on it • all aspects that are required to characterize the direct damage to the floor	• use (e.g., storage, production, shop, office, etc.) • maximum water depth inside the building • damage to: coating/plaster, windows and doors, floor, plants
FORM C: Damage to machinery, production plants, equipment, furniture, store, archive, and mobile goods		
1. Damage to store and archive	Includes aspects for describing damage to store and archive	• damage to stock (raw material, intermediate products, finished products) • damage to paper documents (accounting books, client registers, etc.)

(Continued)

Table 6.5 (Continued)

Section	Description	Aspects
2. Damage to machinery, production plants, equipment, and furniture	Includes aspects for describing damage to machinery, production plants, equipment, and furniture	• damage to machinery (electrical appliances at work/standby, mechanical appliances on work/standby, thermal appliances on work/standby, etc.) • damage to production plants (goods lift, pumps, hydraulic plants, specific plants) • damage to equipment • damage to informatics equipment • damage to furniture
3. Damage to mobile goods	Includes aspect for describing damage to company vehicles	• number of damaged vehicles • types of damaged vehicles (motorbikes, car, van, trucks, forklift, etc.)
FORM D: Recovery/mitigation costs		
1. Recovery costs	Includes aspects for defining recovery costs	• clean-up costs (private expenditure vs. public costs) • clearing and disposal of mud, debris, and flooded material
2. Mitigation costs	Includes aspects for identifying mitigation action taken before and after the flood and their effectiveness	• previous experience (yes/no) • mitigation actions before the event (documents back-up, insurance, etc.) • mitigation actions during the alarm phase (suction pumps, flood shields, stock movements, evacuation, power interruption, etc.) • mitigation actions after the event (archive/store relocation to upper floor, plants/machinery rising above the flood level, use of waterproof material)
FORM E: Indirect damage/reimbursements		
1. Indirect damage	Includes aspects that are important for identifying indirect damage	• lack of usability (days) • activity disruption (days) • missed orders • unemployment (number of people and days) • damage due to humidity
2. Reimbursements	Includes aspects to quantify reimbursement	• private reimbursement (from insurance) • public reimbursement

---

Box 6.3 Stress tests related to data gathering.

Several meetings were organized with data owners in order to understand the procedures they currently implement and the data they collect/own, and to define protocols for data sharing.

The process revealed to be difficult so that, as already discussed in section 6.3, a formal agreement on data sharing has not been reached at present with most of the data owners, but it is the responsibility of the Expertise Centre to negotiate with them every time a flood occurs. Nevertheless, meetings provided, on the one hand, an understanding of stakeholders' needs and constraints for data sharing, which facilitates the negotiation process. On the other hand, most of the required data for the production of a complete event scenario were collected during the meetings, regarding the floods that hit the Umbria region in 2012 and 2013 (see Chapter 11 for the description of scenarios). Still, lack of formal agreements is a limitation for the efficiency of the procedure that must be overcome in the future.

Impediments in data sharing were claimed mostly by private owners (i.e., utilities companies) being related to sensitive information, bureaucratic reasons (especially true for big/national companies with a strong hierarchical structure), and the lack of earning for the company in sharing data. The last point was overcome by providing examples on how information on damage to other interconnected systems can be useful for both managing the emergency phase and for speeding up the recovery phase, for a specific system/sector. Lack of resources was another reason of concern for both public and private subjects. Therefore, it is important that sharing protocols incorporate as many existing practices and procedures as possible to minimize efforts required to implement RISPOSTA. Defining such protocols is the objective of a second set of (ongoing) meetings, aiming at integrating existing practices for data collection and management into the procedure also by means of the ad hoc developed ICT system (see Chapter 15).

Bureaucratic impediments were also found for certain sectors, like agriculture and environment. This is because, for these sectors, both the emergency and the damage compensation phases are not managed by the RCPA but by specific regional divisions so that no links existed among the authorities before the development of the procedure.

Table 6.6 Scheme of the RISPOSTA Procedure for Data Gathering: Activities to be Performed, Times of Actions, and Responsible Actors (Extract from Table 6.1).

ACTIVITIES	ACTOR			ACTION		SECTOR					
	RCPA	Expert. Centre	Survey	Gathering	Coordination	Farms	Infrastructures	Public items	Emergency	People	Environ./cultural
Acquisition of pre-existing knowledge on exposure and vulnerability	X			X		X	X	X	X	X	X
Event											
Acquisition of damage data from the Regional Emergency Room (SOUR)		X		X		X	X	X	X	X	X
90 days											
Acquisition of damage data from the responsible stakeholders (optional)		X		X		X	X	X	X	X	X
Acquisition of monetary damage data		X		X		X	X	X	X	X	X
6 months											
Acquisition of damage data from the responsible stakeholders		X		X		X	X	X	X	X	X
Acquisition of monetary damage data		X		X		X	X	X	X	X	X
12 months											

Activities to be performed, sector by sector, consist in detail of the following:

1. Acquisition of already existing knowledge on the exposure and vulnerability of potentially affected items by extracting information from existing databases (e.g., thematic maps) or by acquiring data directly from data owner(s). Such data (like location and types of strategic buildings, roads, electric lines, etc.) are used as the basic information for the development of complete event scenarios (see Chapter 11).

2. Acquisition of damage data from the Regional Emergency Room (SOUR). As introduced in section 6.2, most of data owners/collectors must communicate (by law) to the RCPA (and to the SOUR in particular) significant information for the management of the emergency phase. This information typically include direct damage to infrastructures, strategic buildings, and people.

3. Acquisition of damage data from the responsible stakeholders. Information collected during the previous step is completed (when required) with information supplied by subjects/authorities responsible for data collection. This action can be performed twice, a few months and several months after the event to get information on both direct and longer-term damage.

4. Acquisition of monetary damage data. As for field survey, information on damage is collected primary in physical units. The monetary values of damage are acquired from the Accounting Division of the RCPA, whereby the latter is in charge of acquiring all reimbursement

requests made by public and private subjects (when eligible). Monetary values can also be acquired directly from data owners.

6.6. DATA COORDINATION

Data coordination is performed in the Umbria region by means of an ad hoc developed Information System (IS), which is described in detail in Chapter 15. The system allows acquiring, storing, and analyzing all the data acquired by means of the procedure.

Activities to be performed under this action are described in Table 6.7 and consist of the following:

1. Inputting and validating data. Data acquired by means of the procedure are first validated to verify their quality and completeness. When required, data are integrated by means of additional field surveys (see section 6.4) or

by asking data owners (see section 6.5). Then data are input into the system database. The timeline for such activities is sector specific.

2. Data analysis. Collected data are elaborated with two main aims: 1) to produce maps/data that are required for the different activities of the procedure (e.g., maps for the survey of the flooded buildings) and 2) to extract all the information to be included in the complete event scenarios. Because such elaborations are required at different steps along the development of the procedure, data analysis is performed at different times.

3. Data sharing. Collected data are shared with actors identified in section 2. According to data of interest and security restrictions, specific subsets of data are shared with the different actors that can query the database at any time before and after the occurrence of the flood event.

Table 6.7 Scheme of the RISPOSTA Procedure for Data Coordination: Activities to be Performed, Times of Actions, and Responsible Actors (Extract from Table 6.1).

ACTIVITIES	ACTOR		ACTION		
	RCPA	Expert. Centre	Survey	Gathering	Coordination
Set-up and management of the IS		X			X
Data sharing		X			X
Event					
Data sharing		X			X
2–3 days					
Data analysis (field survey of flooded areas)		X			X
Data validation (physical event)		X			X
Inputting data (physical event)		X			X
Data sharing		X			X
20 days					
Data analysis (field survey of residences and business)		X			X
Data validation (residences/businesses/SOUR)		X			X
Inputting data (residences/businesses/SOUR)		X			X
Data sharing		X			X
90 days					
Data validation (all sectors except businesses/monetary damage data)		X			X
Inputting data (all sectors except businesses/monetary damage data)		X			X
Data analysis (field survey of residences, complete event scenario)		X			X
Data sharing		X			X
6 months					
Data validation (all sectors except residences/monetary damage data)		X			X
Inputting data (all sectors except residences/monetary damage data)		X			X
Data analysis (field survey of businesses, complete event scenario)		X			X
Data sharing		X			X
12 months					

6.7. CONCLUSIONS

This chapter has presented RISPOSTA a new procedure for the collection of damage data in the aftermath of floods. The procedure can be included in the general effort, made at the European and international level, to standardize ways of collecting, storing, and analyzing disaster data (see Chapter 1 and Chapter 2). The procedure was developed and implemented according to the specific characteristics of the Umbria region, in Italy, as a support tool for managing emergency and recovery phases.

Integration of data and procedures is one of the main achievements of RISPOSTA. The procedure is conceived to furnish a common platform through which all data relevant to the definition of ex-ante and ex-post risk mitigation strategies are gathered and stored and all relevant data owners and users are identified and coordinated. This way consistent and reliable data can be furnished to decision makers, practitioners, risk analysts, and so on. In particular, the multi-usability of collected data is achieved, which is a key requirement, given the efforts required by data collection.

Transferability is another strength of the procedure. Its test implementation after the floods that hit the Umbria region in 2012 and 2013 led to the definition of complete event scenarios as deeply described in Chapter 11, proving the robustness of the general structure of the procedure and of its main distinctive features, such as the sector-based approach, the collection of data on damage and their explicative variables, and the integration of existing practices and responsible stakeholders. On the other hand, stress tests discussed in the boxes corroborated the feasibility of individual activities included in the procedure. Therefore, we expect it to be well transferable to different juridical and physical contexts. The general framework of the procedure may remain substantially unchanged, although some specific aspects would certainly require revision and adaptation to the system under investigation.

For example, authorities responsible for data gathering may change from one juridical context to another. Collecting methods may be adapted to the phenomena characterizing the specific region so that, in the case of riverine floods lasting for several days over wide areas, survey of the flooded area can be avoided because of the existence of reliable satellite images. Collecting methods may change also because of different institutional contexts. In particular, field surveys may be required for sectors other than private residences and commercial/industrial units. In this case, the experience described here highlights the need to design specific survey modalities and tools for each sector concerned, by taking into account the features of exposed items and expected damage.

It may also happen that data survey/collection cannot be performed for certain sectors (for example, because of a lack of resources or because of bureaucratic impediments, that is, private companies do not consent to data sharing) or cannot be carried out at reasonable costs. The RISPOSTA procedure is flexible with respect to this, as the collection/gathering of data for a certain sector is not affected by the availability or unavailability of data from other sectors.

In the case of data missing for certain sectors, the procedure supplies a partial vision of damages occurred. This must be explicitly reported when collected data are presented to decision makers to avoid confusion between "lack of data" and "unaffected sectors" [Ballio et al., 2015].

Previous considerations are equally valid when transferability to other hazards is considered. In this case, however, particular care must be paid also in adjusting time of activities according to the evolution of the physical phenomenon under investigation, as well as in collecting information on those variables that are explicative for the damage it causes.

Future developments must be focused, on the one hand, to achieve a better integration with data owners, making them "active" actors in the procedure. To this aim, participatory meetings are already planned in the next months in the Umbria region with the main objective of defining specific protocols for data sharing. On the other hand, efforts should focus on extending the applicability of the procedure to other physical (e.g., flash floods) and juridical contexts (e.g., other countries) as well as to other hazards (e.g., landslides).

ACKNOWLEDGMENTS

RISPOSTA has been developed as part of the Poli-RISPOSTA project (stRumentI per la protezione civile a Supporto delle POpolazioni nel poST Alluvione) within the Poli-SOCIAL funding scheme of Politecnico di Milano.

The authors acknowledge with gratitude all those involved in the Poli-RISPOSTA project, the Regional Civil Protection Authority of the Umbria region, students, and volunteers who actively took part in the stress tests.

REFERENCES

Associated Programme of Flood Management (APFM) (2013), Conducting flood loss assessment, Integrated flood management tools, Series n. 2 [online]. Available at: http://www.apfm.info/?page_id=787.

Ballio, F., D. Molinari, G. Minucci, M. Mazuran, C. Arias Munoz, S. Menoni, F. Atun, D. Ardagna, N. Berni, and C. Pandolfo (2015), The RISPOSTA procedure for the collection, storage and analysis of high quality, consistent and reliable damage data in the aftermath of floods, Journal of Flood Risk Management, on-line.

Emergency Management Australia (EMA) (2002), Disaster Loss Assessment Guidelines, Australian Emergency Manuals Series, on-line. Available at: https://www.em.gov.au/Documents/Manual27-DisasterLossAssessmentGuidelines.pdf.

Global Facility for Disaster Risk Reduction (GFDRR) (2013), Post Disaster Needs Assessment, Volume A, on-line. Available at: http://www.recoveryplatform.org/assets/publication/PDNA/PDNA%20Volume%20A%20FINAL%20for%20Web.pdf.

Ledoux, B. (2000), Guide pour la Conduite des Diagnostics des Vulnérabilités aux Inondations pour les Entreprises Industrielles–Rapport Final, Ministère de l'aménagement du territoire et de l'environnement, Direction de la prevention des pollutions et des risqué, Sous-direction de la prévention des risques majeurs,

Menoni, S. (2001), Chains of damages and failures in a metropolitan environment: some observations on the Kobe earthquake in 1995. Journal of Hazardous Material, *86* (1–3), 101–19.

Molinari, D., S. Menoni, G. T. Aronica, F. Ballio, N. Berni, C. Pandolfo, M. Stelluti, and G. Minucci (2013), Ex-post damage assessment: an Italian experience. Nat. Hazards Earth Syst. Sci., *14*, 901–916.

7

Data Collection for a Better Understanding of What Causes Flood Damage–Experiences with Telephone Surveys

Annegret Thieken[1], Heidi Kreibich[2], Meike Müller[3], and Jessica Lamond[4]

ABSTRACT

Starting after the severe flood event of August 2002, computer-aided telephone interviews (CATI) have repeatedly been used in Germany to remotely collect large standardized data sets on flood losses and possible influencing factors by questioning affected residents and company owners. In comparison to on-site surveys, the data cover a larger number of affected properties in different environmental and socioeconomic settings, as well as a variety of factors that potentially influence the amount of damage. In this chapter, sampling strategies, questionnaires, and problems encountered when questioning residents and business owners are outlined. Further, the method (CATI) will be compared to other survey techniques. The chapter is complemented by a reflection of approaches in the United Kingdom (UK).

7.1. INTRODUCTION

With the shift from flood defense policies to integrated flood risk management strategies, there is a growing demand for models that are capable of estimating flood losses, particularly direct economic losses on the local and regional scale. For example, structural flood defense schemes and other flood risk reduction measures are increasingly evaluated by means of cost-benefit analyses or multi-criteria analyses, with the aim of finding the best option and optimizing investments on an economic basis

[e.g., *US Army Corps of Engineers (USACE)*, 1996; *Al-Futaisi and Stedinger*, 1999; *Ganoulis*, 2003; *Penning-Rowsell et al.*, 2005]. Along the same lines, flood hazard maps have been supplemented by information on potential losses to enhance risk awareness. For example, the European Union (EU) Floods Directive (EU/2007/60) demands the determination of economic activities that could be affected by flooding. Such analyses are also essential for the comparison of the impacts of different perils [*Jonkman*, 2005; *Grünthal et al.*, 2006]. Besides meteorological, hydrological, and hydraulic investigations, these flood risk analyses require models to estimate flood impacts, such as direct losses.

Furthermore, the (re-)insurance industry needs financial loss modeling for risk appraisals, especially for the estimation of probable maximum losses (PML), in order to guarantee their solvency [e.g., *Kron and Willems*, 2002; *Changnon*, 2003]. Direct insurance underwriters may also need detailed information on flood losses for setting up incentives for property-level mitigation measures or

[1] *Institute of Earth and Environmental Science, University of Potsdam, Potsdam, Germany*

[2] *Section Hydrology, GFZ German Research Centre for Geosciences, Potsdam, Germany*

[3] *Deutsche Rückversicherung NatCat-Center, Düsseldorf, Germany*

[4] *Architecture and the Built Environment, University of the West of England, Bristol, UK*

Flood Damage Survey and Assessment: New Insights from Research and Practice, Geophysical Monograph 228,
First Edition. Edited by Daniela Molinari, Scira Menoni, and Francesco Ballio.
© 2017 American Geophysical Union. Published 2017 by John Wiley & Sons, Inc.

different premiums for different building types (such as buildings with or without a basement) as suggested by *Müller and Kreibich* [2005]. In the case of a flood, first loss estimates or collection of first losses may also provide a basis for decisions about the kind and amount of disaster relief assistance [*Downton and Pielke*, 2005]. Currently, such estimates are highly uncertain [e.g., *Changnon*, 1996; *DKKV*, 2015] and highlight the need to improve loss estimation models.

To develop reliable predictive loss models or loss functions, factors and processes that influence flood damage have to be analyzed, understood, and finally quantified. Accurate, comparable, and consistent empirical data on flood impacts and potentially influencing factors, gathered on the level of individual flood-affected properties serve as a good basis for this purpose. Commonly, depth-damage curves are used for estimating direct flood losses at the property level [*Merz et al.*, 2010]. These curves link the property loss to the maximum flood water level at that property (i.e., above surface level). Different types of buildings might be considered by different curves. An alternative to derive such curves from empirical data on actual events are so-called "what if"-analyses. Thereby, potential losses are estimated by building surveyors for several potential flood scenarios and resulting water levels in the building under study [see *Penning-Rowsell et al.*, 2005].

Currently available loss data tend to be aggregated either on a regional or national level, or they are gathered on the property scale foremost for the purpose of loss compensation. Such data sets often fail to link the property or event loss to hazard characteristics or characteristics of the affected property or society; that is, only a few, if any, explanatory variables such as the water level are available in such data sets. Hence, they cannot be used to derive depth-damage curves or more complex loss models.

Furthermore, data analyses revealed that loss data show large scatter [*Blong*, 2004; *Merz et al.*, 2004]. Water depth and building use, the most frequently considered parameters in depth-damage curves, only explain a proportion of the data variance [*Merz et al.*, 2004]. However, if additional information about the affected object, flood warning, precaution, etc. is missing, data variability cannot be further explained. Consequently, loss modeling remains insufficient. To fill this gap, detailed surveys have been performed among flood-affected residential and commercial properties [e.g., *Ramirez et al.*, 1988; *Joy*, 1993; *Gissing and Blong*, 2004; *Thieken et al.*, 2005; *Zhai et al.*, 2005; *Kreibich et al.*, 2007; *Lamond et al.*, 2009; *Joseph et al.*, 2011; *Bhattacharya and Lamond*, 2014; *Joseph et al.*, 2015; *Kienzler et al.*, 2015; *DKKV*, 2015].

Since an extreme flood hit Central Europe in August 2002, recurrent cross-sectional surveys have been conducted in Germany as CATIs after each big river flood and some pluvial flood events [see *Thieken et al.*, 2005; *Kreibich et al.*,

2007; *Kienzler et al.*, 2015; *DKKV*, 2015]. These data sets have been used to successfully derive and validate the model family called the Flood Loss Estimation MOdel (FLEMO) on the micro as well as on the meso scale [*Büchele et al.*, 2006; *Thieken et al.*, 2008; *Apel et al.*, 2009; *Wünsch et al.*, 2009; *Elmer et al.*, 2010; *Kreibich et al.*, 2010; *Seifert et al.*, 2010; *Thieken et al.*, 2015]. *Cammerer et al.* [2013] successfully adapted the model to Austrian conditions. Recently, more sophisticated models were derived by data-driven approaches such as Bayesian networks or regression trees [*Merz et al.*, 2013; *Schröter et al.*, 2014]. This book chapter aims to provide some insights into the assumptions and approaches that were used in the German surveys to enable the transfer of the approach to other regions (see section 7.3). This elaboration is contrasted by a discussion of the UK experience, where the distribution of flood losses may make such an approach more challenging to apply (see Box 7.1). To facilitate the comprehensibility of this chapter, a general introduction to survey methods and sampling strategies is provided in the next section.

7.2. SURVEY METHODOLOGIES AND SAMPLING STRATEGIES

For scientific flood damage analyses, surveys are undertaken that cover a wide range of parameters, i.e., not only the damage is monitored but also the characteristics of the flood situation. For example, water level, flow velocity, contamination, flood duration or sediment load are monitored, as well as the characteristics of the affected object, such as building type and use, precautionary measures in place, early warning, and emergency measures undertaken. Commonly, the objective of such surveys is the identification of important damage influencing parameters in order to better understand damaging processes. This knowledge and data may be used to support flood risk assessment and management in developing and validating reliable predictive loss models (see Introduction) as well as in suggesting effective precautionary and emergency measures.

In contrast to the collection of data for loss adjustment or disaster relief that cover the whole affected area, in most of the scientific surveys, only a representative sample is investigated because an investigation of the total population, for example, all damaged (residential/commercial) properties, is too expensive. Therefore, a representative sample is aimed for, which shall cover the whole range of affected cases, e.g., different regions, flood situations, and building types. Commonly, property-specific representative random samples of all affected households, companies, etc., from all over the flooded area are created [*Thieken et al.*, 2005; *Kreibich et al.*, 2007]. In particular cases, lumped samples are preferred, i.e., some representative municipalities are chosen (randomly) for the investigation, where a census of all damaged properties is then undertaken [*Buck*

Box 7.1 Experiences from the United Kingdom

Large-scale CATI surveys have not been carried out in the UK for the purpose of damage modeling. English national risk assessments are based on the probability loss curves developed for the Middlesex tables [for example, *Flood Hazard Research Centre*, 2010]. These have been developed based on damage data collected in the early stages of analysis coupled with synthetic damage assessments conducted by experts.

The damage estimates have hardly been subjected to validation, and according to *Penning-Rowsell* [2015], they may lead to inaccuracy in estimation of risk for England and Wales. Conversely, data from insurers is hard to access and ambiguous in meaning. Usually such data represent financial loss, i.e., replacement of old items with new ones. Typically, these sources produce much higher average event damages [*Black and Evans*, 1999]. However, the purpose of such analyses has traditionally been to evaluate the costs and benefits of large-scale flood defenses or expected national losses rather than to assess the detail of property scale precautionary measures. Therefore, detailed surveys have neither been required nor carried out. This is partially a function of the UK insurance regime, whereby government is not involved in the reinstatement of damage to property and insurers may collect, but rarely collate and share detailed damage data in relation to property and flood characteristics. In addition, the properties at risk in England and Wales are dispersed geographically with an equal amount of properties at risk from surface as from river and coastal flooding, rendering repeat surveys of major river basins less applicable in the UK. Furthermore, information regarding properties flooded or at risk, has not generally been available to researchers. Therefore, typically, surveys of damage within the UK have been undertaken on smaller scales as part of diverse research projects and not in a consistent fashion. With the lack of a national database, these surveys have tended to be lumped samples, focusing on areas identified as badly affected.

A growing body of work designed to improve loss modeling within UK-based studies, builds on the synthetic data derivation approach that is part of the underlying database for the Middlesex depth-damage curves [*Messner et al.*, 2007; *Thurston et al.*, 2008; *Association of British Insurers*, 2009; *Adaptation Sub Committee*, 2012; *Consultants*, 2012]. Expert assessment of the damage and loss prevented by precautionary measures is beginning to be brought into the standard risk assessment methodology and cost-benefit analysis, as UK government agencies pursue devolution of management of flood risk and wish to compare alternative strategies for investment on a like for like cost-benefit basis.

Empirical data to validate the desk-based estimates are a critical necessity if the credibility of such estimates is to be maintained. However, the low level of uptake of measures in the UK is an existing barrier to empirical data collection.

A comparison of actual data with modeled damage repair data was provided by the *Association of British Insurers* [2009]. This report is based on damage data from the largest recent UK flood event of 2007 and illustrates the difficulty with using real damage data to build comprehensive databases of damage for all property types, as 4 out of 10 defined UK property construction types were not represented in the damage data. However, comparison of the modeled against actual repair costs also revealed that the modeled costs (based on a "typical" property) consistently underestimated the actual repair costs (obtained from insurance company records) averaged across the properties actually repaired after the 2007 event. Part of this uplift was attributed to the increased costs of reinstatement experienced in the aftermath of a major event when labor and materials are at a premium due to demand factors. Other possible causes of differences include the complexity of housing stock in terms of different sizes, layouts, and quality of internal fittings and contents. For example, differences in the cost of a fitted kitchen, one of the most expensive components to replace, can dominate the reinstatement cost. Furthermore, *Lamond* [2008] observed limitations in householders' recall of direct costs of damage repair following the 2000 flood event. Therefore, in their recent UK survey of households, *Joseph et al.* [2015] used actual damage data taken from insurance databases as the basis for a hybrid data set adding estimates of indirect and intangible impacts derived from census surveys of the same flood affected population. The use of financial, rather than economic, damage data being justified by the purpose of the research targeted at informing costs and benefits at a household level where, in practice, the majority of reinstatement models replace old with new.

A recent survey of commercial property occupiers, also based on the 2007 flood event in the UK [*Bhattacharya-Mis et al.*, 2015], revealed relatively low levels of direct damage from flooding and higher concerns about indirect damage and disruption. In making these assessments (in line with the

EU Flood Directive), it is more difficult to rely on expert judgment alone. The complexity of the building fabric is multiplied hugely by the variety of operational assets and business aspects specific to commercial sector, customer base, and supply chain factors.

This excursus illustrates that the ability to implement a method to collect loss data continuously not only depends on methodological and financial considerations but also on governance aspects and the role governmental institutions, insurer and scientists have in flood risk management.

and Merkel, 1999]. This approach is particularly advantageous for on-site expert surveys or personal interviews since it reduces the necessary traveling time for the experts or interviewers. When different affected regions can be clearly distinguished by a certain characteristic, e.g., socio-economic or flood event characteristics, stratified samples can also be considered as shown in *Thieken et al.* [2007].

In principle, different methods are available for data collection: on-site expert surveys or (semi) structured interviews with affected residents, company owners, etc. via face-to-face or telephone interviews or by a written or online survey.

On-site expert surveys are based on standard forms containing all items on which information is to be collected by the experts (see Chapter 6). It is vital for the data usability and quality that all information items are documented with the same level of detail. A high degree of standardization as proposed, for example, by *Molniari et al.* [2014], tries to minimize the subjectivity of different experts and has the advantage of a homogenous loss assessment for all objects and events. However, detailed information of individual cases, particularities, and extremes blur. Good training and experience of the experts are decisive for the quality of the data, as are standards, e.g., common price and service lists, and other well-documented and commonly applied methods [see *Henderson et al.*, 2009; *Molniari et al.*, 2014]. *Henderson et al.* [2009] further stress the importance of surveyors who are familiar with the region and the local culture. The comparatively high cost is a general disadvantage of on-site expert surveys. To minimize costs, representative areas are chosen and surveyed as a lumped sample (see above); this approach was conducted for the old German flood loss database HOWAS [*Buck and Merkel*, 1999].

The methods for surveying affected households, companies, and so on differ in their level of commitment (cooperation of the interviewees), the amount and type of possible questions, the size of the surveyed area as well as in their (unit) response rates and costs. With most survey methodologies, it is possible to cover large areas at relatively low costs. However, the survey environment is not static. Particularly, response rates of telephone surveys are changing. *Greenberg and Weiner* [2014] report that response rates of telephone surveys dropped from about 70–80% to about 20–25% over the last years in the United States, going along with a rapid replacement of landline telephones by cell phones. The representativeness of samples from telephone surveys might thus be challenged. Good sampling and high response might also be difficult to achieve by online surveys.

Survey answers depend on the subjective valuation of the interviewed people and may therefore vary from person to person, which might be a problem when collecting information on quantitative items such as water levels or damage costs. Thus, it is necessary, although difficult, to collect the best quality of data that is possible, to check the reliability of the answers and to judge the data quality. To collect most accurate answers and to avoid a strategic response bias, questionnaires should not only contain one question about an important fact, e.g., the flood damage, but various questions characterizing an information item from different perspectives. Additional control questions may also be advantageous to avoid data inconsistencies. For instance, a questionnaire may contain detailed questions addressing not only the total damage but also the area affected per story, the damage ratio, the type and amount of the most expensive item damaged, and the type and costs of all building repairs and all expensive domestic appliances affected [*Kreibich et al.*, 2011]. The interviewee could be asked to look up the amount of costs for repair works in the respective bills or to check what he or she has claimed for loss compensation from government funds or from their insurer. This is, however, easier to be implemented in written or online surveys than during telephone interviews. A written advance notice could overcome this problem and might also increase response rates [*Hüfken*, 2000]. The effect of the survey methodology on quality of loss data and accompanying information is not well known. However, a comparison of surveyed damage data collected after the 2002 flood by computer-aided telephone interviews with official damage data from the Saxon Bank of Reconstruction, which was responsible for administering governmental disaster assistance after the 2002 flood in the federal state of Saxony, confirmed good agreement of loss information [*Thieken et al.*, 2005]. (See section 7.3.3.)

An alternative approach to the collection of data about a real flood event is the collection of synthetic damage data by experts (buildings surveyors) for different water levels and building types [*Penning-Rowsell et al.*, 2005], mainly with the aim to develop flood loss models. Synthetic damage surveys or "what if"-analyses estimate the damage, which is expected in case of a certain flood situation, for example, "Which damage would you expect if the water depth was 2m above the building floor?" Examples of this approach are included in the damage functions for the United Kingdom [*Penning-Rowsell et al.*, 2005]. In comparison with the collection of empirical damage data, the synthetic approach has advantages and disadvantages [*Merz et al.*, 2010]. This approach does not rely on actual flood events and can therefore be applied to any area at any time [*Smith*, 1994], which is advantageous. It is particularly recommended for asset types and regions for which empirical damage data is not available. In each building, damage information for various water levels can be retrieved [*Penning-Rowsell and Chatterton*, 1977], and commonly a higher level of standardization and comparability of damage estimates is achieved. A disadvantage of the synthetic approach is the high effort necessary to develop detailed databases (inventory method) or undertake large surveys (valuation survey method) to achieve sufficient data for each category, i.e., building type [*Smith*, 1994]. "What if"-analyses may be more subjective, resulting in uncertain damage estimates [*Gissing and Blong*, 2004; *Soetanto and Proverbs*, 2004]. Premises within one classification can exhibit large variations in susceptibility that are not reflected

by the data [*Smith*, 1994], e.g., due to precautionary measures that are commonly not taken into account in synthetic surveys [*Smith*, 1994] or by assumptions on their effectiveness [e.g., *Escuder-Bueno et al.*, 2012].

7.3. LOSS DATA COLLECTION IN GERMANY AFTER SEVERE FLOOD EVENTS

In August 2002, a severe flood event hit Central Europe causing EUR 11,600 million of damage and 21 fatalities in Germany [*Thieken et al.*, 2006]. This flood event has triggered many changes at all levels of flood risk management in Germany (See *DKKV*, 2015 for an overview). It can also be seen as a starting point for new loss data collection initiatives. As outlined by *Thieken et al.* [2005] and *Kreibich et al.* [2007], extensive standardized surveys were conducted among flood-affected households and companies, respectively. Similar large-scale damage surveys based on CATIs were carried out after river floods in August 2005, April 2006, August 2010, January 2011, and June 2013 in Germany as well as after urban flooding in 2010 and 2014 in the cities of Osnabrück and Münster, respectively. (See Table 7.1 and Table 7.2.) The main aim of the surveys was to identify and quantify factors that influence flood damage to improve flood loss estimation models. It is obvious from Table 7.1 and Table 7.2 that the timing of the surveys differed very much, which was mainly due to the availability

Table 7.1 Chronological Overview of Flood Events and Related Household Surveys in Germany. In the Column "Timing," the Number of Months Between the Damaging Event and the Data Collection is Given.

Flood event (River flood)	Number of interviews	Field time	Timing	Sampling
August 2002	1697	8 Apr to 10 Jun 2003	8–10	Building-specific random sample, stratified by flood type, previously experienced flooding, and socio-economy: 1. The River Elbe and the lower Mulde River 2. The Erzgebirge (Ore Mountains) and the River Mulde in Saxony 3. The Bavarian Danube catchment
August 2005	305	20 Nov to 21 Dec 2006	15–16	Full coverage
April 2006	156	20 Nov to 21 Dec 2006	7–8	Full coverage
August 2010	349	16 Feb to 20 Mar 2012	18–19	Full coverage
January 2011	209	16 Feb to 20 Mar 2012	13–14	Full coverage (Municipalities upon the River Elbe were excluded.)
June 2013	1652	18 Feb to 24 Mar 2014	8–9	Building-specific random sample, stratified by flood type, previously experienced flooding, and socio-economy: 1. Residents in Baden-Wurttemberg, Lower-Saxony, and Thuringia 2. Residents in the Elbe catchment 3. Residents in the Bavarian Danube catchment
Urban flooding				
Osnabrück, August 2010	91	16 Feb to 20 Mar 2012	18–19	Full coverage
Münster/Greven, July 2014	510	Nov and Dec 2015	16–17	Full coverage

Table 7.2 Chronological Overview of Flood Events and Related Surveys Among Flood-Affected Companies in Germany. In the Column "Timing," the Number of Months Between the Damaging Event and the Data Collection is Given.

Flood event (River flood)	Number of interviews	Field time	Timing	Sampling
August 2002	307	Oct 2003	14	• Site-specific random sample in Saxony
	+108	May 2004	21	• Large-scale companies in Saxony
	+64	19 to 30 Oct 2006	50	• Companies upon the River Elbe downstream of Saxony
August 2005	102	19 to 30 Oct 2006	14	Full coverage
April 2006	61	19 to 30 Oct 2006	6	Full coverage
August 2010	60	May to Jul 2013	33-35	Full coverage
January 2011	58	May to Jul 2013	28-30	Full coverage
June 2013	557	12 May to 17 Jul 2014	11-13	Full coverage

of resources to finance the surveys. In general, loss data collection should allow a timeframe of at least 3 to 6 months after the event [*Queensland Government*, 2002] since indirect and intangible damage might occur later in time. The timeframe should, however, not exceed five years.

7.3.1. Sampling Strategies

Lists of inundated streets and zip codes served as a basis for the selection of telephone numbers of all potentially affected residents or companies. These lists were generated for the survey after the 2002 flood on the basis of information that had been provided by the affected communities and districts upon request. In successive surveys, these lists were compiled based on flood reports or press releases, as well as with the help of flood masks derived from satellite data (DLR, Center for Satellite-Based Crisis Information, www.zki.dlr.de). For the surveys of companies, the lists were complemented by company names that were explicitly mentioned in flood reports. For the flood of 2002, the Saxon chamber of industry and commerce provided 183 addresses of affected large-scale companies that were surveyed separately [*Kreibich et al.*, 2007]. (See Table 7.2.)

Telephone numbers were generally retrieved from the public telephone directory or the commercial telephone directory (yellow pages). The subcontracted pollster, SOKO GmbH, Bielefeld or explorare GmbH, Bielefeld, retrieved the numbers and performed the sampling. For the household survey of the 2002 flood, a building-specific random sample of households was generated per stratum (see Table 7.1) in order to include a broad variation of hydrological and socio-economic conditions [*Thieken et al.*, 2007]. A similar sampling was applied in the survey of the 2013 flood [*DKKV*, 2015]. Since the other floods affected a smaller area and were not as severe as the floods of 2002 and 2013, the surveys started with a random sampling, which was later replaced by a full inclusion of all retrieved telephone numbers. Besides refusal, difficulties conducting a larger number of inter-

views resulted from the fact that approximately 40% of households called had not been affected by flooding [*Kienzler et al.*, 2015]. It is unknown how many non-affected people are included in the group who could not be reached during the field time or who refused to participate, therefore, (unit) response rates cannot be calculated reliably. This underlines the fact that more precise spatial information about the maximum flood extent would help to improve the lists of affected streets that are the basis for the sampling process [*Kienzler et al.*, 2015].

7.3.2. Contents of the Questionnaire

The damage of a building is influenced by many more factors than just the water level. As proposed in Figure 7.1, it is dependent upon the load on the structure on the one hand and its resistance on the other hand [*Thieken et al.*, 2005].

Following this theoretical model, a questionnaire with the following topics was developed [*Thieken et al.*, 2005]:

• Characteristics of the flood: water level, flood duration, flow velocity, type of flooding (e.g., sudden groundwater rise, overflow of the sewer system, river inundation, flash flood), contamination of the flood water

• Flood warning: flood warning source, flood warning information, lead time

• Emergency measures (short term): undertaken emergency measures, number of people involved in emergency measures, time spent on emergency measures, perceived effectiveness of the undertaken measures

• Information on the building: building type, construction material, number of stories, total living area, type of basement, utilization of the basement

• Precautionary measures (long term): type of measure, time of realization, perceived effectiveness of self-protective measures

• Flood experience: number of previously experienced events, date of last experienced flood event, knowledge about living in a flood-prone area

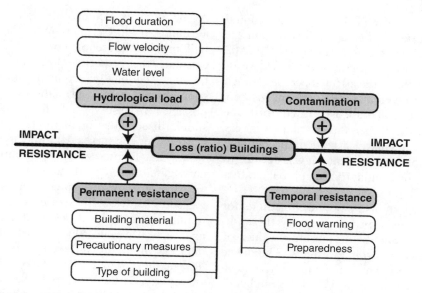

Figure 7.1 Underlying theoretical model used to operationalize the contents of the flood loss questionnaire in Germany [adapted from *Thieken et al., 2005*].

• Loss: repair costs (building) and replacement costs (household contents)

• Further topics: evacuation, clean-up work and recovery, aid and financial compensation, socio-demographic information

Altogether, the questionnaire for affected private households contained about 180 questions. The duration of an interview was on average 30 minutes.

Analogue to the household surveys, flood-affected companies were interviewed. For these studies, the questionnaire was adapted and shortened to about 90 questions so that the average duration of an interview did not exceed 20 minutes. In particular, questions were omitted regarding the characteristics of the affected building, which were questioned in the household survey with great detail in order to create a data basis that allowed reliable estimation of the asset value of the building and the household contents. With regard to companies, only the general setting of the premise was addressed by one question. Since it was assumed that company owners are more familiar with their assets than residents, the interviewees were directly asked to estimate the asset values of buildings, equipment, etc. Further, the socio-demographic questions of the household questionnaire were replaced by questions about the company (e.g., branch, number of employers, total revenue, number of customers and suppliers in the region). Questions about business disruption were added, and detailed questions on warning, response, precaution, and recovery were left out to save some time.

For most questions in both the household and the company surveys, a list of possible answers was given with either a single answer or multiple answers possible.

In a few cases, open-ended questions were posed. Furthermore, interviewees were asked to assess qualitative or descriptive items on a rank scale from 1 to 6, where 1 described the best case and 6 described the worst case. The meanings of the end points of the scales were given to the interviewee. The intermediate ranks could be used to graduate the evaluation.

Owing to the available funding, the household surveys were conducted 7 to 19 months after the event under study; often, affected companies were investigated later, up to four years after the event. (See Table 7.1 and Table 7.2.) The household surveys contained up to almost 1700 cases per event; surveys among companies were considerably smaller. (See Table 7.2.)

In general, surveys among companies are more difficult to perform than household surveys, since company owners are often not willing to participate in long surveys. Therefore, the number of questions had to be restricted (see above). Furthermore, company environments/settings are far more heterogeneous than residential homes in that they might range from a small store on a building's ground level to industrial premises with several buildings. Consequently, the questionnaire must be designed in a way that all possible settings are covered. As a further consequence, the collected information is often less detailed than it is for household surveys.

7.3.3. Post-processing and Quality of the Data Collected

Since each topic of the questionnaire was addressed by a number of questions (see also section 7.2), it was possible to undertake some validity checks of the answers, especially with regard to answers about the affected and total area, affected stories, and estimates of the damage and the asset values. After the first survey of the 2002 flood, it turned out that the reliability of the damage estimates was high since most of the affected people claimed their losses from governmental funds or from insurers and thus had a good idea of the costs. In contrast, the quality of estimates of the property asset values was very low; the reported loss regularly exceeded the denoted property value (data not shown). Hence, it was decided to estimate asset values with a standardized approach rather than asking residents about asset values.

Because the insurance industry provides methods for estimating the asset values of private property, these values were calculated as follows: the absolute values of buildings were estimated according to the VdS guideline 772 1988-10 [*Dietz*, 1999], which is commonly used in the insurance sector in Germany. It provides mean building values in "Mark 1914" per square meter (m²) living area for different building types. The building type and the living area of a building were determined with the help of the answers concerning the total floor space of the building, the number of stories, the basement area, and the roof type. The mean building values were upgraded or degraded depending on the quality and equipment of the building, e.g., the heating system [*Dietz*, 1999]. The resulting insurance sum in "Mark 1914" can be transferred to a replacement value of any given year by the price index for buildings published by the German Federal Statistical Agency.

The values of the household contents were calculated by multiplying the m² living area with EUR 650 in the case of the survey on the August 2002 flood. In subsequent surveys, the square meter price of EUR 650 was adapted to the then-current price level by consumer price indices published by the German Federal Statistical Agency. This is a commonly used method of the insurance industry in Germany. With the estimated values of buildings and household contents, it was then possible to calculate the respective loss ratios, i.e., the relation between the building/content damage and the corresponding value. The calculation of asset values was not possible for the surveys of affected companies since such standardized valuation methods are missing.

Moreover, the survey data were used to create indicator values that summarize the information on flow velocity, contamination of the flood water, flood warning, (short-term) emergency measures, (long-term) precautionary measures, previously experienced floods and socioeconomic status. (See *Thieken et al.*, 2005.) With this approach, it was possible to verify the theoretical model depicted in Figure 7.1, except for flood warning. (See *Thieken et al.*, 2005.) Particularly the data of the 2002 survey were further used to derive and validate the model family FLEMO [*Büchele et al.*, 2006; *Thieken et al.*, 2008; *Apel et al.*, 2009; *Wünsch et al.*, 2009; *Elmer et al.*, 2010; *Kreibich et al.*, 2010; *Seifert et al.*, 2010; *Thieken et al.*, 2015]. The model is currently updated and adapted to different flood types and biophysical environments.

For the development of loss models, data on water levels and losses are of particular importance. Therefore, Table 7.3 summarizes non-response rates and averages of the reported water levels as well as the reported repair/replacement costs for buildings and household contents damaged by river floods in Germany. It can be seen that the item non-response rate is the lowest for the water level, followed by the damage costs to household contents, and finally the damage costs to the building. A correlation analysis revealed that, except for household contents, for which the correlation between the time of the interview and the item non-response rate amounted to –0.35, the severity of the event and the sample size have much greater

Table 7.3 Sub-Sample Sizes, Item Non-Response Rates, and Average Values for Flood Water Levels as well as Damage Costs for Buildings and Household Contents in the Household Surveys after River Floods in Germany. (See also Table 7.1).

Event	2002	2005	2006	2010	2011	2013
Average time between event and interview [months]	9	15.5	7.5	18.5	13.5	8.5
Surveyed households (N)	1697	305	156	349	209	1652
Item non-response: water level	1.5%	6.6%	5.8%	3.4%	4.8%	7.0%
Mean water level (above surface) (cm)	64.2	–19.4	18.8	58.3	–19.5	53.5
Median water level (above surface) (cm)	64.5	–26.0	15.0	61.0	–20.0	40.0
Cases with damage to the building	1340	222	126	331	202	1299
Item non-response: monetary damage to building	19.5%	27.0%	31.7%	32.3%	41.1%	27.1%
Mean building damage (EUR; price level as of 2013)	52,680	23,043	29,840	46,833	11,369	55,086
Median building damage (EUR; price level as of 2013)	30,037	7650	10,950	21,436	2112	30,000
Cases with damage to the household contents	1489	218	80	264	76	1172
Item non-response: monetary damage to contents	14.5%	27.5%	30.0%	14.4%	30.3%	18.7%
Mean content damage (EUR; price level as of 2013)	20,538	14,253	11,526	18,553	7298	19,044
Median content damage (EUR; price level as of 2013)	9535	4566	1687	10,560	1345	8500

influence on the item (non-)response rate than the timing of the interview. A similar pattern emerged for the company surveys (data not shown). It is concluded that more, and probably better, loss information can be collected for severe flood events like the ones in 2002, 2010, and 2013 than for more frequent flood events that occurred, for example, in 2005, 2006, and 2011; the timing of the survey seems to be less important.

7.3.4. Comparison of Flood Losses from Different Data Sets

As a further quality control measure, the reported losses were compared to other loss data sets that were provided by the insurance sector or by public authorities in charge of managing governmental disaster relief.

Figure 7.2 shows the mean building loss based on different data sets. The mean flood loss of residential buildings in Saxony caused by the flood in 2002 is 25% lower in the CATI data set than in the (official) SAB data set. However, an analysis of the cumulative distributions of the SAB and the CATI data by *Thieken et al.* [2010] revealed an overall good agreement of the two data sets. In the first 50% of the data, which account for 86% of the total loss, the differences were a bit higher than in the second half. It appears that high building losses are underrepresented in the CATI data. A building loss of more than EUR 280,000 was only assigned to 2.1% of the records in contrast to 3.4% of the SAB data. Very small losses also tended to be underrepresented in the CATI

data. These systematic errors can be explained by the sampling procedure that was used in the CATI survey; affected households were determined on the basis of maps showing the inundated areas and lists of affected streets, which were provided by the municipalities or were found in flood reports. Small losses due to heavy rain or rising groundwater might occur outside of inundated areas and were thus probably not included in the sample. Completely destroyed objects could not be recorded as a result of damaged and missing telephone landlines. Both kinds of buildings are, however, included in the SAB data set resulting in a higher mean loss value (Figure 7.2) but a lower median. In total, the similarity of the distributions for Saxony and the 2002 event is satisfactory.

In Bavaria, where the CATI data were compared to the HOWAS data from two flood events in 1985 and 1988, respectively, larger discrepancies occur (Figure 7.2) [*Thieken et al.*, 2010]. Apart from the differences in the surveying methods (CATI versus on-site building surveyors) two other things might explain the differences. First, the CATI data contain a (building) representative sample that is spread all over the affected area in Bavaria in 2002, whereas the HOWAS data are concentrated in three municipalities (lumped sample as explained in section 7.2). Therefore, the HOWAS data do not cover as much variability as the CATI data: the coefficients of variation amount to 97% (HOWAS, Inn 1985), 72% (HOWAS, Danube 1988), and 160% (CATI, Bavaria 2002). Secondly, the time between the flood events of the HOWAS database on one hand and the CATI data on the

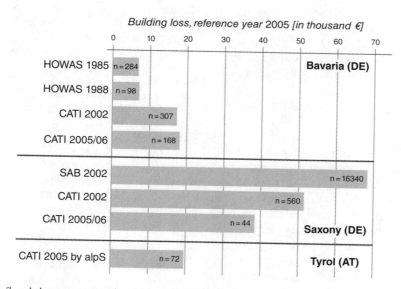

Figure 7.2 Mean flood damage at residential buildings in the German federal states Bavaria and Saxony as well as in the Austrian province Tyrol for some flood events. (All prices are indexed to the reference year 2005. HOWAS data for the floods in 1985 and 1988 were taken from *Buck and Merkel* [1999]. SAB data for the flood 2002 were provided by the Saxon Relief Bank (Sächsische Aufbaubank), and data for the 2005 flood in Tyrol, Austria, were provided by the alpS GmbH, Innsbruck. CATI interviews as listed in Table 7.1.)

other hand is almost 20 years. It has to be assumed that in the meantime more assets were accumulated in the buildings and that newer buildings might be more valuable. The mean building damage has been increasing by approximately 60% every 10 years.

Finally, the difference in the mean damage values in Saxony and Bavaria reflect the different flood intensities. Although the flood of 2002 exceeded the 100-year flood discharge in many places in Saxony, this was not the case in Bavaria in 2002, nor in 2005 or 2006. In regions with similar flood intensities in 2005 and similar construction types, such as Tyrol (Austria) and Bavaria (Germany), the mean losses derived from data collected by telephone surveys are also similar (Figure 7.2) underlining the validity of loss data collected by CATI. This is further supported by the investigations of *Cammerer et al.* [2013] who successfully validated loss models derived from Bavarian data in a Tyrolian study site.

7.4. CONCLUSIONS

Altogether, the experience in Germany with collection flood loss information by CATI and the analysis of the damage data lead to a number of conclusions concerning the collection of damage data. CATI is, in principle, a suitable method for collecting valid loss data although item non-response rates might be high for technical or very specific questions. Very low as well as extremely high damage records tend to be underrepresented in CATI data sets due to the sampling procedures. However, the average losses and damage patterns are well covered in such data sets, making them extremely valuable for detailed damage analyses and loss modeling. They allow insights into damage-causing processes and effective countermeasures, since far more parameters, e.g., about the flood impact or the characteristics of the affected structure, are gathered than during the loss adjustment procedures (for examples, see *Kreibich et al.*, 2005; *Thieken et al.*, 2005, 2006). In order to take advantage of data sets recorded for loss adjustment, a few additional parameters, such as building characteristics, water level, and precautionary measures, should be commonly recorded.

Finally, the increase in average damage demonstrates that continuous efforts to collect damage data are needed. It has to be doubted whether depth-damage curves that are based on data that were gathered 10 or 20 years ago are still valid today. *Johnson et al.* [2007] show for the UK that damage potential in the residential and commercial sector has considerably risen over the past 15 years. The case of the UK (see Box 1) also indicates that in countries with well-established loss assessment guidelines and appraisal procedures and a well-developed flood insurance market, new (empirical) approaches are more difficult to establish than in Germany where until 2002 only very little work on loss estimation had been undertaken.

REFERENCES

Adaptation Sub-Committee (2012), Climate change–Is the UK preparing for flooding and water scarcity? London: Committee for Climate Change. *Progress Report 2012*.

Al-Futaisi, A., and J. R. Stedinger (1999), Hydrologic and economic uncertainties and flood-risk project design. *Journal of Water Resources Planning and Management*, *125*, 314–324.

Apel, H., G. T. Aronica, H. Kreibich, and A. H. Thieken (2009), Flood risk assessments–How detailed do we need to be? *Natural Hazards*, *49*(1), 79–98.

Association of British Insurers (2009), Resilient reinstatement–The cost of flood resilient reinstatement of domestic properties, Research Paper, London: Association of British Insurers.

Bhattacharya-Mis, N., and J. Lamond (2014), An investigation of patterns of response and recovery among flood affected businesses in the UK: Case study in Sheffield and Wakefield, Flood Recovery Innovation and Response, 2014 Poznan, Poland, WIT Press.

Bhattacharya-Mis, N., R. Joseph, D. Proverbs, and J. Lamond (2015), Grass-root preparedness against potential flood risk among residential and commercial property holders. *International Journal of Disaster Resilience in the Built Environment*, *6*, 44–56.

Black, A. R., and S. A. Evans (1999), Flood Damage in the UK. New insights for the insurance industry. University of Dundee Geography Department.

Blong, R. (2004), Residential building damage and natural perils: Australian examples and issues. *Building Research & Information*, *32*, 379–390.

Büchele, B., H. Kreibich, A. Kron, A. Thieken, J. Ihringer, P. Oberle, B. Merz, and F. Nestmann (2006), Flood-risk mapping: contributions towards an enhanced assessment of extreme events and associated risks. *Nat. Hazards Earth Syst. Sci.*, *6*(4), 485–503.

Buck, W., and U. Merkel (1999), Auswertung der HOWAS Schadendatenbank, (Analysis of the HOWAS data base), Institut für Wasserwirtschaft und Kulturtechnik der Universität Karlsruhe, HY98/15.

Cammerer, H., A. H. Thieken, and J. Lammel (2013), Adaptability and transferability of flood loss functions in residential areas. *Nat. Hazards Earth Syst. Sci.*, *13*, 3063–3081.

Changnon, S. A. (1996), The Great Flood of 1993: Causes, Impacts and Responsibilities, Westview Press, Boulder, CO.

Changnon, S. A. (2003), Shifting Economic Impacts from Weather Extremes in the United States: A Result of Societal Changes, not Global Warming. *Natural Hazards*, *29*, 273–290.

Consultants, J. (2012), Establishing the Cost Effectiveness of Property Flood Protection: FD2657.

Dietz, H. (1999), Wohngebaeudeversicherung Kommentar, 2nd ed., VVW Verlag Versicherungswirtschaft GmbH: Karlsruhe, Germany.

DKKV (Ed.) (2015), Das Hochwasser im Juni 2013: Bewährungsprobe für das Hochwasserrisikomanagement in

Deutschland, DKKV-Schriftenreihe Nr. 53, Bonn, ISBN 978-3-933181-62-6, 207 pp.

Downton, M.W., and R.A. Pielke Jr. (2005), How accurate are Disaster Loss Data? The Case of U.S. Flood Damage. *Natural Hazards, 35*, 211–228.

Elmer, F., A. H. Thieken, I. Pech, and H. Kreibich (2010), Influence of flood frequency on residential building losses. *Nat. Hazards Earth Syst. Sci., 10*, 2145–2159.

Escuder-Bueno, I., J. T. Castillo-Rodríguez, S. Zechner, C. Jöbstl, S. Perales-Momparler, and G. Petaccia (2012), A quantitative flood risk analysis methodology for urban areas with integration of social research data., *Nat. Hazards Earth Syst. Sci., 12*, 2843–2863.

Flood Hazard Research Centre (2010), The Benefits of Flood and Coastal Risk Management: A Handbook of Assessment Techniques, London, FHRC.

Ganoulis, J. (2003), Risk-based floodplain management: A case study from Greece, *International Journal of River Basin Management, 1*, 41–47.

Gissing, A., and R. Blong (2004), Accounting for Variability in Commercial Flood Damage Estimation. *Aust. Geogr., 35*(2), 209–222.

Greenberg, M. R., and M. D. Weiner (2014), Keeping Surveys Valid, Reliable, and Useful: A Tutorial. *Risk Analysis, 34*(8), 1362–1375.

Grünthal, G., A. H. Thieken, J. Schwarz, K. S. Radtke, A. Smolka, and B. Merz (2006), Comparative risk assessments for the city of Cologne–Storms, floods, earthquakes. *Natural Hazards, 38*(1–2), 21–44.

Henderson, T. L., M. Sirois, A. C. C. Chen, C. Airriess, D. A. Swanson, and D. Banks (2009), After a disaster: Lesson in Survey Methodology from Hurricane Katrina. *Popul. Res. Policy Rev., 28*, 67–92.

Hüfken, V. (2000), Kontaktierung bei Telefonumfragen. Auswirkungen auf das Kooperations- und Antwortverhalten. Methoden in Telefonumfragen, Westdeutscher Verlag: Wiesbaden, p. 11–31.

Johnson, C., E. C. Penning-Rowsell, and S. Tapsell (2007), Aspiration and reality: flood poliy, economic damages and the appraisal process. *Area, 39*(2), 214–223.

Jonkman, S. N. (2005), Global Perspectives on Loss of Human Life Caused by Floods, *Natural Hazards, 34*, 151–175.

Joseph, R., D. Proverbs, and J. Lamond (2015), Assessing the value of intangible benefits of property level flood risk adaptation (PLFRA) measures. *Natural Hazards, 79*, 1275–1297.

Joseph, R., J. Lamond, D. Proverbs, and P. Wassell (2011), An analysis of the costs of resilient reinstatement of flood affected properties: a case study of the 2009 flood event in Cockermouth. *Structural Survey, 29*, 279–293.

Joy, C. S. (1993), The cost of flood damage in Nyngan. *Climatic Change, 25*, 335–351.

Kienzler S., I. Pech, H. Kreibich, M. Müller, and A. H. Thieken (2015), After the extreme flood in 2002: changes in preparedness, response and recovery of flood-affected residents in Germany between 2005 and 2011. *Nat. Hazards Earth Syst. Sci., 15*, 505–526.

Kreibich, H., I. Seifert, B. Merz, and A. H. Thieken (2010), Development of FLEMOcs–A new model for the estimation of flood losses in the commercial sector. *Hydrological Sciences Journal, 55*(8), 1302–1314.

Kreibich, H., M. Müller, A. H. Thieken, and B. Merz (2007), Flood precaution of companies and their ability to cope with the flood in August 2002 in Saxony, Germany. *Water Resources Research, 43*, doi: 10.1029/2005WR004691.

Kreibich, H., S. Christenberger, and R. Schwarze (2011), Economic motivation of households to undertake private precautionary measures against floods. *Nat. Hazards Earth Syst. Sci., 11*(2), 309–321.

Kron, W., and W. Willems (2002), Flood risk zoning and loss accumulation analysis for Germany, in: International Conference on Flood Estimation (M. Spreafico and R. Weingartner, eds). CHR, Berne, pp. 549–558.

Lamond, J., D. Proverbs, and F. Hammond (2009), Accessibility of flood risk insurance in the UK–confusion, competition and complacency. *Journal of Risk Research, 12*, 825–840.

Merz, B., H. Kreibich, A. Thieken, and R. Schmidtke (2004), Estimation uncertainty of direct monetary flood damage to buildings. *Nat. Hazards Earth Syst. Sci., 4*(1), 153–163.

Merz, B., H. Kreibich, and U. Lall (2013), Multi-variate flood damage assessment: a tree-based data-mining approach. *Nat. Hazards Earth Syst. Sci., 13*, 53–64.

Merz, B., H. Kreibich, R. Schwarze, and A. Thieken (2010), Review article "Assessment of economic flood damage." *Nat. Hazards Earth Syst. Sci., 10*(8), 1697–1724.

Messner, F., E. C. Penning-Rowsell, C. Green, V. Meyer, S. M. Tunstall, and A. van der Veen (2007), Evaluating flood damages: guidance and recommendations on principles and methods, Floodsite, Oxfordshire, UK: H R Wallingford.

Molinari, D., S. Menoni, G.T. Aronica, F. Ballio, N. Berni, C. Pandolfo, M. Stelluti, and G. Minucci (2014), Ex post damage assessment: an Italian experience. *Nat. Hazards Earth Syst. Sci., 14*, 901–916.

Müller, M., and H. Kreibich (2005), Private Vorsorgemaßnahmen können Hochwasserschäden reduzieren. Schadensprisma, *2005*(1): 4–11.

Penning-Rowsell, E.C. (2015), A realistic assessment of fluvial and coastal flood risk in England and Wales. *Transactions of the Institute of British Geographers, 40*, 44–61.

Penning-Rowsell, E. C., and J. B. Chatterton (1977), The benefits of flood alleviation: A manual of assessment techniques. Gower Technical Press, Aldershot.

Penning-Rowsell, E., C. Johnson, S. Tunstall, S. Tapsell, J. Morris, J. Chatterton, and C. Green (2005), The Benefits of Flood and Coastal Risk Management: A Manual of Assessment Techniques. Middlesex Univ. Press, UK.

Queensland Government (2002), Disaster loss assessment Guidelines. Queensland Department of Emergency Services and Emergency Management Australia. Written by Handmer, J. C. Read, and O. Percovich.

Ramirez, J., W. L. Adamowicz, K. W. Easter, and T. Graham-Tomasi (1988), Ex Post Analysis of Flood Control: Benefit-Cost Analysis and the Value of Information. *Water Resources Research, 24*, 1397–1405.

Schröter, K., H. Kreibich, K. Vogel, C. Riggelsen, F. Scherbaum, and B. Merz (2014), How useful are complex flood damage models? *Water Resources Research, 50*, 3378–3395, doi: 10.1002/2013WR014396.

Seifert, I., H. Kreibich, B. Merz, and A. H. Thieken (2010), Application and validation of FLEMOcs–A flood loss estimation model for the commercial sector. *Hydrological Sciences Journal*, 55(8), 1315–1324.

Smith, D. I. (1994), Flood damage estimation–A review of urban stage damage curves and loss functions. *Water SA*, 20(3), 231–238.

Soetanto, R., and D. G. Proverbs (2004), Impact of flood characteristics on damage caused to UK domestic properties: the perceptions of building surveyors. *Struct. Survey*, 22(2), 95–104.

Thieken, A. H., A. Olschewski, H. Kreibich, S. Kobsch, and B. Merz (2008), Development and evaluation of FLEMOps–A new Flood Loss Estimation MOdel for the private sector. Flood Recovery, Innovation and Response. Proverbs, D., C. A. Brebbia, E. Penning-Rowsell, Eds.), WIT Press, p. 315–324.

Thieken, A. H., H. Apel, and B. Merz (2015), Assessing the probability of large-scale flood loss events–A case study for the river Rhine, Germany. *Journal of Flood Risk Management*, 8(3), 247–262.

Thieken, A. H., M. Müller, H. Kreibich, and B. Merz (2005), Flood damage and influencing factors: New insights from the August 2002 flood in Germany. *Water Resources Research*, 41(12), W12430, doi: 101029/2005WR004177.

Thieken, A. H., T. Petrow, H. Kreibich, and B. Merz (2006), Insurability and mitigation of flood losses in private households in Germany. *Risk Analysis*, 26(2), 383–395.

Thieken, A., R. Schwarze, V. Ackermann, and U. Kunert (2010), Erfassung von Hochwasserschäden–Einführung und Begriffsdefinitionen, Chapter 2 in Hochwasserschäden–Erfassung, Abschätzung und Vermeidung. Thieken, A.H., I. Seifert, and B. Merz, eds., Oekom-Verlag: Munich, pp. 21–49.

Thieken, A. H., H. Kreibich, M. Müller, and B. Merz (2007), Coping with floods: preparedness, response and recovery of flood-affected residents in Germany in 2002. *Hydrological Sciences Journal*, 52(5), 1016–1037.

Thurston, N., B. Finlinson, R. Breakspear, N. Williams, J. Shaw, and J. Chatterton (2008), Developing the evidence base for flood resistance and resilience. Environment Agency.

US Army Corps of Engineers (USACE) (1996), Risk-based analysis for flood damage reduction studies. EM 1110-2-1619, Washington DC.

Wünsch, A., U. Herrmann, H. Kreibich, and A. H. Thieken (2009), The Role of Disaggregation of Asset Values in Flood Loss Estimation: A Comparison of Different Modeling Approaches at the Mulde River, Germany. *Environmental Management*, 44(3), 524–541.

Zhai, G., T. Fukuzono, and S. Ikeda (2005), Modeling Flood Damage: Case of Tokai Flood 2000. *Journal of the American Water Resources Association*, 41, 77–92.

8

Utilizing Post-Disaster Surveys to Understand the Social Context of Floods–Experiences from Northern Australia

David King[1] and Yetta Gurtner[2]

ABSTRACT

Floods are an integral part of the Australian landscape and psyche. They destroy and refresh. People have long acknowledged their inevitability and their role, but increasingly urban populations are less accepting of loss, while at the same time urban development has increased vulnerability. Analysis of the impacts of floods in regional and metropolitan Australia reveals patterns of vulnerability countered by resilience, mitigation, and adaptation to the possibility of increasing threats of flood induced by climate change. Issues include relocation and abandonment of flood hazardous places. Local government and public administration authorities benefit directly from research following floods because it deals directly with vulnerability, enhancement of resilience, land use planning, and mapping within a broader policy context of flood mitigation. Valuable data and insights into community behavior and experiences during and following disastrous floods have been derived from post-disaster studies that were carried out in flooded communities. The contributions of this type of research to post-disaster research methodology and to disaster risk reduction policies and strategies are illustrated, and methods are examined.

"I love a sunburnt country,
A land of sweeping plains,
Of ragged mountain ranges,
Of droughts and flooding rains."
– Dorothea Mackellar, My Country

8.1. THE MYTHOLOGIZING OF FLOODS AND NATURAL DISASTERS IN AUSTRALIAN CULTURE

It is both a fact and a myth that Australia is a land of "drought and flooding rains." The fact is that Australia is arid and tropical/subtropical. Like similar climate zones on earth, it is subject to extreme seasonal variations and very great differences in rainfall from one year to another. Much of the rain in subtropical and tropical zones is brought by annual monsoonal systems and is frequently driven by tropical cyclones.

Indigenous residents of Australia had adapted their lifestyle and resource uses to such seasonal variations of wet and dry over many millennia. Aboriginal groups were predominantly nomadic as a response to an environment that gave seasonal abundance within specific locales. Indigenous knowledge of their environment and its seasons, and the variability of climate from one year to another was established and passed on from generation

Centre for Disaster Studies
[1]Centre for Tropical Urban and Regional Planning, School of Earth and Environmental Sciences, James Cook University, Townsville, Queensland, Australia
[2]Centre for Disaster Studies, James Cook University, Townsville, Queensland, Australia

Flood Damage Survey and Assessment: New Insights from Research and Practice, Geophysical Monograph 228,
First Edition. Edited by Daniela Molinari, Scira Menoni, and Francesco Ballio.
© 2017 American Geophysical Union. Published 2017 by John Wiley & Sons, Inc.

to generation. It took the Europeans settling in Australia a long time to understand and learn from aboriginal knowledge. An example of a catastrophic flood in small town of Gundagai in 1851 records more than 40 people being rescued by a local aboriginal man using his canoe, and a further 89 of the 250 residents drowned [*Coates*, 1999; *Carbone and Hanson*, 2012]. In 1916 Queensland, the regional town of Clermont was destroyed, killing 64 people, despite local aboriginal people warning the settlers that the place where they had established the town was dangerously flood prone. The town of Nyngan, which severely flooded in the 1990s, literally means "a place of flood" in the local aboriginal language.

However, as European Australians progressively learned of aboriginal knowledge of climate and of the place or country, this knowledge became almost mythologized into European Australian thinking that has tended to see aborigines as people living in harmony with nature, people who had an understanding and acceptance of the variability of its wet and dry seasons. This thinking imbues indigenous people with an innate understanding of the environment that is portrayed in direct contrast to Western exploitation and misuse of the environment, and a subsequent failure or refusal of European settlers to adapt appropriately. Such a perception is not entirely empathetic or representative of the circumstances.

Early European settlers in Australia had predominantly come from Britain and Ireland where rainfall patterns and storms are mostly less extreme. They brought an idea of hot and cold seasons, as summer and winter, but were less aware of wet and dry, which are the predominant seasonal contrasts of Australia. Also many nineteenth century settlers came from newly industrializing cities where climate had become less significant to their livelihoods. As such, immigrants and pioneers learned of the Australian extremes of wet and dry, they adapted through endurance and stoicism, which have become core values of the Australian spirit and especially of the mythological outback spirit.

Thus, by the end of the nineteenth century, Mackellar's poem captured the nature of this adaptation by expressing a love for a land of extremes, especially the wet and dry. Many more poems, art, and music have celebrated the beauty of this dramatic landscape and its storms and droughts that leave people both in awe and in a state of dependence. In a real sense, community acceptance of the climatic extremes of this country underscores the resilience and stoicism of Australians when confronted by the powerful forces of nature and natural disaster.

There is a further side of this acceptance that mythologizes natural hazards as something that people can survive but never overcome. This is an especially powerful attitude and pervasive view in the north of Australia. Anyone who has lived in "the North" for a couple of decades or more, is likely to have a personal story of hard-

ship associated with the challenges of confronting the perilous impacts of Mother Nature. Each and every story features the actors as enduring survivors, bringing people together in the memory of that shared experience. This is particularly the case for past floods and cyclones. Droughts are less forgiving in nature, destroying many lives and livelihoods, contributing to business and farm failures, suicide, and mental health impacts.

The mythologizing of such natural disasters gives the events meaning and imbues the survivors with knowledge and experience upon which they draw the next time. However, the next time is never the same; people are older, family and community situations have changed, even the scale and size of the event is usually different. Preparations for the "next" flood or cyclone however remain based on memories of prior events that have been mythologized and reimagined to give them meaning and emotion, in which everyone was okay, which was summed up in the classical Australian philosophy of "she'll be right."

These courageous attitudes and values of stoicism and fortitude, coupled with concern and help shown between neighbors, create a powerful basis for resilience. The problem is that being based on endurance through past events and experiences, they serve to filter government and institutional messages and changes in warnings, advice, and practice. Given the imminent threat of a natural hazard, people are apt to minimize preparations, take unnecessary risks, and leave it to the last minute to take recommended actions such as provisioning and evacuation because "everything will be okay" (like last time), because it is not culturally acceptable to make a fuss or to show too much emotion, and because a likely flood or severe storm are part of the regional pattern of the seasons and its extremes.

This mythologizing of natural hazards presents an even more problematic situation as we factor in climate change. Lessons learned in the past may not be a guide to events such as more severe floods and cyclones in the future. This is particularly evident given that historical experiences are already an inaccurate gauge for the changing context of subsequent natural hazards, let alone potentially worse events in the future, regardless of climate change. Even with such uncertainty, it is the experience of past floods and cyclones and an acceptance of the inevitability of natural disasters that on the one hand makes these people resilient and stoic in the face of loss, and on the other hand constrains active or innovative adaptation.

8.2. NATURAL HAZARDS IN AUSTRALIA

From millennia of traditional indigenous systems of resilience and self-reliance in the face of natural hazards, patterns of European settlement and the increasing urbanization of Australia saw a concomitant growth in levels of public administration, laws, and systems of governance. With the formal establishment of sovereignty and taxation systems, government authorities inherited a greater role, responsibility, and expectation to maintain and protect the safety of the general populous. Continued population growth, expansion, and associated environmental degradation also resulted in greater numbers of people inhabiting vulnerable areas that are exposed and susceptible to the impacts of natural hazards.

Apart from the duty of care to the people and their communities, the government is expected to address the political and economic costs of natural disasters. Table 8.1 below shows the death rate from natural disasters in Australia from the point of first settlement by Europeans in 1788. Just two years later came the first deaths in a flood at Sydney. Up until 2014, floods had caused the highest number of deaths of all recorded natural hazards in Australia except heatwaves (heatwave/extreme heat is the biggest single estimated cause of death from natural hazard but is defined inconsistently in public policy). In the wet season of 2010/2011, 35 people died in Queensland floods prompting a Commission of Inquiry in which each individual death was recorded in detail. In 2015, a further eight lives were lost in eastern Australia from vehicles swept away during torrential rainfall flash flooding events. Deaths prompt a far greater political reaction than economic losses.

The total number of flood events since 1788 is difficult to pin down because places that were flooded when the population was very small, or areas were sparsely inhabited, subsequently became more flood prone or hazardous as the population moved in and increased. Examples of serious floods and their impacts [*Coates*, 1999; *Carbone and Hanson*, 2012] include 89 deaths in Gundagai in 1851 and 64 deaths in Clermont in 1916, with both towns subsequently relocated out of the flood zone. In 1893, Ipswich in Queensland experienced 35 deaths with 300 injured. In 1927, 47 people died in floods in Brisbane and Cairns with extensive infrastructural damage as well. In 1929, there were 22 deaths in northern Tasmania with

1000 homes destroyed. In 1934, 36 deaths occurred in South Gippsland with 6000 people made homeless. In 1955, there were 24 deaths in the Hunter Valley with 5200 houses flooded and 40,000 people evacuated from 40 settlements. The 1974 Brisbane floods resulted in 14 deaths, and 56 homes destroyed and 6000 damaged. Then in 1986, the Hawkesbury and Georges rivers left six dead and 10,000 homes damaged. The 2010 and 2011 wet season flooding in Queensland resulted in three-quarters of the council areas of the state declared disaster zones, affecting at least 90 towns, with the forced evacuation of over 200,000 people and the tragic deaths of 35 people [*Queensland Flood Commission of Inquiry*, 2012].

Table 8.2 records the annual financial (insured) costs of natural disasters in Australia from 1970 to 2013. In addition to the fatalities, the wet season of 2010/2011 in Queensland added an enormous financial toll with extensive flooding in the state capital of Brisbane, as well as in many other urban areas. Damages were estimated at $2.38 billion. The financial burden of natural disasters to the economy of Australia is extremely high, with impacts from hydrological hazards such severe storm, cyclones, and floods foremost on the list. This particularly reflects the frequency and scale of events in Queensland in the last decade, where at times most of the state has been flood declared.

Although part of the economic burden has been upon infrastructure and is thus borne by different levels of government, damage to residential housing comprises the largest component of economic impact. Costs to residential housing are borne by individuals and householders, with some of the financial burden having gone to insurance companies. However, many people who do not have adequate flood insurance are reliant upon central and state governments to provide assistance in flood recovery. Better preparedness by residents may reduce the financial burden to householders, insurance companies, and the government. This provides further justification for attempts to understand how people prepare for natural hazards and prioritize household strategies for disaster risk reduction.

Table 8.1 Summary of Deaths in Natural Hazards in Australia: 1788–2014[a].

Hazard	First Recorded Death	Number of Deaths	% Total Deaths
Earthquake	1902	16	0.2
Landslide	1842	107	1.7
Bushfire	1850	913	14.3
Thunderstorm	1824	787	12.3
Tornado	1861	52	0.8
Cyclone	1839	2165	33.9
Flood	1790	2352	36.8
Tsunami		0	0.0
Total		6392	100.0

[a]Excludes complex/latent hazards such as heatwave, drought, and epidemic.
Revised and updated from Blong [2005], p. 6, with Australian Emergency Management Knowledge Hub [2015].

Table 8.2 Insurance Losses by Natural Hazard in Australia 1970–2013[a] (Millions of Dollars [2011 Dollars])[b].

Event Type	Bushfire	Cyclone	Flood	Storm	Hail	Earthquake	Total $
New South Wales	527	36	965	2747	4856	1657	10,788
Victoria	1650	-	400	2439	294	–	4783
Queensland	–	3329	3630	1376	949	–	9283
South Australia	189	–	–	47	92	–	327
Western Australia	96	486	24	1232	0	15	1852
Tasmania	100	–	51	34	86	–	271
Northern Territory	–	1529	123	–	–	–	1652
Australian Capital Territory	440	–	–	–	–	–	440
AUSTRALIA	3002	5379	5129	7874	6277	1652	29,395
% Total	10.2	18.3	17.4	26.8	21.4	5.6	100

[a]Where events were recorded as impacting multiple states, costs have been divided evenly across those states; nil or rounded to zero.
[b]Indicative of direct insured costs only, does not account for indirect intangible losses.
Based on Productivity Commission data, Natural Disaster Funding Arrangements [2014], p. 281.

8.3. RESEARCH IN THE POST-DISASTER CONTEXT

Despite common acceptance of the seasonality and regularity of floods, the most common response from people after a disastrous flood is that it had never been this high or extensive before. People have tended to contrast the natural disaster flood to other minor floods or threats that they had experienced in previous years. Their expectation of flooding was surpassed by the extreme event that brought significant damage and widespread destruction. For most of the floods analyzed in the identified post-disaster surveys, the specific event had not been the worst ever. Rather it had been the worst in the individual respondent's experience. Previous catastrophic events had very likely occurred decades earlier so that only small numbers of residents could remember these events, and in many instances, the maximum probable flood had occurred over 100 years prior. Indigenous knowledge and quaternary records suggest probable maximum floods had occurred far back in the past, before European settlement.

Consistently, the community disaster subculture based in memory of past events does not accurately prepare existing or new members of the community for what may happen in the future. Education campaigns for preparedness and awareness of risk, coupled with improved warnings and an upgraded understanding among the community of how to respond to warnings, are essential parts of disaster risk reduction. Significant knowledge and recommendations are developed through learning from direct experiences and behavior during related events.

The objective and purpose of post-disaster research is to gather information from people who recently have been involved in a natural disaster, that is, residents, victims, and respondents, in order to understand what happened to individuals and households, how people had prepared for the event, how they behaved during its passage, sources and effectiveness of warnings and assistance, actual impact on households and structures, and any issues specific to that event. Often sponsored by government authorities or agencies, research and information acquired is generally widely disseminated to influence enhanced strategies, management, and public policy.

8.4. CENTRE FOR DISASTER STUDIES RESEARCH

The Centre for Disaster Studies of James Cook University has conducted post-disaster studies since its inception in the 1970s following two severe cyclones that devastated Townsville and Darwin. The United Nations International Decade for Natural Disaster Reduction during the 1990s prompted research support from the Bureau of Meteorology and the Queensland Department of Emergency Services, sponsoring the conduct of immediate rapid response post disaster studies that would contribute to institutional debriefing and analysis. From the mid-1990s onward, researchers in the Centre have conducted post-disaster studies by rapid appraisal methods to complement broader research on vulnerability (and subsequently resilience), hazard awareness, preparedness and understanding of warnings, and the roles and experiences of emergency managers, local government, planners, private enterprise, and nongovernmental organizations (NGO).

The focus of research for the Centre for Disaster Studies has been primarily on social impact; however, it has also included collaboration with psychologists as well as engineers from the Cyclone Testing Station, whose disciplinary emphases complement social impacts, but who have carried out independent post-event analyses. After most natural disasters, research results have been brought together by a range of researchers in reporting to federal,

Table 8.3 Post-Disaster Studies Following Floods, Australia.

Year	Place	Type of Flood	Impact	Issues
1997	Cloncurry, N Qld	River	Small town, houses, belongings	Warnings, response
1998	Gulf of Carpentaria, N Qld	River and floodplain	Rural, small towns	Health, isolation
1998	Townsville, N Qld	Extreme rainfall inundation	Urban, housing, businesses, infrastructure	Insurance, building heights, behavior
1999	Innisfail, N Qld	Heavy rain, river, Cyclone Rona generated	Rural town, housing, businesses	New developments and overland flow, opening of sluices
1999	Wujal Wujal, N Qld	Cyclone Rona generated, river	Small indigenous community, isolation	Warnings, supply, recovery
2008	Mackay, N Qld	Flash flood, extreme rainfall	Urban, inside levee, housing, businesses, evacuation	Warnings, evacuation
2008	Charleville, S Qld	River/creek	Rural town, inside levee, isolation, housing	Warning, safe development paradox
2010	Ingham, N Qld	River	Isolation, housing, businesses	Warnings, supply, recovery
2010	Emerald, Central Qld	River	Medium urban, housing, businesses	Warnings, insurance
2011	Brisbane	River	Multiple suburbs, housing, businesses, infrastructure, evacuation	Politics, human intervention opening dam, insurance, response
2011	Donald, Victoria	River floodplain	Rural town, from drought to flood	Evacuation center for surrounding rural communities
2011	Tully and Hull Heads, N Qld	Cyclone storm surge	Beachside townships, catastrophic damage, evacuation	Rebuilding, trauma
2011	Lockyer Valley, S Qld	Flash flood, river floodplain	Rural towns, catastrophic damage, deaths, inquiry, evacuation	Inquiry, relocation

Notes: Qld – State of Queensland; CDS – Centre for Disaster Studies.
Emerald, Brisbane, and Donald flood studies, and site visit to Lockyer Valley were carried out by Risk Frontiers, Macquarie University, in collaboration with CDS, funded by the National Climate Change Adaptation Research Facility. All other flood studies were carried out by CDS, and most were funded by the Bureau of Meteorology and Emergency Management Australia.

state, and local government departments and institutions. Over a 20-year period, the Centre for Disaster Studies researchers predominantly visited impacted communities following cyclones and floods.

Apart from three miscellaneous disasters (tsunami, civil war, and terrorism), 11 cyclones in 19 separate towns and communities have been covered. Table 8.3 lists the 13 flood events that were also investigated. Three of the events in Table 8.3 were conducted as part of associated cyclone studies, that is, floods that were a consequence of a cyclone or storm surge that came with a cyclone. The table indicates the type of flood that occurred in each event: river, floodplain, extreme rainfall event, flash flood, and storm surge. The impact refers to the type of settlement that was the focus of the study, and specific locales within it, such as inside the levee. Figure 8.1 provides reference to the diverse geographic locations and hazard exposure for each place. All studies involved household interviews and questionnaires, but some (as indicated) included surveys of affected businesses.

8.5. FLOOD TYPOLOGY AND IMPACTS

The temporal distribution, scale, and subsequent impacts of flood hazards in Northern Australia are heavily influenced by region and topography. Although flood can be generally defined as an overflow of water onto land that is normally dry, the post-disaster research studies demonstrate that the cause, consequences, and individual experiences can be variable between both inland and coastal locations.

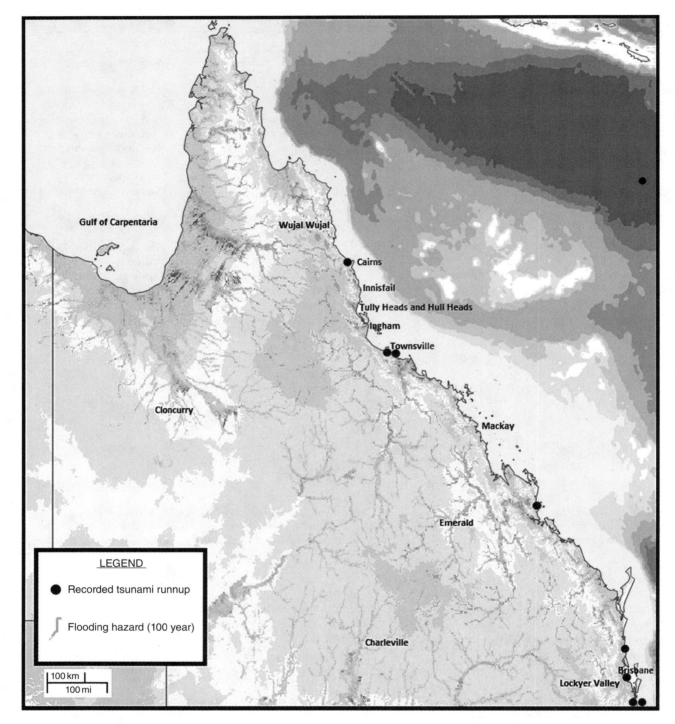

Figure 8.1 Flooding hazards and CDS post-disaster research study areas in Queensland *Map generated from Global Risk Data Platform 2015.*

8.5.1. Riverine–Heavy Rain Over River Catchments

River floods are monitored by gauges in many parts of the catchment so that flood heights and timings of expected floods can be predicted as the surge moves downstream. Evacuation advice can be provided well ahead of the flood, enabling people to remove or raise belongings. Flood warnings were issued in the Cloncurry, Ingham, and Emerald floods, where efforts were made to protect personal belongings, although floodwaters still rose much higher than people expected. Such river floods may be controlled by artificial structures such as levees, banks, dams, and sluice gates. These have given many people a false sense of security and even created extreme controversy where human intervention has occurred to reduce risk by releasing dammed or restricted floodwaters. This occurred at Innisfail in 2000 and in Brisbane in 2011. Such interventions target some communities (or a part of the community) for overall disaster risk reduction, but such decisions made under pressure have subsequently been subject to enquiries such as the Queensland Flood Inquiry.

8.5.2. Storm Surge–Coastal Inundation/Flooding

Storm surges accompany tropical cyclones, bringing seawater inundation to low-lying areas adjacent to the coast. Warnings to evacuate are issued before cyclone landfall, but four levels of uncertainty act as constraints to appropriate community response. First, the track of the cyclone is highly variable. Forecasting in recent years has become much more precise as a consequence of atmospheric modeling, but landfall is frequently within a 100-km zone where maximum storm surge is highly concentrated at the center of the storm. Second, surge depth overland varies according to the state of the tide. A third uncertainty is the speed of passage of a cyclone; a slow-moving system may push a surge over the period of the tidal range and inundate communities for a much longer period, thereby adding to wave damage. The fourth uncertainty is the severity of a cyclone because they often strengthen on approaching the coast, but weaken quite rapidly once they make landfall. These uncertainties translate into an unwillingness of people to evacuate if they judge it to be unnecessary, regardless of official warnings and information. At Hull Heads and Tully Heads in North Queensland, the storm surge in 2011 cyclone Yasi was predicted, the cyclone category was severe, and houses in these communities were built on an extremely vulnerable beach ridge. All of the population evacuated safely and destruction of the communities was severe.

Storm surges are also an increasing probability quite independent of cyclones occurring in the immediate vicinity. Storms far out in the ocean have generated surges that have inundated coastal communities, causing widespread flooding and damage such as what occurred along the northern coast of Papua New Guinea and its islands during 2010. Sea level rise as a consequence of climate change will probably increase the risk of surges and see some beachside communities become unsustainable.

Although often preceded by an earthquake event, coastal inundation from tsunami run-up is also associated with similar levels of uncertainty. Despite the increased use of offshore buoys to detect and measure any rapid fluctuations in sea levels, the speed, height, distance, and subsequent force of tsunami run-up on individual coastal communities is influenced by a large number of variables including local bathymetry and topography. Although the historical incidence of tsunamis in Australia is relatively small and low impact, there is still the risk of a high magnitude event impacting coastal communities on either the east or west coast of the continent. Local and regional tsunami inundation mapping can be used to identify vulnerable areas and populations.

8.5.3. Flash Floods/Extreme Rainfall (Heavier Precipitation Events)

Associated with severe storm systems, the least predictable floods are flash floods. These are often part of much larger flood systems where a channeling of floodwaters creates a localized catastrophe. This occurred in Toowoomba and the Lockyer Valley in Queensland during the 2011 floods that inundated Brisbane. Nineteen people died in the Lockyer Valley and the small town of Grantham when catastrophic floodwaters poured off the dividing range into the headwaters and lowlands of the Lockyer River. The suddenness and severity of the flash flood made warnings virtually impossible, and the depth of floodwaters swept away houses, infrastructure, cars, and people.

8.6. SPECIFIC ISSUES IDENTIFIED IN NORTHERN AUSTRALIA

Just as each flood hazard type can have different impacts, the post-disaster surveys conducted reveal that specific issues varied from one flood to another. These were either incorporated into the survey instrument or they emerged as significant issues from open-ended questions and interviews.

Isolation and loss of access to services and supplies are critical issues for remote communities, where sometimes people have been cut off for several weeks. This was a particular issue for the aboriginal community of Wujal Wujal, at the base of Cape York Peninsula, where rivers to the south and north of the settlement cut road access for several weeks causing shortages of food and fuel. The Gulf of Carpentaria river floods in 1998 similarly isolated mostly indigenous communities in small towns and outstations, creating similar problems of shortages of supplies. Both of these flood events added additional health hazards from pollution of drinking water and from disruption and overflow of sewerage systems. At Wujal Wujal, the settlement is above river flood waters, but on the Gulf of Carpentaria Lowlands, slow-moving floodplain waters inundated many houses and eroded the limited access roads.

Extreme rain events occurred in Townsville, Charleville, and Mackay resulting in extensive inundation of the urban areas. The 1998 rainfall event in Townsville dumped 770 millimeters (mm) in six hours from a decayed cyclonic system. Passive floods also contained pockets of fast moving water, where constraints such as levees were broken or floodwaters channeled. The local river channeled floodwaters through a beachside settlement at Black River causing extensive destruction of property and houses. In the 2008 floods in Charleville and Mackay, inundation occurred inside the levee, forcing evacuation of residents to other parts of the town. In many of the towns of North Queensland, older fibro and wooden houses raised a couple of meters on stilts, have been built under to form apartments. These residential units suffered extensive damage. Increasing the height of habitable floor levels was a direct recommendation implemented following the 2011 Brisbane floods.

Consequently, the type and location of a flood exert a strong influence on response and recovery and the issues that need to be explored in post-disaster surveys. Neither warnings nor impacts are equal. Responses and disaster risk reduction strategies have to be community and event specific. The primary purpose of post-disaster surveys is to gather experiences and information to provide immediate feedback to emergency management debriefings to provide advice for education, awareness, and policy developments of local and state governments.

8.7. IMPACT OF FLOODS

In 2002, the Centre for Disaster Studies was commissioned by Emergency Management Australia to analyze and summarize post-disaster studies in order to contribute to the development of methodologies for rapid response post disaster surveys [*King*, 2002]. This began with an analysis of the findings from the events that had previously been studied by Centre for Disaster Studies researchers. These studies summarized seven groups of impacts and issues. More than a decade later, further post-disaster studies have reviewed the impacts of experiences of many more floods, cyclones, and other types of disasters. The seven impact areas remain as core summary findings. With the recent emphasis on climate change, two further groups of issues may be added. The initial issues and impacts identified were 1) unequal distribution of the impact within the community, 2) the loss of services during the event, 3) a lack of expectation of the severity of the impact, 4) late or minimum preparation, 5) strong community or neighborhood response, 6) a level of confusion concerning warnings and the role of the media, 7) a strong level of resilience, and with the additional climate change related issues are 8) constraints to adaptation for future disasters, and 9) retreat and relocation.

Most of these issues were identified immediately after a disaster, and they linked directly to the vulnerability and exposure of the community. In terms of flooding events, the unequal distribution of the flood often meant that only a portion of the community experienced severe loss or impact. People generally expected that they will lose power and water for a while during a natural disaster, but a loss of services has often been exacerbated by a lack of capacity of emergency services and local government to respond as rapidly as people would like. Following most floods, people expressed surprise and disbelief at the rapidity and height of the inundation. Due to this lack of expectation, people had frequently not prepared adequately, or their preparations occurred at the last minute.

In all communities, people showed great care for their neighbors and family and provided valuable assistance. Emergency services rely upon the media to pass on warnings, information, and advice, so people have frequently found fault with the tendency of the media to exaggerate and politicize the disaster story. Since the initial analysis in 2002, enormous advances in communications technology have improved the capacity of organisations and individuals to provide information and warnings, but this information has also increased gossip, hearsay, and misleading information. Strong levels of resilience have been encountered in all communities; however, this varies in style and character from one place to another.

After flood events, people have generally expressed a willingness to modify their homes and properties to reduce risk in future events. There is an underlying acceptance of the reality of climate change, that is, people are mainly divided over the extent to which this is human generated, and there is consequently a willingness to adapt to the threat of further disasters. However, many people described limitations and constraints to being able to do much. This leads to the related issue of relocation. A small but significant proportion of the community is likely to leave in the face of repeated or worsening floods. This is in itself a hazard adaptation for those individuals and families, but its effect will be to reduce the overall resilience of communities that are at risk of disaster.

These issues that emerge from post-disaster studies reinforce the need to continue to explore vulnerability, resilience, and adaptive capacity in order to enhance disaster risk reduction. The issues become constructs to test in future studies, where knowledge of previous events and community experience forms the basis for the next generation of post-disaster surveys. Analysis of such studies therefore prompts the development of methodology.

8.8. METHODOLOGY AND OBJECTIVES OF POST-DISASTER SURVEYS

Analysis of post-disaster studies in 2002 further examined 130 reports available at the libraries and websites of the Australian Emergency Management Institute (AEMI), Mount Macedon and the Natural Hazards Research Applications and Information Center at the University of Colorado [*King*, 2002]. Both organizations operated funding schemes for post-disaster studies at that time. Floods comprised 21 of the studies and were the second largest single disaster type after cyclones and hurricanes, which in many instances also brought flood impacts. Of the research methods employed, 23 studies were post-trauma related conducted by psychologists or psychiatrists. These are specialist types of studies with a well-established method and structure. For social scientists, such as in the Centre for Disaster Studies, a lack of specialized training and expertise required for this style of research meant that it was not a viable post-disaster survey option. The 2002 review indicated that post-disaster surveys mostly used questionnaires, interviews, and secondary sources [*King*, 2002]. Such methods have broadly continued up to the present time with the addition of emphases on climate change adaptation and innovative approaches and modeling that are made possible by changes in technology.

Social and community impact analysis of natural disasters received significant attention during the United Nations International Decade for Natural Disaster Reduction, which set targets for disaster risk reduction by understanding and tackling vulnerability, and through identifying and enhancing community resilience. At the same time during the 1990s, there was a noticeable shift in concepts and practice of development following the end of the Cold War. Disaster relief, a subset of development aid, had grown rapidly in the 1980s and 1990s [*Lindahl*, 1996; *Duffield*, 1994]. Emerging ideas in ecologically sustainable development influenced the linking of relief and especially disaster risk reduction to development. Furthermore, with the end of superpower rivalry by the beginning of the 1990s, peacekeeping operations targeted complex humanitarian disasters. Aid agencies realigned their priorities and programs [*Lindahl*, 1996; *Duffield*, 1994], and some personnel, both development specialists and researchers, shifted emphasis from development to disaster risk reduction as well as a combination of both. Development researchers brought disaster analysis methods from the developing world, such as rapid rural appraisal. This type of method was especially suited to post-disaster studies.

The most common approach for these rapid appraisal studies is to visit a disaster-impacted community within a week or two of the event (when access is open and flood waters have receded), interviewing people face to face. Although there may be a significant level of trauma and stress, people have usually been very receptive and willing to talk to researchers. For many, the interview and related discussion acts as something of a debriefing and often entails heightened emotions and fervor for retelling the story. Frequently, the adrenalin of the event experience is still present, and it is not yet perceived as a depressing process. The researcher usually records the responses immediately; however, in some circumstances where the length and complexity of the survey instrument have been longer than is appropriate for a face-to-face interview, a drop off questionnaire has been used. The questionnaire is then picked up the next day. In some instances, the interviews were supplemented with longer openended discussion that enabled issues to be identified that had not been considered before the event. Drop off and pick up surveys also enabled people to go into more detail in the open-ended questions. Since 2005, we have used online software such as *Survey Monkey* as a supplemental or alternate instrument.

Until 2000, a number of post-disaster surveys were also conducted as telephone surveys. Up to that period, most households still had landlines that were recorded in telephone directories, enabling a random sampling technique to be used. Otherwise face-to-face and drop off/pick up surveys were either systematic structured samples to ensure that all areas of the community were covered, or the survey was focused entirely on the primary area of impact, such as river or beach side. After 2000, the rapid expansion of mobile phones, coupled with invasive phone advertising, reduced the reliability and viability of telephone surveys. The phone survey was additionally restricted by the need to use a small number of brief, short and simple questions.

The post-disaster survey instruments used have been primarily qualitative, although some answers could be quantified using non-parametric methods. Mostly the data were expressed as indicators rather than representative samples, although a number of surveys have generated data that employed sophisticated quantitative analysis methods. Key informants and focus groups have provided specific qualitative data that put events into perspective and identified event-specific issues. Table 8.4 below summarizes the range of data-gathering techniques that have contributed to post-disaster studies in Australia and that have been utilized by researchers in other parts of the world.

The approaches identified in the table are neither mutually exclusive nor singular. A post-disaster study will usually encompass several approaches, for example, a survey instrument that includes both quantifiable questions and open-ended comments and explanations, alongside key informant interviews, secondary data, and observation. Well being and livelihoods approaches involve specific types of questions and methodology. Likewise action research has a specific methodology and is especially suited to response and recovery phases where the researcher learns from participation in the process and the community. Grounded research remains based in a variety of methods intended to assist to understand individuals, households, and whole communities from their experiences, needs, behaviors, actions, and reactions, over a specified period of time. It is evident from the literature and experience that there is no singular, specialized post-disaster survey technique or methodology. Field-oriented data collection practices should be adaptable and relevant to the context of the specific disaster impact environment and available resources.

The post-disaster study should address issues that contribute to mitigation as well as to an understanding of community experience in a disaster. The key mitigation themes are vulnerability, resilience, and adaptation. These are identified through constructs that define those elements that contribute to each theme. These are then developed as indicators that can be derived from specific questions and data at a range of levels such as individuals, households, or community characteristics. Some data may be derived from the census, but in trying to understand the relationship between demographic and socio-economic characteristics, attitudes, behaviour and experience, the demographic and socioeconomic data are incorporated into a survey instrument that is primarily concerned with issues that pertain to the disaster that has occurred.

A suggested list of characteristics that may be used to generate indicators of vulnerability, resilience, and adaptation may include the following areas. These data may be gathered through a number of approaches that are identified in Table 8.4.

Indicators of vulnerability, resilience, and adaptation are provided below. (Note: This is not an exhaustive list.)

• Physical: geography/geology, environment, climate, critical infrastructure and facilities, utilities plant, technology, hazardous industries, cultural, environmental and historically significant assets, transport, buildings resource base

• Demography: population (numbers, growth), density, age/structure, gender distribution, minorities, isolated/marginalized, special needs, mobility

• Health: medical/emergency services and facilities, health status, disabilities/mobility, nutrition, air/water quality, sanitation, disease

• Economy: income, production and productivity, trade, employment, livelihoods, insurance, individual assets/savings, capital, investments, opportunity, agriculture, livestock, sectoral diversification, social safety net, financial support, economic status

• Communications: public education, awareness, information and warning systems, patterns and transference, media, equipment, credibility, censorship

• Emergency Management: plans, resources and equipment, access, trained personnel, knowledge, experience, services/agencies, response and recovery capability

• Psychological: disaster experience, awareness/knowledge, disaster subculture, stress, acceptance, risk aversion, bravado, attitude, motivation, preparation, independence, flexibility, resilience, adaptive capacity

• Organized community structure/social aggregation: government, NGO services and social institutions, politics, logistics, policies, planning processes, management, law, authority structure, mitigation, extra family/community ties and networks, safety and security, external networks

• Societal/Cultural: coping strategies/mechanisms, cohesion, language, leaders, beliefs, resilience, lifelines, adaptation, innovation, skills, traditions, improvisation,

Table 8.4 Research Approaches to Post-Disaster Studies.

Type of Data and Approach	
Quantitative	Qualitative
Vulnerability assessment	Rapid appraisal type methods
Well-being	Well-being
Livelihoods	Livelihoods
Spatial information and analysis	Action research (volunteer/employee)
Needs assessment	Community participation and consultation
Individual and household questionnaires	Focus groups
Post-traumatic assessment	Post-traumatic assessment
Economic and infrastructural inventory	Observation (photographic records)
Longitudinal (any of above)	Longitudinal (any of above)
Text analysis	Stories, art, self-expression, meaning
	Key informant interviews
	Community interviews

values, ideologies, religion, ethnicity, integration, membership, participation, social activism, transition

Such indicators of vulnerability, resilience, and adaptation may be derived as pre-event vulnerability and resilience assessments to inform awareness and preparedness, warnings education and disaster risk reduction strategies. The same indicators also form a key element of the post-disaster survey, where they are combined with information relating to the actual disaster event. Many of these data may be provided by key informants and officials, but there is a powerful value in deriving perceptions and information from community members as well.

Indicators of disaster impact follow:

• Event-specific Issues: shock, trauma, sense of community
  • Media coverage and interpretation
  • Level of response
  • Political issues
  • Loss and Impact
  • Death and injury
  • Victims: numbers, involvement, impact, post trauma
  • Emergency management personnel: numbers, involvement, impact, trauma
  • Service providers: numbers, involvement, impact, post trauma
  • Volunteers and researchers: numbers, involvement, impact, post trauma
  • Relocation and loss of community members: temporary, long term, permanent
  • Economic impact and loss: sectors
  • Insurance: availability, level, policy coverage, efficiency
  • Loss of structures and infrastructure
  • Loss of land
  • Loss of livelihoods
  • Loss of vegetation, amenity, and symbols
  • Loss of information, communications, and history
  • Resource availability

Finally, there are data and indicators that pertain to the phases of impact and recovery. Each of these has a defining influence on the kinds of approaches to data collection and issues that emerge after the event. In most cases, the post-disaster study takes place very shortly after the event, but sometimes the study is delayed for reasons of access, safety, and evacuation of residents. There is also great value in carrying out longitudinal studies during the period of recovery. These timing issues qualify the types of approaches and indicators.

Timing and phases issues follow:

• Event: immediate post, six months post, one year, five years
• Impact, recovery, rebuilding, sense of loss, depression, moving on

The post-disaster studies have looked specifically at what happened to people and communities during the passage of a natural hazard. The dominant theme of all the surveys involved is that exposure to the source of the hazard is the primary level of impact. Receipt and appropriate response to hazard warnings is critical at this point of the disaster. Some socioeconomic and demographic factors play a role in processing such warnings, but the impact on the community is primarily exposure to the hazard. Mitigation of social factors of vulnerability and enhancement of resilience and adaptive capacity play very important roles in pre-disaster awareness and preparedness campaigns, and especially in the long process of recovery after a disaster.

Although there are inherent limitations and ethical concerns associated with conducting such rapid social assessment and research in the post-disaster context, from a response, governance, and strategic planning perspective, these are often outweighed by the need for timely, candid, local knowledge and understanding. Issues and experiences identified in the immediate post-impact environment can be used to better inform the direction of recovery, mitigation, and future preventative planning. Ideally, social assessment and community monitoring remains a proactive, iterative, process premised in integrated disaster risk reduction and management.

8.9. SUMMARY OF FINDINGS FROM FLOOD POST-DISASTER SURVEYS

8.9.1. Vulnerability

It has been shown elsewhere that vulnerability is a composite of a cluster of factors, such as social class, ethnicity, gender, age, socioeconomic status, and education [*Anderson-Berry and King*, 2005]. In the extensive literature on vulnerability, emphasis has been placed on the social factors listed above, in order to identify areas of vulnerability that could be targeted and reduced. From analysis of vulnerability, it became clear that most characteristics of vulnerability were structural, or could not be reduced or mitigated in the short term, or in some cases at all. Thus, resilience emerged as a more appropriate area to understand, and to enhance, because it concerned the more positive aspects of society and individuals. However, *Clark et al.* [1998] suggested that vulnerability involved exposure to the hazard and coping ability, which consisted of a combination of vulnerability and resilience. Vulnerability is arguably not the opposite of resilience, but it is largely an independent and parallel set of characteristics.

The emphasis on social and community aspects of vulnerability put exposure to one side as an obvious element of the risk equation that did not need further analysis. Post-flood studies, however, have shown us that exposure to the hazard is the single most significant aspect of vulnerability. Some lower socioeconomic households are located in flood-prone locations, but much wealthier households are also on absolute beach or river frontages, for the amenity of the area. Indigenous people occupy remote places where extensive floods isolate settlements and add health problems for extended periods of the wet season, but they share residence with non-indigenous people who are thus equally vulnerable to the effects of the flood. The 1998 Gulf floods isolated the towns of Normanton, Karumba, Burketown, and Doomadgee for several weeks, impacting all people, although those who were employed had been better enabled to stockpile food. Similarly at Wujal Wujal in 1999, the indigenous population was isolated alongside a predominantly low-income equal number of non-indigenous people, who lived in the same valley.

The flash floods in the Lockyer Valley destroyed houses and drowned people in the wealthier settlements at Murphy's Creek, Postman's Ridge, and Helidon before moving down river and devastating the lower social economic community of Grantham. The Cloncurry River flood of 1997 devastated a new housing area alongside an anabranch of the river. Social economic factors were of minor importance. The critical issue was a lack of awareness of the level of risk and consequent late preparation and response [*Goudie and King*, 1997]. Similarly in the 2008 flood in Mackay, the most significant aspect of vulnerability was a lack of information about floods, as well as community knowledge and perception of the accuracy of flood information, coupled with the extreme rapidity of a high rainfall event. In the same year, the flood vulnerability of residents in Charleville primarily related to low levels of household flood insurance [*Apan et al.*, 2010]. These vulnerability characteristics in both Charleville and Mackay applied also to businesses that were analyzed as part of the same flood study.

The floods that occurred in Brisbane in 2011 [*Bird et al.*, 2013] and Townsville in 1998 [*King*, 1998] impacted households, infrastructure, and buildings that were in low-lying areas, regardless of socioeconomic status. In Brisbane, floods inundated suburbs adjacent to the river. Some communities were wealthy, and others were much less well off. Upstream of the city center, riverside suburbs are expensive and prestigious; close to the central business district and down river there are many more rental properties and households that are adjacent to industrial areas where residents are much less well off. The floods were not socially or demographically selective. Exposure to the riverside determined impact and vulnerability.

Townsville's 1998 flood was an extreme rainfall event that inundated almost all of the low-lying flat floodplain on which the city is built. Three quarters of all residential properties experienced flood inundation on their land, but only 15% had water enter the house or apartment. On the edge of the city, two local creeks flooded adjacent peri-urban housing, causing significant damage in com-

munities that were socioeconomically and demographically representative of the average characteristics of the city. Within the main urban area, the worst impact was in older parts of the city, especially in old river and creek courses. Single-story apartments built flat on the ground, or the built under apartments of high set houses, suffered especially badly. Many of these were low-income rental properties occupied by elderly and other low-income residents, including students. Because 1000 people had participated in the post-disaster survey, it was possible to map locations of flood-impacted properties and discern the pattern of flooding of the 150 households that had experienced inundation inside their buildings.

This mapping indicated the low-lying old river courses as being especially vulnerable. However, there was virtually no inundation in all of the newer suburbs, despite most houses being surrounded by floodwaters up to a meter in depth. The reason was very simple. In all the newer suburbs, houses are constructed on a raised pad 600 millimeters above ground height. The roads are designed to carry run-off and large drains 2 to 3 metres deep above ground, are landscaped for recreational purposes, and are part of the suburban landscape design. The outer suburbs are typically occupied by low to middle income families, although they include a significant number of retirees. However, these areas were not vulnerable to the flood simply because post-1970 suburbs are designed to mitigate the runoff of flash flooding. Thus, exposure to the flood was not related to socioeconomic or demographic vulnerability.

8.9.2. Resilience and Adaptation

Awareness, preparedness warnings, and resilience have constituted elements of all post-disaster studies. Since 2005, climate change adaptation has been an additional focus of flood studies, but survey instruments have been careful to avoid direct reference to climate change for fear of polarizing responses. Rather, adaptation has been an extension of aspects of resilience, prompting people to consider how they might prepare for similar hazards in the future. This has included questions on attitudes toward relocation in the event of future floods.

Relocation is a disaster adaptation strategy that does nothing to enhance the resilience of the flood-impacted community. In most instances, out-migration will consist of the younger economically active members of the community whose loss reduces overall community resilience and often contributes to the economic decline of a town, a suburb, or a region. This is a particular problem for small rural towns whose lack of local opportunities and long-term agricultural decline (especially a reduction in the agricultural workforce) has caused loss of the younger population. Ingham, which is regularly flooded by the

Herbert River, illustrates a pattern over all censuses since 1990 of a net decline in all age groups under 45 years of age [*Boon et al.*, 2013]. Each flood adds a cumulative cost to both the local sugar cane farms and local businesses that prompts further economic decline and out-migration.

Following floods in Mackay in 2008, Charleville in 2008, Ingham in multiple years, inundated suburbs of Brisbane adjacent to the river in 2011, Donald in Victoria in 2011, and Emerald in 2010, household surveys asked people to consider their intentions to relocate, either to another suburb in the same town or to another community in the event of further floods in the future. Other surveys included communities impacted by cyclone, bushfire, and drought, where responses to the relocation questions were consistent, with an overall recording of 1463 responses [*Bird et al.*, 2013; *Boon et al.*, 2013; *Apan et al.*, 2010; *King et al.*, 2013]. With the exception of Donald, where long-term drought had already decimated the community, in each flood-impacted community, between 10 and 20% of respondents expressed a consideration that future events may prompt them to move elsewhere [*King et al.*, 2014]. In cities there are alternative suburbs to absorb migrants, but in small towns, relocation would involve a move to a completely different town. Analyses of demographic and socioeconomic characteristics of respondents who considered relocation indicated a preponderance of younger families on middle incomes, predominantly vocationally qualified core members of the community. Home ownership was a constraint to relocation, and older people had a strong sense of place and commitment to the community. Another finding was that people who displayed strong adaptive capacity in relation to floods were less likely to consider leaving. However, it was clear while carrying out all of these household surveys that some people had already left the community after the flood and had not returned.

Although out-migration potentially reduces resilience and may represent an unwillingness or inability to adapt to future hazard events, all communities showed positive characteristics of resilience. Ingham illustrated the most stoicism. Resilience in this community was based on prior experience with floods, sound governance and leadership, strong and coherent social networks, and a strong sense of place [*Boon et al.*, 2013]. Cloncurry and Charleville residents also demonstrated stoicism and gave support and assistance to one another during their floods. They, like the people of Ingham, interpreted this as the small town/outback spirit where communities come together in a time of crisis. However, the same spirit was recorded in Brisbane and Townsville during and after their disaster, as people came to the help of friends, neighbors, and complete strangers. A volunteer army developed through social media and came together without formal organization or leadership to assist in the immediate clean up in

Brisbane. Cities possess far greater diversity and capacity through their much greater numbers of networks and linkages. Additionally, urban dwellers have closer access to media, institutions, and leaders and were able to articulate criticism and demands. For all that, they also behaved with the same stoicism and acceptance in the suburbs of Brisbane, Mackay, and Townsville as in the remote and rural communities.

All flood surveys recorded high levels of community resilience, but all equally found areas of constraints to adaptive capacity, that is, the capacity of people and communities to move forward in anticipating the next event and to prepare more effectively to reduce the risk of disaster. These constraints included purely physical limitations such as a hazardous location of houses and an inability to develop protective measures for block built houses. Constraints also involved insurance, which discouraged people from "building back better," as well as the escalating costs of insurance and the unsupportive responses of several insurance companies to claims.

As a consequence of climate, environment, and settlement patterns, flood events remain an inevitable reality for people living in Northern Australia. Although scale and physical exposure remain significant indicators of the level vulnerability to direct impacts, the mythologizing of disasters in the Australian ethos has significantly influenced resilience and adaptive capacity. The challenge is to effectively integrate relevant knowledge and experience toward strategic, proactive disaster risk reduction planning.

8.10. CONCLUSION

The research derived from post-flood studies contributes to governance and policy, local government planning responses, flood mapping, communication, and raising awareness. The missing link is routine household preparation for a flood season. Every year in Australia there is a formal beginning to the cyclone season and the bushfire season. This is accompanied by an education and awareness campaign that specifies preparatory actions, explains and reinforces warning messages and expected responses to those messages on the part of individuals and households. *Cottrell* [2002] looked at women's preparation for the wet season in northern Australia and showed that women are aware of the flood hazard and make appropriate preparations. The wet season is a period of heightened hazard risk and inconvenience, yet there is no formal season launch similar to the cyclone or bushfire type. Respondents to the 2001 post-flood surveys in southeast Queensland [*Bird et al.*, 2013] made many statements concerning the lack of communication from authorities about what to do. This is a gap in pre-hazard preparation. The consistent message for public administration from each of the post-disaster flood studies is that households and individuals need clear, consistent advice on how to prepare in advance of a flood, and what to do in the case of a warning. Further progress in the delivery of effective, targeted hazard information, relevant timely planning, and practical actions and response may not necessarily reduce economic costs or the direct impacts of floods and other natural hazards on infrastructure, but it will reduce the disaster risk and subsequently save lives.

REFERENCES

Anderson-Berry, L., and D. King (2005), "Mitigation of the Impact of Tropical Cyclones in Northern Australia through Community Capacity Enhancement." Special issue of Mitigation and Adaptation Strategies for Global Change (2005) 10: 367–392 ed. E. Haque.

Apan, A., D. U. Keogh, D. King, M. Thomas, S. Mushtaq, and P. Baddiley (2010), The 2008 Floods in Queensland: A Case Study of Vulnerability, Resilience and Adaptive Capacity, A Final Report Submitted to NCCARF, Toowoomba, Queensland.

Australian Emergency Management (2015), Australian Emergency Management Knowledge Hub, 10 September 2015 viewed at https://www.emknowledge.gov.au/disaster-information/.

Bird, D., D. King, K. Haynes, P. Box, T. Okada, and K. Nairn (2011), Investigating Factors that Inhibit and Enable Adaptation Strategies Following the 2010/11 Floods: A Report for the National Climate Change Adaptation Research Facility Synthesis and Integrative Research Program, Risk Frontiers, Macquarie University, Centre for Disaster Studies, James Cook University, NCCARF, Griffith University.

Bird, D., D. King, K. Haynes, P. Box, T. Okada, and K. Nairn (2013), Impact of the 2010–11 floods and the factors that inhibit and enable household adaptation strategies, National Climate Change Adaptation Research Facility, Gold Coast. pp.160.

Blong, R. (2005), Natural Hazards Risk Assessment: An Australian Perspective, Issues in Risk Science. Issue 4. Benfield Hazard Research Centre: London.

Bureau of Infrastructure, Transport and Regional Economics (BITRE) (2008), About Australia's regions. Bureau of Infrastructure, Transport and Regional Economics: Canberra.

Carbone, D., and J. Hanson (2012) Floods: 10 of the deadliest in Australian history. Australian Geographic, 8 March 2012. viewed at http://www.australiangeographic.com.au/topics/history-culture/2012/03/floods-10-of-the-deadliest-in-australian-history/.

Chambers, R. (1994), The Origins and Practice of Participatory Rural Appraisal. World Development, Vol. 22, No. 7, 953–969, Elsevier Science.

Clark, G.E., S. C. Moser, S. J. Ratick, D. Kirstin, W. Meyer, S. Emani, W. Jin, J. X. Kasperson, R. Kasperson, and H. E. Schwarz (1998), Assessing the vulnerability of coastal communities to extreme storms: the case of Revere, MA, USA, Mitigation and Adaptation Strategies for Global Change, 3, 59–82.

Coates, L. (1999), Flood fatalities in Australia, 1788–1996. Aust Geogr.; 30:391–408.

Cottrell, A. 2002. Women and the Wet Season in Northern Australia. EMA Project 13/2002. Emergency Management Australia, Canberra and Centre for Disaster Studies, James Cook University.

Cottrell A., and D. King (2010), Social assessment as a complementary tool to hazard risk assessment and disaster planning. The Australasian Journal of Disaster and Trauma Studies, 1.

Duffield, M. (1994), Complex Emergencies and the Crisis of Developmentalism. IDS Bulletin 25.4, 1994. Institute of Development Studies.

King, D. (1998), Townsville Thuringowa Floods January 1998; Post Disaster Household Survey. Report Prepared for Department of Emergency Services, Bureau of Meteorology, Emergency Management Australia. Centre for Disaster Studies, James Cook University.

King, D. (2002), Post Disaster Surveys: experience and methodology. Australian Journal of Emergency Management Vol.17 issue 3.

King, D., D. Bird, K. Haynes, H. Boon, A. Cottrell, J. Millar, T. Okada, P. Box, D. Keogh, and M. Thomas (2014), Voluntary relocation as an adaptation strategy to extreme weather events. International Journal of Disaster Risk Reduction 8, 83–90.

King, D., J. Ginger, S. Williams, A. Cottrell, Y. Gurtner, C. Leitch, D. Henderson, N. Jayasinghe, P. Kim, C. Booth, C. Ewin, K. Innes, K. Jacobs, M. Jago-Bassingthwaighte, and L. Jackson (2013), Planning, building and insuring: Adaptation of built environment to climate change induced increased intensity of natural hazards, National Climate Change Adaptation Research Facility, Gold Coast, 361 pp.

King, D., and D. Goudie (1997), "Breaking Through the Disbelief: the March 1997 Flood at Cloncurry," The Australian Journal of Emergency Management, Vol. 12 No. 4.

Lindahl, C. (1996) Developmental relief? An Issues Paper and an Annotated Bibliography on Linking Relief and Development. Sida Studies in Evaluation 96/3, Department for Evaluation and Internal Audit, Stockholm.

Mackellor, D. (1908), My Country. The Spectator (London), 5 September 1908, p. 329.

O'Keefe, P., and J. Rose (2008), Humanitarian Aid. In V. Desai and R. B. Potter, Eds. The Companion to Development Studies.

Productivity Commission (2014), Natural Disaster Funding Arrangements, Inquiry Report no. 74, Canberra.

Queensland Flood Commission of Inquiry (2012), Queensland Floods Commission Report: Final Report. Queensland Floods Commission of Inquiry, Brisbane.

9

Understanding Crowdsourcing and Volunteer Engagement: Case Studies for Hurricanes, Data Processing, and Floods

Shadrock Roberts[1] and Tiernan Doyle[2]

ABSTRACT

Crowdsourcing is a method for data collection that involves obtaining data from a large number of individuals via the Internet and is one method for supporting information needs in disaster response. However, this phenomenon is often misunderstood or underutilized by traditional humanitarian agencies. Through a survey of current research on crowdsourcing in disasters, and our experience as practitioners, we unpack the concept of crowdsourcing and situate it in a broader history of volunteer engagement. We survey the current state of literature and present five case studies on crowdsourcing for disasters, with an emphasis on flooding and water events, and draw on our own experiences to investigate the concept of crowdsourcing to identify and describe different formulations of "crowd" and emphasize the volunteerism that underpins most crowdsourcing efforts. We examine the role of engagement with the crowd versus passive data collection to more precisely understand its relationship to data quality. Finally, we outline key concepts to consider for crowdsourcing and how each contributes to better engagement with the crowd and results in better data for decision-making.

9.1. INTRODUCTION

Over the last five years, the idea of "crowdsourcing" as a method for data collection during disasters has gone from esoteric, to disruptive, to buzzword. Crowdsourcing is most commonly understood as the process by which information is obtained or digital tasks are broken up and completed by a number of people who are generally not considered to be "experts," typically via the Internet in the form of a large open call. Initially coined in the technology magazine *Wired* to describe an innovative approach emerging in the technology sector, crowdsourcing captured the attention of the international community

during the 2010 Haiti earthquake where it became a method to gather a wide variety of data related to the earthquake. At the same time, the sudden addition of thousands of virtual volunteers and overwhelming amounts of data in unorthodox formats created confusion for established modes of humanitarian data production and positioned proponents of crowdsourcing as both a potential threat and a welcome revolution to individuals within the humanitarian industry.

The results of this nexus of technology and collective action have evolved in a variety of ways: previously nebulous groups of volunteers have become highly organized and structured themselves with clear lines of communication and some traditional humanitarian organizations have begun to incorporate various forms of crowdsourcing into their operations or experiment with it. Despite the recognition of the opportunities for disaster response

[1] Ushahidi, Resilience Network Initiative, Athens, Georgia, USA & Nairobi, Kenya

[2] VOAD and Resilience Networks BoCo Strong Boulder, Colorado, USA

Flood Damage Survey and Assessment: New Insights from Research and Practice, Geophysical Monograph 228,
First Edition. Edited by Daniela Molinari, Scira Menoni, and Francesco Ballio.
© 2017 American Geophysical Union. Published 2017 by John Wiley & Sons, Inc.

that lie latent in the use of both technology and aggregated data, we have found that crowdsourcing and, perhaps more importantly, the various forms by which "non-experts" convene around data creation and sharing during disasters remain underutilized by traditional organizations responsible for disaster response. This overview presents a summary of literature on crowdsourcing in natural disasters, with a focus on flooding and storm surge events, as a step toward best practices.

We draw on our experience as practitioners working within traditional humanitarian organizations and community-based flood relief organizations to situate crowdsourcing in a longer history of citizen engagement and collective action that are critical for effective disaster response and post-disaster recovery. This chapter also examines the role of engagement with the crowd and its constituent volunteers to facilitate greater inclusivity in decision-making and improve the quality of data collected. We present several case studies that illustrate the value of crowdsourcing but also the considerations required to ensure its effective integration to disaster response. Finally, we outline key concepts to consider for crowdsourcing and how each contributes to better engagement with the crowd and improved data collection during disasters.

9.2. UNDERSTANDING CROWDSOURCING

9.2.1. Defining "the Crowd"

In our experience, crowdsourcing is a term that is only partially understood by those who manage response to natural disasters and is often viewed with a range of preconceived notions from extreme distrust to panacea. At its root, any crowd is composed of individuals and how you understand and build relationships with them or "the crowd" to which they belong will have direct impacts on how you are able to leverage the phenomena of crowdsourcing. Put more succinctly, "people's motivations to contribute information have implications for its credibility" [*Flanagin and Metzger*, 2008], and both accuracy and level of engagement can vary depending on the type of data being collected and their interest in the problem for which they are being collected [*See et al.*, 2013; *Swain et al.*, 2015]. This being said, there is no a priori reason to doubt that the quality of crowdsourced data or information can be good: in some cases, notably in the creation of geographic data, it can rival the quality of data created by commercial or governmental organizations [*Haklay et al.*, 2010].

The term crowdsourcing originally described a company or institution outsourcing a function once performed by employees to a large, undefined network of people, but the term has since been used to refer to a wide variety of data creation practices in which those contributing have varying degrees of engagement, motivations, or awareness. In some

cases, crowdsourcing is the act of harvesting data (often via social media) from a large network of individuals who share information for personal reasons and not as part of a collective action [see *Cobb et al.*, 2014; *Herfort et al.*, 2014]. In other cases, it may describe focused collective action to support disaster response for a specific event [see *Shahid and Elbanna*, 2015]. Some literature uses the term "microtasking" and implies that this is a more structured form of crowdsourcing involving defined workflows and Internet platforms designed to maximize time and efficiency on specific tasks [*Morris et al.*, 2012]. However, microtasking appears to have its roots in crowdsourcing tasks done in exchange for pay by individuals in low-income countries [*Grant*, 2010; *Gino and Staats*, 2012], and although the term appears in conjunction with crowdsourcing in studies about its role in disaster response [*Meier*, 2013], a precise definition of how it relates to, or is different from, other forms of crowdsourcing done with specific online platforms remains unclear. The broadness of these terms, therefore, may obscure important details for those who want to operationalize crowdsourcing during floods or other natural disasters. Following *Estellés-Arolas and González-Ladrón-de-Guevara* [2012], we use it here to describe "a type of participative online activity in which an individual, an institution, a non-profit organization, or company proposes to a group of individuals of varying knowledge, heterogeneity, and number, via a flexible open call, the voluntary undertaking of a task". The participants from the crowd fundamentally form a volunteer group that is part of an active engagement process in disaster response, and we encourage a view of crowdsourcing as a form of collective action.

9.2.2. The Many Faces of the Crowd

Although crowdsourcing is a relatively new phenomenon, the formation of new social relationships to meet the unique demands of a disaster is not new. In order to more clearly understand what crowdsourcing is, we situate it within the historical context of volunteerism related to disasters. Writing in 1985, *Stallings and Quarantelli* [1985] describe "emergent citizen groups" (ECG) during disasters. These ECG are loosely organized, tend to have a flat hierarchy, exist for short periods of time, and are composed by a core of continuing members with other individuals participating irregularly (Ibid). These characteristics are strikingly similar to certain group formations of crowdsourcing with the notable exception that ECGs have been historically defined as those directly affected by, or in closest in proximity to, natural disasters.

The phenomenon of crowdsourcing in natural disasters is related to that of ECG. The evolution of technology has resulted in information and communication technologies (ICT) that have lowered the traditional costs and barriers of social activities such as communicating to large audiences and connecting with others who have similar interests. This, in turn, makes it easier to organize groups of individuals and take collective action on issues that they care about, whether formally or informally [*Shirky*, 2008]. When seen in this light, crowdsourcing combines the motives of ECG with a greater capacity to collect, analyze, and disseminate information via technology. In the humanitarian sector, concerned individuals can now formulate new social relationships and approaches to data collection, analysis, and dissemination during natural disasters, often challenging the status quo of information management [*Meier*, 2015; *Crowley*, 2013; *Liu and Palen*, 2010].

The group formation of these individuals can take many forms. In recent years, the term "volunteer and technical communities" (V&TC) has been applied to volunteer-based communities, usually with some defined leadership, who apply their technical skills to support formal responders to natural disasters [*Waldman et al.*, 2013]. V&TC may convene entirely online or be represented by an individual or team at the site of a natural disaster. The online network of individuals who support the work of V&TC or engage in other forms of crisis or natural disaster response using online tools are sometimes referred to as "digital humanitarians" [*Meier*, 2015], digital humanitarian organizations [*Crowley*, 2013], or "crisis mappers," so named after an annual conference at which many of these new concepts were presented and new networks formed (www.CrisisMappers.net) [*Ziemke*, 2012]. Therefore, crowdsourcing can involve loosely affiliated individuals or highly organized groups, community-based organizations or pools of global volunteers who may never meet in person, and groups who self organize to complete a specific action on their own or those who actively support existing forms of disaster response. Their common characteristics are a strong sense of volunteerism and access to technology that enables them to engage with the humanitarian sector in new ways.

9.2.3. Trusting the Crowd: The Role of Engagement for Better Data

Current research regarding crowdsourcing for natural disasters often suggests or investigates its usefulness for situational awareness and, whether implicitly or explicitly, the degree to which it may contain geographical information. In a recent review, *Horita et al.* [2013] found that the research regarding volunteered geographic information (VGI) in disasters focused primarily on its use during the response phase and that floods and fires were the most common disasters in which it was used. The practice of large groups of volunteers using online mapping tools to create new data has been labeled neogeogra-phy [*Turner*, 2006], volunteered geographic information [*Goodchild*, 2007], or crowdmapping [*Shahid and Elbanna*, 2015], and several research agendas for VGI and crowd-sourced data have been suggested [*Elwood*, 2008b; *Zhao and Zhu*, 2014].

The main issues raised concerning the use of VGI are credibility, reliability, and quality [*Flanagin and Metzger*, 2008] because they are produced by non-experts in a context that differs significantly from the highly structured format in which "expert" data are created. Understanding the true value of crowdsourced data may be difficult due to its inherent heterogeneity and attendant absence of systematic "professional" standards [*Feick and Roche*, 2013]. At the same time, several studies have shown that crowdsourced data are credible, reliable, and of generally high quality in a variety of contexts. *Haklay et al.* [2010] found that the spatial accuracy of crowdsourced geographic data was comparable to that generated by governmental authorities. The United States (US) government has crowdsourced reports of seismic activity to such great affect that it is considered a "valuable new data resource for both qualitative and quantitative earthquake studies and has the potential to address some longstanding controversies in earthquake science" [*Atkinson and Wald*, 2007] and provides a real-time earthquake detection system [*Liu*, 2014]. Several other reviews of crowdsourced data in practice suggest the promise it holds for environmental monitoring [*Wiggins and Crowston*, 2011; *See et al.*, 2013] and creating geographic data [*Neis and Zielstra*, 2014]. In a review of crowdsourced data relating to wildfire outbreaks, *Goodchild and Glennon* [2010] found that the benefits of these data can outweigh the risks.

Similarly, the case studies provided below show various measures for assessing the relative quality or value of crowdsourced information. However, we feel that absolute measures of credibility, reliability, and quality are inappropriate outside of specific contexts in which crowdsourced data may be created. Like *Feick and Roche* [2013], we argue that the socio-technological processes that permit individuals and groups, who may not otherwise interact, to create these types of data offer value in themselves. We see the engagement between the crowd and the end-user or requestor who may initiate a crowdsourcing activity as a critical space for exchange that can both foster greater inclusivity at the same time that it provides more useful data.

In the domain of disaster risk reduction, there is an emerging view of knowledge as a convening tool that can allow different stakeholders to build common understanding or risk and the way to cope with it. Like *Menoni et al.* [2015], we believe that co-production of knowledge should "enhance the cooperation and the coordination capacity among different stakeholders and social groups, particularly when they are working in the same geographical context on the same problems." The role then, of a motivated crowd or group of volunteers, is a key variable to success [*Gouveia and Fonseca*, 2008; *Swain et al.*, 2015], and it is "critical that the crowd is treated as a partner in the crowdsourcing initiative, and the needs, aspirations, and motivations of the crowd must remain an important consideration" [*Zhao and Zhu*, 2014].

Crowdsourcing works on multiple scales and is not limited to specific types of tools; therefore, it can be difficult for informal and non-traditional responders to collect and distribute data in ways that are recognized as authoritative or accurate. Grass-roots responders must negotiate the paradox that their work relies on multiple forms of knowledge "but at the same time must tame them and abstract from them" to align with those pre-existing needs and information management structures of formal responders [*Burns*, 2014].

Recognizing that grassroots groups, especially ECGs, will often emerge during a disaster in order to meet a specific need is a first step toward allowing them to function as spontaneous subject matter experts. Both collecting and dispersing data from the affected population, these groups often possess insider knowledge for specific aspects of a disaster as well as access to communication channels and informal networks that may be more effective than official systems during a disruptive event. Emergent groups can rapidly increase the capacity of formal responders by identifying local resource channels and volunteers and have the potential to provide specialized knowledge derived directly from affected populations. As such, they can simultaneously model effective methods of data acquisition for often highly disparate communities and model service delivery to the affected population.

Several efforts exist to focus and streamline crowdsourcing by establishing volunteer networks and formalizing their crowdsourcing practices and protocols. Both *Liu* [2014] and *Crowley* [2013] examine several of these networks including the Standby Task Force (http://blog.standbytaskforce.com), Humanity Road (http://humanityroad.org), and the consortium of similar organizations known as the Digital Humanitarian Network (http://digitalhumanitarians.com), which organizes disaster simulations to streamline collaboration among groups and regularly publishes guidance for both V&TC who would like to support traditional disaster responders [*Waldman et al.*, 2013] and for formal humanitarian organizations that would like to better understand crowdsourcing [*Capelo et al.*, 2012]. As traditional disaster response organizations engage with crowdsourcing and, specifically, these emerging groups, the way in which crowdsourcing becomes a way to negotiate social and political processes around knowledge production will

become increasingly important to improving future forms of engagement [*Burns*, 2014]. Already, we are seeing that this interaction can and does influence both institutional practice and the behavior of these groups [*Shahid and Elbanna*, 2015], and understanding this engagement as a dynamic relationship is an important step in building true participation as well as receiving data that are an appropriate fit for the task at hand.

9.3. CASE STUDIES

The cases presented here are classified by three variables: event type, typology of data processing, and applicable phase in the disaster cycle, the last two being linked (see Table 9.1). Literature about the disaster management process identifies various phases, almost all of which include four basic phases: mitigation, preparedness, response, and recovery [*Hiltz et al.*, 2010]. Different types of information will be required at different phases of response. For example, baseline data and development of scenarios are required for mitigation and preparedness while damage assessments will be required for response. Although crowdsourcing activities have most often been studied during a response to a disaster [*Horita et al.*, 2013], the case studies presented here highlight the use of crowdsourcing in various phases of the disaster management cycle. Finally, these cases illustrate the need for building relationships with the "crowd" and creating partnerships, versus only focusing on the outputs of the "crowd" as a passive generator of data, thus situating crowdsourcing in the context of understanding volunteer engagement and community building.

9.3.1. Crowdsourcing Damage Assessments

A damage assessment is a preliminary onsite evaluation that captures the extent and cause of damage in the aftermath of a disaster. Assessments are often the first step in formulating response and relief operations such as evacuation, search and rescue, and shelter. Both of these cases were in response to cyclones and used crowdsourcing to produce data and analysis for damage assessments during the response phase of the disaster cycle.

9.3.1.1. FEMA Damage Assessments during Hurricane Sandy. In the wake of Hurricane Sandy's landfall on the eastern seaboard of the US in 2012, the US Federal Emergency Management Agency (FEMA), used crowdsourcing to evaluate more than 35,000 global positioning system (GPS)-tagged aerial photographs of damage-affected areas to aggregate geo-located data for situational awareness. FEMA, working with the Humanitarian OpenStreetMap Team (HOT) and the V&TC GISCorps, employed crowdsourcing as a method

Table 9.1 Case Studies.

Section and Case Study	Event Type	Data Processing	Applicable Phase in Disaster Cycle
Section 9.3.1.1 FEMA damage assessments during Hurricane Sandy	Hurricane	Damage assessment	Response
Section 9.3.1.2 American Red Cross damage assessments during Typhoon Haiyan	Typhoon	Damage assessment	Response
Section 9.3.2.1 Crowdsourcing baseline data creation for the US Department of State	Multiple	Baseline data creation	Multiple
Section 9.3.2.2 Data Processing for USAID	Economic development	Baseline data creation/Data processing	Mitigation/Recovery
Section 9.3.4 Floods and ECG: the case of Boulder Flood Relief	Flood	Operational data collection and response coordination	Response/Recovery

to evaluate and rate the level of damage shown in the images via an online system that allowed for three broad ratings: little/no damage; medium damage; or heavy damage. The after-action assessment of this effort used inter-volunteer agreement as a metric for evaluating accuracy. This is common in crowdsourced tasks when the "correct" answer is not known and is predicated on Linus's law that, if there is a large amount of agreement concerning a judgment from multiple volunteers, then it is likely that the shared judgment is the correct one [*Haklay et al.*, 2010].

The analysis was restricted to 17,070 images that had been rated by three or more volunteers. In these cases, the crowd had majority agreement on 93.54% of the images. Additionally, 720 images believed to be the most difficult to rate were also rated by 11 experts from GISCorps using the same process for comparison. Experts generally agreed about how to rate these more difficult images: 81% had an agreement among the experts, compared to just 37% for public volunteers, showing that the volunteers were not as accurate (in terms of inter-annotator agreement) for these images. The biggest divergence of expert opinion from the crowd was on images that the crowd deemed little/no damage but which the experts said showed medium damage. In broad strokes, this matches the findings of the crowdsourcing damage assessment conducted by the American Red Cross (ARC) (section 9.3.1.2) and suggests that crowdsourcing may not be an ideal method for damage assessments due to the subtle distinctions that might not be apparent to an untrained eye. This being said, the report also describes the role that image quality plays in aerial image interpretation and suggests that the crowd

would likely learn from better and more frequent interaction with experts during rating process.

9.3.1.2. American Red Cross Damage Assessments During Typhoon Haiyan. In 2013, the deadliest Philippine typhoon on record killed at least 6,300 people. It was one of the strongest tropical cyclones ever recorded and devastated portions of Southeast Asia. During the response to Typhoon Haiyan (also known as Yolanda), the ARC led a crowdsourcing effort to move beyond the geographic base data typically collected by crowdsourcing such as streets, houses, and farms and create information about building-level damage in areas affected by natural disasters [*ARC*, 2014]. For this effort, ARC chose to use OpenStreetMap (OSM), which is most commonly understood as a free online map of the world. It is, however, actually a combination of elements: a database of geographic data, a website to display portions of the database, and a variety of tools that allow users to interact with the database to download, add, or edit portions of it. It is also a global community that interacts via various web-based communication channels in-person for conferences or activities that facilitate adding to the database. More than just an online map, OSM is a "multi-faceted project that enables distributed work around a common product" [*Soden and Palen*, 2014].

The use of OSM as a spatial data infrastructure for base vector data derived from satellite imagery during disasters is becoming increasingly common [*Crowley*, 2014; *Soden et al.*, 2014; *Campbell*, 2015]. This is due, in large part, to HOT, an organization that emerged out of the response to the 2010 Haiti earthquake specifically to lead crowdsourcing efforts using OSM. HOT combines both globally distributed and localized work for different aspects of data creation and advancement of the social practices that surround the use of OSM [*Soden and Palen*, 2014].

Together, ARC and HOT tasked the crowd with tracing satellite imagery made available via the US State Department's Humanitarian Information Unit. Instructions to volunteers were communicated via existing OSM wikis. list-servs, and the effort was widely disseminated via social media. To test the validity of the damage assessment, paid enumerators assessed building damage of 1,343 structures, in the field, in randomly selected municipalities that were most highly affected by the typhoon. The results show that the crowd did a "reasonably good job of identifying affected buildings but overestimated the number of buildings completely destroyed by the typhoon and underestimated the number of buildings that were majorly damaged." ARC found that the quality of the imagery available directly affected the crowd's ability to perform the assessment especially because the orthographic nature of the imagery might conceal "partial" damage to the sides or insides of buildings. Furthermore imagery interpretation by the crowd was likely affected by three factors:

1. The resolution of the imagery used was too low to allow the crowd to reliably differentiate between destroyed and merely damaged buildings.

2. Buildings with major damage in particular may be mistaken for destroyed; habitable buildings with heavily damaged roofs can appear destroyed at a 1 square meter pixel resolution.

3. The time required to plan and implement ground truthing allowed for repairs and reconstruction to have taken place on some structures.

The inability of the crowd to more accurately infer damage from remotely sensed imagery shouldn't be surprising considering the manual interpretation, which is considered to be both a "science and art" that takes time and experience to perform well [*Lillesand et al.*, 2008]. Moreover, existing literature regarding the spatial accuracy of crowdsourced data in OSM suggests that volunteers do very well when identifying easily distinguishable objects. In other words, although OSM may not be the best method for collecting damage assessments, it remains useful for the collection of baseline geographic data. Indeed, ARC suggests that greater investment in baseline geographic data, specifically detailed building data layers within OSM prior to disasters, would improve the crowd's ability to spot missing or damaged buildings.

Finally, this case study lauds the "the continued responsiveness and diligence of the crowd," which mapped and validated an entire municipality within 48 hours upon receiving the request. It is noticeable that this occurred three weeks after Typhoon Haiyan made landfall when media attention of the typhoon was minimal and public interest had faded. In our experience, formal response organizations are often skeptical that volunteer efforts are a reliable enough workforce to merit any organizational investment in crowdsourcing, but this case shows otherwise.

9.3.2. Working the Crowd for Baseline Data Collection and Processing

Although crowdsourcing during the response phase of the disaster cycle is most prevalent in literature, it is important to consider how it might be used during the less acute phases. The length of time required to build an appropriate relationship with a given body of volunteers may be better matched to mitigation and preparedness phases. The cases listed below show the value of crowdsourcing for the creation of baseline data or data processing and provide examples of what could be applied to a variety of event types.

9.3.2.1. Crowdsourcing Baseline Data Creation for the US Department of State. The increasing value of OSM as resource, tool, and work force to create geographic data during disasters, and the consistent need to create baseline data before disaster strikes, especially in areas at high risk of disasters, led to an initiative of the US State Department's Humanitarian Information Unit (HIU) known as MapGive. The purpose of MapGive is to increase the amount of free and open geographic data. MapGive combines the "cognitive surplus" [*Shirky*, 2010] of volunteers with the power of the US government to provide updated high-resolution commercial satellite imagery to volunteers for vetted humanitarian purposes. MapGive has been used in several major disasters and confirms that crowdsourcing is a sustainable resource.

Launched in 2014, MapGive is one element of a larger ecosystem and has successfully implemented several steps (outlined below) to harness the power of crowdsourcing. Despite its name, MapGive did not begin with a map but with a multi-year effort to establish the legal, policy, and technical framework for sharing commercial satellite imagery, purchased by the US government with volunteers [*Campbell*, 2015]. HIU developed a geographic computing infrastructure built from open-source software to publish updated satellite imagery as web services that can be quickly and easily accessed via the Internet, allowing volunteers to trace the imagery to extract visible features such as roads and buildings. HIU also reached out to existing organizations to help organize crowdsourcing efforts and has built a significant relationship with HOT to increase the number of volunteers by providing outreach, education, and training materials.

Finally, MapGive is accompanied by a thoughtful communications strategy to give maximum visibility to their initiative and garner the maximum amount of volunteers possible when needed. When taken in sum, these sets have established a repeatable, sustainable mechanism for the US government to help catalyze and direct volunteer mapping efforts [Ibid].

MapGive has played an active role in crowdsourcing geographic information for refugee camp mapping in the Horn of Africa, risk mapping and strengthening community resilience in Uganda, disaster risk reduction in Kathmandu, Nepal, disaster response to Typhoon Haiyan in the Philippines, and humanitarian planning in the Democratic Republic of Congo. It is cited as a model for governmental use of crowdsourcing [Crowley, 2013] and is estimated to have leveraged between $1.5 and 2 million dollars worth of imagery to crowdsourcing efforts [Campbell, 2015].

9.3.2.2. Data Processing for the US Agency for International Development.

Similarly, the US Agency for International Development (USAID) began piloting crowdsourcing initiatives in 2012 to test the sustainability of using online volunteers to process data regarding international development and to investigate the overall quality of the resultant data. A data set regarding loans made as part of USAID's Development Credit Authority (DCA) was identified as a potential target for crowdsourcing. This was an ideal way to test the viability of crowdsourcing for a variety of development and humanitarian actions in a non-crisis environment. The DCA database was originally structured to capture information regarding the amount, sector, and purpose of each loan in accordance with the guarantee agreement, and it paid less attention to the geographic specificity of each loan. Users who entered data were given a single field marked "City/Region," and all geographic information was stored as free-form text in a single column in the database. Typically, databases have detailed geographic information collected in separate fields that are machine-readable. The DCA database, on the other hand, did not originally envision a demand for mapping its data and did not separate these fields. Moreover, there was no standardization given for how to enter various pieces of information (e.g., spelling of place names, abbreviations to use, separation of discrete pieces of information by commas). This unstructured, non-standard input translated into a column of information containing only partial geographic information that could not be automated for mapping.

Working with the private companies Socrata and Esri, USAID created a website that allowed volunteers to "check out" loan records and mine them for clues to the appropriate administrative unit, such as a county or municipality, to which the loan record should be geo-referenced. Because of incomplete data, not all records could be processed, and it was important to allow volunteers to flag such records as "bad data." Working with GISCorps and the Standby Task Force, both of whom are highly visible V&TC, USAID broke new ground by engaging the public, for the first time, in processing to map and open USAID data. The project attracted the attention of more than 300 volunteers worldwide, 145 of whom far exceeded all expectations by processing almost 10,000 records in roughly 16 hours: less than one-third of the time anticipated for this task. Of the records processed, 7,085 contained useful enough information to derive the needed geographic information. An accuracy assessment, using a separate group of expert volunteers, found that the selected subset of 322 records has been processed at 85% accuracy. After adjusting for a common transcription error, the accuracy level reached 90% [Roberts et al., 2012].

As with MapGive, the USAID project was preceded by significant time spent on policy issues such as establishing an appropriate technological environment, building strong relationships with organizations that would ensure a "core" group of volunteers around which a wider crowd could coalesce, and developing a robust communication strategy to ensure transparency and visibility for all stakeholders. Indeed, the accuracy assessment was conceived as a necessary step to dispel persistent myths that crowdsourcing is not a viable method for quality data processing. It is notable that relatively few official USAID maps contain any form or accuracy assessment.

Together, these studies show that the productive cooperation between formal organizations, in this case governmental, and the crowd is not only possible, but can improve data and the methods by which they are collected and processed if enough consideration is given to the entire process. The advantages of crowdsourcing also extended well beyond the defined tasks, that is, DCA saw a 20% increase in Twitter followers and increased Facebook friends by 15%, thus ensuring that their work was more widely known and supported by the public. The project is widely considered a success by other experts in the field.

9.3.3. Floods and ECG: The Case of Boulder Flood Relief

The case of Boulder Flood Relief differs somewhat from the previous cases in that it is an ECG and was never presented as a crowdsourced project. The activities of the organization, however, do fit the definition of crowdsourcing in that they engaged in participatory online activity in which individuals and groups of individuals of varying knowledge voluntarily undertook given tasks. The organization began by collecting various forms of operational data for their own response needs, and over time, they engaged in a wide range of data collection and management practices aimed at better coordination among responders and activities for long-term recovery.

Boulder Flood Relief (BFR) is an ECG that provided organizational infrastructure to quickly mobilize volunteers for community disaster relief during the 2013 Colorado floods. *Doyle* [2015] describes how official communications channels meant to direct or manage volunteer efforts were often not sufficient. Official websites, such as the state of Colorado's http://helpcoloradonow. org, were overwhelmed and unavailable during the floods as "the demand for information became greater than the technology could supply." Using social media and a variety of tools associated with crowdsourcing to formulate community-based response, BFR functioned at the nexus of technology and volunteerism by filling gaps in situational information and acting as a convening point for volunteerism that found little direction or support through official channels.

As exponentially increasing numbers of community volunteers outstripped BFR's initial data management solutions, the ECG reversed the top-down approach of official channels, choosing instead to share data as openly as possible within the shifting roster of volunteer office staff. Using freely available collaborative tools such as Google Docs enabled a more open form of knowledge production. As volunteers rotated hourly or daily through the office, open data sharing was paramount in order to compensate for the lack of centralized updating mechanisms. Methods of project development and logistics remained highly flexible dependent on the number of volunteers available, homeowner requests, jurisdictional restrictions, and onsite leadership. Information sharing was the best solution to all of these issues and was relied upon to the point that the act of knowledge production became its own form of update. The use of sharing technology allowed volunteer dispatchers and office staff to collaborate effectively as events unfolded.

The contrast between "top-down" and open-source modes of knowledge production is a regular feature in the conceptual landscape where ad hoc volunteerism abuts highly regimented governmental procedures and is often a source of concern for those wishing for a more tightly controlled flow of information. However, consistent research has shown that open-source modes of production can, in fact, produce widely useful results [*Gouveia and Fonseca*, 2008; *Goodchild and Glennon*, 2010; *Haklay et al.*, 2010]. This approach has long been trusted for open-source software development. In describing this process, *Raymond* [2001] likens the top-down processes to that of building a cathedral: guarded, structured, and integrated, whereas the open-source approach resembled a bazaar: chaotic, rapid, and iterative. Raymond describes how the inclusion of collaborative volunteers, connected via the Internet, produce better software because the code can be reviewed and corrected by a wider audience at a greater speed than with a top-down process.

BFR used this approach to successfully share situational information and try to match volunteers to specific tasks: dispatching over 1,300 volunteers to meet more than 300 requests for assistance. However, *Doyle* [2015] points out that, although flexible, this mode of working presented other challenges such as ensuring adequate privacy, which were dealt with via peer oversight to the best of the organization's ability. BFR also used a "bottom-up" approach for sharing information externally by leveraging social media to deploy emerging terms the public was using to search for information. By watching these terms emerge, and adapting to them, BFR promoted rapid diffusion of the information they had on hand. BFR also became adept at structuring "packaged tasks" or well-defined needs with requests for specific types of help that could be distributed via social media. This combination of internal and external information management was well suited to the self-organized nature of BFR and the pace at which the general response was moving. It was simply "impractical to wait for updates from a centralized location."

BFR serves as an important example that the technique of crowdsourcing may already be in practice without explicitly being stated or advertised as such. Indeed, the flexibility of ECGs suggests that, when available, they will take advantage of ICT to help achieve their goals. Unlike crowdsourcing efforts that direct a global pool of remote volunteers to collect or process data to support decision makers, which may be in headquartered on an entirely different continent from where the disaster is happening, groups like BFR are collecting situational information specifically to make decisions in their community. They may represent the affected population, real-time data collection, and a valuable resource for disaster response in situ all at once.

9.4. HARNESSING THE POWER OF THE CROWD

Whatever your interest in crowdsourcing, you will need to devote some time to defining how it can be beneficial to your organization and developing the appropriate resources and relationships to accomplish this. Although not necessarily requiring additional financial resources, engaging with crowdsourcing will likely require time in the form of staff who can interface with volunteer groups or technical analysts who can process the multiple types of data coming from the crowd. We list here several recommendations that we have directly experienced as being critical for effective crowdsourcing.

9.4.1. Technology

Because crowdsourcing is facilitated through ICT, either as a tool for organizing and communicating tasks to the crowd or as a method of data collection, crowdsourcing is often conflated with ICT and other technological terms, such as "social media" and "big data". It is critical to understand these as distinct elements if one is to have a clear picture of how crowdsourcing operates and the role of technology within that.

Crowley [2013] separates crowdsourcing into several distinct elements, including social media channels and the suite of hardware and software tools that enable the completion of crowdsourced tasks. The term social media refers to the various channels through which an individual, volunteer or otherwise, may provide data or information during a disaster. These include short text messages such as Twitter, websites for sharing photographs and video, or online platforms that combine multiple forms of media, such as Facebook. The social media outlet may be a channel through which social media data are collected. See *Cobb et al.* [2014] and *Herfort et al.* [2014] for examples. Social media channels may also be used to communicate with the crowd and to help organize specific actions. See *Roberts et al.* [2012] and *Doyle* [2015] for examples.

Furthermore, crowdsourcing efforts generally use a suite of hardware and software tools to perform a given function such as using OSM for mapping road networks. Some tools have been explicitly built with crowdsourcing in mind, such as the open-source software Ushahidi (https://www.ushahidi.com) that allows users to collect reports from affected communities via a range of social media channels or directly through the software itself. The use of open-source software and tools, that is, those for which the original computer code is freely available and may be redistributed and modified without licensing restrictions, allows one to continually refine and adapt them and further invites volunteers to improve on the design of the tool itself. However, many ECG or V&TC use a wide array of freely accessible ICT such as Google Docs (https://www.google.com/docs/about) and develop practices to align with the capabilities and constraints of the technology they are using [*Liu*, 2014].

It is also important to distinguish the companies or nonprofit organizations that develop these tools from organizations that organize crowdsourcing efforts or respond to natural disasters. As *Crowley* [2013] notes, these organizations may donate their time to provide support in specific instances, but they are not disaster response organizations, and their mission is to build the ecosystem around the software that drives social value, a revenue model (profit or nonprofit), or both.

When considering technological issues, it is critical to coordinate with ECG or V&TC to understand the technology that they have found most useful and how it may interact with the technology being used by other stakeholders. Engaging with volunteers over the method by which data exchange will take place can foster greater trust among stakeholders and is another critical piece of successful crowdsourcing. The digital infrastructure used to collect and manage data should not be seen as separate from, but complementary to, the act of crowdsourcing since the crowd may both contribute improvements concerning how to collect or manage data, but they may also wish to access and use the data they have helped create. The increasingly blurred line between those who collect data, those who manage it, and those who use it suggests the need for new approaches to data exchange [*Budhathoki et al.*, 2008], which can harness the use of ICT to promote effective collaboration as well as data collection [*Gouveia and Fonseca*, 2008].

Even though ICT undoubtedly facilitates aspects of crowdsourcing, it also excludes those who may not have equal access to it. And although there is potential for crowdsourcing to foster greater participation and equity among stakeholders, the manner in which it is carried out can also be exclusionary. The rate of Internet penetration and the geographical distribution of digital data reflect a "digital divide" and the uneven development levels of our world [*Sui et al.*, 2013]. Critical reflection about the social and technological processes surrounding crowdsourcing is an important part of understanding which social and political interactions are supported and promoted through it [*Elwood*, 2008a].

9.4.2. Interoperability

Because technology is a fundamental aspect of crowdsourcing, it is vital for an organization to ensure that its information systems can work together within and across organizational boundaries to consume, share, and disseminate information. Many V&TC collect, analyze, and share data using open-source tools composed of software that can be freely used, changed, and shared (in modified or unmodified form) by anyone and publish their data in open standards. Understanding the way that data coming into your organization must be structured and how this may differ from the standards in use by volunteer groups or the crowd is an important starting point. It is possible that your organization may want to process data itself. However, most government and public service entities do not possess the computing power necessary to process the massive amount of data that can be generated by a large crowd. In this way, partnering with V&TC or volunteer groups who can help process and clean data can be very valuable.

The use of open-source tools and compliance with open-data standards can greatly improve interoperability in cases where an organization would like a free flow of information between volunteer organizations or "the crowd." In the United States, some government data standards are controlled by an ecosystem of vendors, whose platforms may not support open-data standards [*Crowley*, 2013]. Organizations who contract the development of information management technology should include clear language in the terms or reference or other legal documents that ensure tools will be built with open standards in mind. The US Department of State, which has implemented an open-source and open-data approach to crowdsourced mapping has built strong institutional relationships with volunteer organizations and have reaped the rewards of crowdsourcing to create large amounts of open data [see *Liu*, 2014; *Campbell*, 2015].

9.4.3. Data: Share and Share Alike

Most formal responses to a natural disaster begin with data to help prioritize and guide response [*National Research Council*, 2007; *Campbell*, 2015; *Verjee*, 2007]. Data such as known human settlements, elevation models, and critical infrastructure serve as a baseline to understand the possible scope of a disaster and the resources that exist to respond to it. Incoming data from damage assessments, surveys, and eyewitness reports can be combined with these baseline data to continually update situational awareness. Although VGI is often viewed as a singular data source, it does show value as a helpful validator when combined with other data [*Herfort et al.*, 2014] and a variety of cases support crowdsourcing as a way to supplement data in a natural disaster [*Goodchild and Glennon*, 2010; *Horita et al.*, 2013; *Liu*, 2014].

Ensuring your organization publishes open data, that is, data that can be freely used, re-used, and redistributed, is an excellent way to avoid duplication of effort, such as collecting data that are already acquired, and to identify gaps in data or areas that need to be validated. For data to serve decision makers across a society, data need to be fully open. This means that data must be:

• Technically open: Many organizational data sets are published in formats that can only be read by proprietary software (and sometimes hardware, like obsolete magnetic tape back-up drives). The data must be released in ways that allow any device or software to read it.

• Legally open: The data sets must be licensed in such a way that they may be used and shared widely.

Beyond data analysis, simply understanding how certain administration information must be structured, such as a request for assistance, can present a challenge for ECT, V&TC, or individuals trying to share information. As traditional humanitarian organizations each maintain their own data management systems, it can be difficult for grassroots organizations to not only engage in information sharing but for them to establish baseline data to work from. BFR found that multiple agencies responded to the same areas. Individuals affected by the disaster had multiple damage assessments performed without knowing who or when anyone would come to their aid. Though data were collected in iterative streams, it was impossible to tell where records became duplicative between agencies. Residents requesting assistance accepted whatever assistance arrived first, including the mass deployment of disaster response groups into neighborhoods. As a result, some residents were the subject of scams, and the overall response was scattered and non-comprehensive. When grassroots and faith-based groups began sharing their information across a single platform in an agreed upon format, the information from these informal responders was used to create a much more comprehensive picture of the damage and ensure that resources were more effectively deployed to those in need.

By refusing to share data or even collaborate on common information structures, traditional humanitarian organizations will perpetuate a narrow field of disaster relief that is clearly limited by organizational capacity, information sources, and mutable notions of authority. Though most ECGs begin as ad hoc efforts, the very learning process that they go through can be of benefit to traditional responders. ECGs are not only gathering data but also refining their methods for using it and streamlining their own interface with crowdsourcing. By engaging with and using the volunteer nature of these crowds, traditional responders can help foster innovation, flexibility, and learning in these social groups that will enhance the data that they provide as well as preventing the "ritualistic behavior" of bureaucratic response [*Majchrzak, Jarvenpaa, and Hollingshead*, 2007]. Choosing to increase interoperability, and share information structures can motivate volunteer and grassroots responders to protect, verify, and aggregate their data streams; enhance direct relief processes within communities; and increase community participation in the long-term recovery process. In recognition of this, open data created via crowdsourcing or as part of ECG efforts plays an increasing role in disaster risk reduction programming [*Crowley*, 2014; *Soden et al.*, 2014] and crisis response more generally [*Campbell*, 2015].

9.4.4. Read the Fine Print: Understand Policy Implications for Crowdsourcing

A wide variety of actions must be sequenced to formulate and implement a response to floods or other natural disaster. These actions are held together by an intricate web of rules and policies that govern the process that is rooted in legal regimes at a variety of scales (city, state, federal, international) and political decisions that are esoteric or even unknown to a large portion of the public. It is, perhaps, for this reason that the policy implications do not garner as much attention as they deserve. Although they have been the subject of some study [*Crowley*, 2013; *Liu*, 2014; *Campbell*, 2015], they are far outstripped by literature relating to the technological aspect of crowd-sourced data. Policy specifics will change according to the actors involved, and a complete review of all possible policy issues is beyond the scope of this chapter. However, we find that they can be broadly thought of in two categories: policies about people that govern relationships and policies about information that define what types of information can be shared and how.

Many policies exist to govern sharing and dissemination of information. These may include non-disclosure agreements that limit how much information volunteers can share; the management of sensitive or personally identifiable information; or rules that hold final data to a certain standard of quality. Other times, policies may limit how organizational data may be shared. In order to distribute satellite imagery for MapGive, the US Department of State invested considerable time to help establish the legal frameworks for how this would be accomplished, which in turn defined the technological and operational parameters necessary to implement the sharing of imagery [*Campbell*, 2015].

Because a complete and detailed view of all policies of a given organization is unlikely even by those working within that organization, many advocates of crowdsourcing within governmental organizations cite the importance of finding a legal advocate within the organization itself to help accomplish what needs to be done within the bounds of existing law and create new precedents [*Crowley*, 2013b].

9.4.5. Build Relationships with the Crowd

As we have shown, crowdsourcing as an approach to data collection or analysis involves individual human beings organized to accomplish a shared goal. Whether creating an open call to gather real-time information from the ground, organizing online volunteers to help process data, or asking for local assistance in debris removal, the most effective crowdsourcing efforts will be those that are part of a longer-term strategy of inclusiveness. Maintaining emphasis on the individual nature of crowd members is the most effective way to increase organizational and response capacity through crowdsourcing. This opens the door to innovative problem solving through the individual skill sets and unique backgrounds of your crowd members while also increasing the effectiveness and level of participation in shared workflows.

The crowd is a resource, and crowdsourcing should be understood as a project that requires adequate time dedicated to management and a considerable amount of communication among partners to ensure a mutually beneficial experience and positive outcomes. Any organization planning to engage with crowdsourcing should build this management capacity into their organization. V&TC, ECG, or other groups that help direct this work should be viewed in the same light as a business partner. In our experience, all parties are working toward a shared goal with, generally, limited resources and under stressful timelines, so clear communication is paramount. Volunteer coordinators must fully understand the task, workflow, and potential pitfalls that volunteers may encounter to help resolve problems during the project. In our work harnessing V&TC and various formulations of "the crowd," we found that time spent in advance to refine workflow, communications, and test applications was vital to the success of our effort [*Roberts et al.*, 2012]. The case of BFR shows that open calls for assistance can be highly productive when tasks are clearly defined and packaged with specific requests to directly maximize the efficiency of a volunteer's time [*Doyle*, 2015]. Questions of appropriate technology, interoperability, open data, and policies can all be used as starting points for discussions with the crowd, whether through V&TC or ECG, as shown in Figure 9.1. Having a shared understanding of the technological and organizational environment you are working in will greatly facilitate working with the crowd and help build a trusting relationship.

Beyond communicating for the sake of efficiency, remember that people volunteer to make a difference and that connecting with others often becomes a profoundly important aspect of the volunteer experience. In all of our projects, we have found that fostering these connections can create a sense of community that helps improve outcomes during the course of an effort. BFR found that residents who were assisted during initial relief efforts became themselves some of the longest lasting and motivated participants in volunteer activities once their own situation was secure. Many volunteers also developed relationships with the people they were helping, and they began self-deploying to assist their new community rather than relying on an external organization for project creation. After building strong, visible partnerships for its first crowdsourcing effort, USAID found that volunteers would suggest better ways of working or make introductions to individuals who had important information to share or who could facilitate both data collection and dissemination throughout a community. Volunteers came forward with these suggestions because they felt that they were truly contributing to a social good and that they were valued as a partner in the process. In short, they felt invested. BFR found that this form of data brokerage can

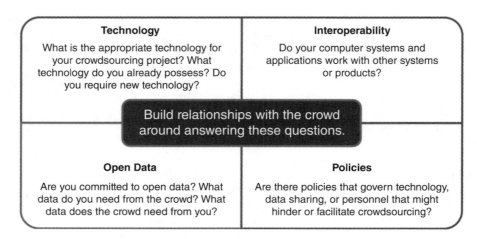

Figure 9.1 Technological and Organization Environment.

be very specific to different community contexts and will change over time, so finding and maintaining relationships with those who understand how to appropriately collect and share data throughout a community can be critical.

Communication is also important to better understand the desires of the crowd or the volunteers who are helping to manage it. We have found that decision makers often see crowdsourcing as a way to harvest data, specifically for "situational awareness." Although the advent of portable and personal technology now allows for "citizens as sensors" to capture and disseminate a wide range of geographic information [*Goodchild*, 2007], it is a mistake to focus solely on harvesting data from individuals or groups in a one-way flow of information. During a flood or other disaster, the affected population may also be your data provider. Alienating them by lack of engagement means losing access to the data they collect and reducing their participation in their own recovery. As much as possible, use the crowd as a way to disseminate information about the response itself as part of a two-way communications strategy. *Doyle* [2015] found that when flood response activities were shared broadly on social media, via local radio, and disseminated through volunteers, they inspired greater community participation and helped identify new opportunities for partnerships. Conversely, Doyle notes that a lack of updated official information or official information that contradicted the lived experience of the affected population created more impetus for communities to seek out knowledge and support for themselves versus complying with official directives.

Finally, building relationships with the crowd or volunteer organizations can significantly increase public support for or interest in those entities leading the formal response. Improved public relations can help promote your work, build political capital, and help align your organization with the needs of those for whom it stands

to serve. By investing in open dialogue with the crowd and V&TC, USAID's DCA saw a significant increase in social media followers during the crowdsourcing effort. Use of social media boosted Twitter followers by 20% and increased Facebook friends by 15%, thus ensuring that their work was more widely known and supported by the public.

9.5. SUMMARY

Contemporary technology facilitates the organization of volunteers to collect or transmit their observations about natural disasters and to synthesize them into a variety of outputs to support traditional humanitarian response. And although crowdsourcing is a relatively new phenomenon, it is best understood in the larger context of volunteer efforts that spring up in times of disaster. The fundamental question raised by this chapter follows: How can one best engage volunteers, in any number of social groupings, to provide effective assistance to traditional humanitarian responders?

We situate crowdsourcing along a continuum of volunteerism in disasters and advocate for it as an avenue for greater inclusivity during the act of data collection and decision-making. As our experience and the studies presented here show, it can be an effective method for increasing efficiency and gaining new insight if appropriately implemented under a philosophy of greater inclusion. We find several technological and data-related elements can increase productivity and relative quality of crowdsourced data, including the quality of existing baseline data, clearly defined tasks, and regular communication with volunteers. We also find that crowdsourcing may not be effective for certain forms of subjective categorization such as categorizing damage from satellite images but that it may also provide insider or highly specialized knowledge. On the basis of our experience and the studies presented, we advocate for appropriate technology to be adopted in conjunction with the crowd, greater inter-operability, and open data to maximize transparency and coordination in order to produce the most appropriate forms of data for responders. Finally, we see crowdsourcing as an opportunity to foster inclusivity into the decision-making process around natural disasters. When seen as a space of exchange, the process of crowdsourcing can support greater participation, particularly in reinforcing the transparency and responsiveness of disaster response.

REFERENCES

ARC (2014), *Groundtruthing OpenStreetMap Damage Assessment Review*, Washington D.C.

Atkinson, G. M., and D. J. Wald (2007), "Did You Feel It?" Intensity Data: A Surprisingly Good Measure of Earthquake Ground Motion, *Seismol. Res. Lett.*, *78*(3), 362–368, doi:10.1785/gssrl.78.3.362.

Budhathoki, N. R., B. Bruce, and Z. Nedovic-Budic (2008), Reconceptualizing the role of the user of spatial data infrastructure, *GeoJournal*, *72*(3-4), 149–160, doi:10.1007/s10708-008-9189-x.

Burns, R. (2014), Moments of closure in the knowledge politics of digital humanitarianism, *Geoforum*, *53*, 51–62, doi:10.1016/j.geoforum.2014.02.002.

Campbell, J. S. (2015), Imagery to the Crowd, MapGive, and the CyberGIS: Open Source Innovation in the Geographic and Humanitarian Domains, University of Kansas.

Capelo, L., N. Chang, and A. Verity (2012), *Guidance for Collaborating with Volunteer & Technical Communities*, Geneva.

Cobb, C., T. McCarthy, A. Perkins, A. Bharadwaj, J. Comis, B. Do, and K. Starbird (2014), Designing for the deluge: understanding & supporting the distributed, collaborative work of crisis volunteers, in *Proceedings of the 2014 ACM Conference on Computer Supported Cooperative Work*, pp. 888–899.

Crowley, J. (2013), *Connecting Grassroots and Government for Disaster Response*, Washington D.C.

Crowley, J. (2014), *Open Data for Resilience Initiative Field Guide*, Washington D.C.

Doyle, T. (2015), *Boulder Flood Relief 2013 Report: An Occupation of Love and Shovels in a Time of Disaster*, Boulder.

Elwood, S. (2008a), Volunteered geographic information: future research directions motivated by critical, participatory, and feminist GIS, *GeoJournal*, *72*(3), 173–183, doi:10.1007/s10708-008-9186-0.

Elwood, S. (2008b), Volunteered geographic information: key questions, concepts and methods to guide emerging research and practice, *GeoJournal*, *72*(3), 133–135.

Estellés-Arolas, E., and F. González-Ladrón-de-Guevara (2012), Towards an integrated crowdsourcing definition, *J. Inf. Sci.*, *38*(2), 189–200, doi:10.1177/0165551512437638.

Feick, R., and S. Roche (2013), Understanding the Value of VGI, in *Crowdsourcing Geographic Knowledge: Volunteered Geographic Information (VGI) in Theory and Practice*, edited by D. Sui, S. Elwood, and M. Goodchild, pp. 15–29, Springer Netherlands, Dordrecht.

Flanagin, A., and M. Metzger (2008), The credibility of volunteered geographic information, *GeoJournal*, *72*(3), 137–148.

Gino, F., and B. R. Staats (2012), The microwork solution, *Harv. Bus. Rev.*, *90*(12), 92–97.

Goodchild, M. (2007), Citizens as sensors: the world of volunteered geography, *GeoJournal*, *69*(4), 211–221.

Goodchild, M. F., and J. A. Glennon (2010), Crowdsourcing geographic information for disaster response: a research frontier, *Int. J. Digit. Earth*, *3*(3), 231–241, doi:10.1080/17538941003759255.

Gouveia, C., and A. Fonseca (2008), New approaches to environmental monitoring: the use of ICT to explore volunteered geographic information, *GeoJournal*, *72*(3), 185–197.

Grant, T. (2010), Microwork is the New, New Buzzword in Global Outsourcing, *Globe Mail*, 10 March.

Haklay, M., S. Basiouka, V. Antoniou, and A. Ather (2010), How Many Volunteers Does it Take to Map an Area Well? The Validity of Linus Law to Volunteered Geographic Information, *Cartogr. Journal,*, *47*, 315–322.

Herfort, B., J. de Albuquerque, S.-J. Schelhorn, and A. Zipf (2014b), Exploring the Geographical Relations Between Social Media and Flood Phenomena to Improve Situational Awareness, in *Connecting a Digital Europe Through Location and Place SE - 4*, edited by J. Huerta, S. Schade, and C. Granell, pp. 55–71, Springer International Publishing.

Hiltz, S. R., B. van de Walle, and M. Turoff (2010), The Domain of Emergency Management Information, in *Information Systems for Emergency Management*, edited by S. R. Hiltz, M. Turoff, and B. van de Walle, pp. 3–20, Routledge, Armonk, NY.

Horita, F., L. Degrossi, L. Assis, A. Zipf, and J. Porto de Albuquerque (2013), The use of Volunteered Geographic Information and Crowdsourcing in Disaster Management: a Systematic Literature Review, *Proc. Ninet. Am. Conf. Inf. Syst.*, 1–10.

Lillesand, T. M., R. W. Kiefer, and J. W. Chipman (2008), *Remote sensing and image interpretation*, 6th ed., Wiley, New York.

Liu, S. B. (2014), Crisis Crowdsourcing Framework: Designing Strategic Configurations of Crowdsourcing for the Emergency Management Domain, *Comput. Support. Coop. Work*, *23*(4), 389–443, doi:10.1007/s10606-014-9204-3.

Liu, S. B., and L. Palen (2010), The New Cartographers: Crisis Map Mashups and the Emergence of Neogeographic Practice, *Cartogr. Geogr. Inf. Sci.*, *37*, 69–90.

Meier, P. (2013), Human Computation for Disaster Response, in *Handbook of Human Computation*, edited by P. Michelucci, pp. 95–104, Springer New York, New York, NY.

Meier, P. (2015), *Digital Humanitarians: How Big Data Is Changing the Face of Humanitarian Response*, CRC Press.

Menoni, S. et al. (2015), *Enabling Knowledge For Disaster Risk Reduction and its Integration into Climate Change Adaptation*.

Morris, R. R., M. Dontcheva, and E. M. Gerber (2012), Priming for better performance in microtask crowdsourcing environments, *IEEE Internet Comput.*, *16*(5), 13–19, doi:10.1109/MIC.2012.68.

National Research Council (2007), *Tools and methods for estimating populations at risk from natural disasters and complex humanitarian crises*, National Academies Press, Washington, D.C.

Neis, P., and D. Zielstra (2014), Recent Developments and Future Trends in Volunteered Geographic Information Research: The Case of OpenStreetMap, *Futur. Internet*, 6(1), 76–106, doi:10.3390/fi6010076.

Raymond, E. S. (2001), *The Cathedral and the Bazaar: Musings on Linux and Open Source by an Accidental Revolutionary*, O'Reilly Media, Sebastopol, CA.

Relations Between Social Media and Flood Phenomena to Improve Situational Awareness, in *Connecting a Digital Europe Through Location and Place SE - 4*, edited by J. Huerta, S. Schade, and C. Granell, pp. 55–71, Springer International Publishing.

Roberts, S., S. Grosser, and D. Ben Swartley (2012), *Crowdsourcing to Geocode Development Credit Authority Data : A Case Study*, Washington D.C.

See, L., A. Comber, C. Salk, S. Fritz, M. van der Velde, C. Perger, C. Schill, I. McCallum, F. Kraxner, and M. Obersteiner (2013), Comparing the Quality of Crowdsourced Data Contributed by Expert and Non-Experts, *PLoS One*, 8(7), e69958.

Shahid, A. R., and A. Elbanna (2015), The Impact of Crowdsourcing on Organisational Practices: The Case of Crowdmapping, 1–16.

Shirky, C. (2008), *Here comes everybody: the power of organizing without organizations*, New York: Penguin Press.

Shirky, C. (2010), *Cognitive surplus: creativity and generosity in a connected age*, New York: Penguin Press.

Soden, R., and L. Palen (2014), From crowdsourced mapping to community mapping: The post-earthquake work of openstreetmap haiti, in *COOP 2014-Proceedings of the 11th International Conference on the Design of Cooperative Systems*, edited by C. Rossitto, L. Ciolfi, D. Martin, and B. Conein, pp. 27–30, Springer International Publishing, Nice.

Soden, R., N. Budhathoki, and L. Palen (2014), Resilience-Building and the Crisis Informatics Agenda: Lessons Learned from Open Cities Kathmandu, in *Proceedings of the 11th International ISCRAM Conference*, pp. 1–10, University Park (Pennsylvania).

Stallings, R. A., and E. L. Quarantelli (1985), Emergent Citizen Groups and Emergency Management, *Public Adm. Rev.*, 45, 93–100, doi:10.2307/3135003.

Sui, D. Z., M. F. Goodchild, and S. Elwood (2013), Volunteered Geographic Information, the Exaflood, and the Growing Digital Divide, in *Crowdsourcing Geographic Knowledge: Volunteered Geographic Information (VGI) in Theory and Practice*, pp. 1–12, Dordrecht; New York: Springer.

Swain, R., A. Berger, J. Bongard, and P. Hines (2015), Participation and Contribution in Crowdsourced Surveys, *PLoS One*, 10(4), e0120521.

Turner, A. (2006), *Introduction to Neogeography*, O'Reilly, Sebastopol, CA.

Verjee, F. (2007), An Assessment of the Utility of GIS-Based Analysis to Support the Coordination of Humanitarian Assistance, The George Washington University, Washington D.C.

Waldman, A., A. Verity, and S. Roberts (2013), *Guidance for Collaborating with Formal Humanitarian Organizations*, Geneva.

Wiggins, A., and K. Crowston (2011), From conservation to crowdsourcing: A typology of citizen science, in *System Sciences (HICSS), 2011 44th Hawaii international conference on*, pp. 1–10.

Zhao, Y., and Q. Zhu (2014), Evaluation on crowdsourcing research: Current status and future direction, *Inf. Syst. Front.*, 16(3), 417–434, doi:10.1007/s10796-012-9350-4.

Ziemke, J. (2012), Crisis Mapping: The Construction of a New Interdisciplinary Field?, *J. Map Georg. Libr. Adv. Geospatial Information, Collect. Arch.*, 8(2), 101–117, doi:10.1080/15420 353.2012.662471.

Part IV
Data Analysis

10

After the Flood Is Before the Next Flood: The Post-Event Review Capability Methodology Developed by Zurich's Flood Resilience Alliance

Michael Szoenyi[1], Kanmani Venkateswaran[2], Adriana Keating[3], and Karen MacClune[2]

ABSTRACT

As part of Zurich Insurance's flood resilience alliance, the post-event review capability (PERC) provides analysis and independent reviews of large flood events, while providing accessible, consistent, and generalizable insights. Research has established the need to build resilience in infrastructure, services, and agents' capacity and livelihood systems if risk is to be proactively reduced. A consistent, practice-based and transdisciplinary disaster forensic analysis methodology will help push disaster risk management and resilience building out of their traditionally sector-focused realms. Building future resilience is dependent upon consistently capturing and reviewing lessons of the past, both within and across disciplinary boundaries, especially in a dynamic environment of urbanization and climate change.

This chapter explains the rationale and approach of the PERC methodology and presents consolidated findings from studies conducted so far. PERC seeks to answer questions related to flood resilience, flood risk management, and catastrophe intervention. It looks at what worked well (identifying best practice) and opportunities for improvements (providing actionable recommendations). Since 2013, PERC has analyzed flood events in locations from Western Europe to Nepal to Morocco, while engaging in dialogue with authorities, affected people, and actors in the disaster risk management space. Knowledge collected in these analyses is being consolidated and made available to help build resilience. Despite the flexibility of the PERC approach, which encourages context-specific analysis without predetermining outcomes, upon consolidating our findings and lessons learnt from the PERC studies conducted to date, we find there are profound similarities in the points of failure, success stories, and capacities during floods across a wide context.

10.1. INTRODUCTION: DISASTER RESILIENCE, DISASTER FORENSICS, AND POST-EVENT REVIEW CAPABILITY

The PERC methodology has been designed as part of the Zurich flood resilience alliance, a program launched in 2013 under Zurich Insurance Group's corporate responsibility strategy. PERC evaluates what happened before, during, and after a disaster, identifies the critical

[1] *Zurich Insurance Group, Zürich, Switzerland*
[2] *Institute for Social and Environmental Transition-International, Colorado, USA*
[3] *International Institute for Applied Systems Analysis, Laxenburg, Austria*

Flood Damage Survey and Assessment: New Insights from Research and Practice, Geophysical Monograph 228,
First Edition. Edited by Daniela Molinari, Scira Menoni, and Francesco Ballio.
© 2017 American Geophysical Union. Published 2017 by John Wiley & Sons, Inc.

Table 10.1 The Five Cs (capitals) that Make up a Set of Measurable Indicators.

Capital	Description
Human	The education, skills, and health of the people in the system
Social	Relationships and networks, bonds that aid cooperative action, and links to exchange and access ideas and resources
Natural	The natural resource base, including land productivity and actions to sustain it, as well as water and other resources that sustain livelihoods and well-being
Physical	The things produced by economic activity from "other" capital, such as infrastructure, equipment, and improvements in crops and livestock
Financial	The level, variability, and diversity of income sources and access to other financial resources that contribute to wealth and help with financial risk transfer

gaps and successes in the overall disaster risk management (DRM) system, and presents actionable recommendations in the context of a changing risk landscape. Individual studies are collected in a central, publically accessible repository so they can be searched and mined for future use across scales and contexts, including consolidation of findings to identify generalizable lessons.

"Resilience" is increasingly becoming a core concept in the disaster and climate change adaptation fields, as evidenced by a surge in resilience-focused publications [e.g., United Nations Office for Disaster Risk Reduction (UNISDR), 2011; National Research Council (NRC), 2012; Asian Development Bank (ADB), 2013] and initiatives (e.g., the Global Resilience Partnership[1]; 100 Resilient Cities[2]). PERC is informed by the Zurich flood resilience alliance's conceptualization of resilience [*Keating et al.*, 2014], and is based on a combination of two complementary frameworks: Institute for Social and Environmental Transition-International's (ISET) Climate Resilience Framework (CRF) [*Tyler and Moench*, 2012] and the alliance's community flood resilience measurement framework [*Zurich Insurance Group*, 2015a].

Firstly, ISET's CRF provides the conceptual framework for simplifying and analyzing the relationships within the community in question that led to development and risk outcomes. This analysis looks at the relationships between people, the physical environment, and the rules and norms that govern behavior as a means to identify entry points for building resilience and reducing vulnerability. The CRF identifies the underlying drivers that mediate vulnerability and resilience to shocks and stresses across scales in a given context.

Secondly, in the Zurich flood resilience alliance's resilience framework, community disaster resilience is defined as "the ability of a system, society or community to pursue its economic and social development and growth objectives, while managing its risk over time in a mutually reinforcing way" [*Keating et al.*, 2014]. This includes both the ability to learn from the disturbance and to incorporate risk into decisions about future investment across multiple sectors. As resilience declines, the magnitude of a shock from which the system can recover gets smaller and smaller, whereas a resilient system is forgiving of external shocks. This perspective on resilience has brought to PERC a strong consideration for the underlying development goals of the community in question and in particular how the pursuit of these goals is influencing the evolution of disaster risk.

More recently, the evolutions of both the flood resilience alliance's framework and PERC have brought them closer together, moving toward the development of a coherent framework and approach for perspectives of both pre-event "sources" of resilience, and post-event resilience "outcomes." The alliance's framework is described as the "5C-4R community-based flood resilience measurement framework." The 5C model uses the five capitals that originated in the United Kingdom's Department for International Development (DFID) sustainable livelihoods framework (SLF) [*DFID*, 1999]. These "5Cs" characterize the assets a community has access to and utilizes. They are complementary resources that, if used well, can increase personal and collective wealth, provide a sense of security, enhance environmental stewardship, and sustain and improve the overall communities' well-being (Table 10.1). The 4R model was developed by the *Multidisciplinary Center for Earthquake Engineering Research (MCEER)* [2007] at the University of Buffalo in the United States and postulates that a system has four properties that determine resilience: robustness, redundancy, resourcefulness, and rapidity. Collectively known as the 4Rs, they were originally used in a built environment (one that comprises physical infrastructure), but we believe they apply equally to all assets in a system, both tangible physical assets and less tangible ones (Table 10.2).

The alliance has developed and is testing a Community Flood Resilience Measurement Tool based on these five capitals. This tool is designed for disaster and development practitioners to help determine where in the local context resilience can be built pre-event to reduce the loss of lives and assets; to measure if and how outcomes of resilience have manifested during and after a hazard

---

[1] *http://www.globalresiliencepartnership.org/*
[2] *http://www.100resilientcities.org/*

Table 10.2 The Four Properties of a Resilient System.

Type of R	Description
Robustness	The ability to withstand a shock, for example, housing and bridges built to withstand a flood
Redundancy	Functional diversity, for example, having many evacuation routes
Resourcefulness	The ability to mobilize when threatened, for example, a group within a community that can quickly mobilize to convert a community center into a flood shelter
Rapidity	The ability to contain losses and recover in a timely manner, for example, quick access to sources of financing to support recovery

event; and to evaluate if and how community-based initiatives and risk management strategies are delivering on their promise of building resilience. It is critical that PERC provides the post-event insight into how resilience was built by the elements within the 5C-4R framework or how they contributed to it.

The Zurich alliance's and ISET's frameworks are built on the premise that resilience needs to go beyond simply recovering from a shock to the pre-shock state. In particular, bouncing back to a previous "stable" state is problematic if that state was vulnerable to begin with. Resilience means ensuring human well-being by bouncing forward and building back better in such a way that future shocks have a lesser impact. Ultimately, disaster resilience is about living and thriving in the face of disaster risk and uncertainty, especially the uncertainty of the trends in the environment (natural and built environment) of that community.

Assessing and understanding how resilience was and/or was not present in the event of a shock is essential for the operationalization of the concept. The methodologies of disaster forensics provide an important opportunity for enhancing the understanding of resilience. Forensics originated in the field of criminal investigation and has evolved to various fields, including disaster analysis, as a way of consistently working from an event to its full analysis and understanding the root causes leading to a disaster. Conducting a forensic post-event study is similar to detective work in that it needs an experienced team of experts, a consistent and iterative meta-structure, and guidance on how to pursue leads and new information. The research team needs to be open to any new insights, and must apply a diligent yet flexible way of working to uncover the important aspects of an event and enable learning for the future.

A review of the disaster forensic approaches currently available (summarized in Table 10.3) shows that they target different aspects of the event and its root causes, are conducted in different time frames, and are produced for different stakeholder audiences.

An examination of the characteristics of these disaster forensics methodologies shows that they can be grouped into three categories: 1) those focused on assessing the immediate response needs and direct damages for informing government-led response and recovery such as

CEDIM [2015], Damage and Loss Assessment (DaLA) [*Worldbank*, 2010] and post-disaster and needs assessment (PDNA) [GFDRR, 2013]; 2) those with an engineering scope like the *Earthquake Engineering Field Investigation Team (EEFIT)* [n.d.] or the post-event validation of vulnerability and catastrophe models, which are more narrowly focused on understanding the engineering performance of a physical structure in the event of an earthquake or the performance of an applied cat model; and 3) long-term analysis such as the Forensic Investigations of Disasters (FORIN) project [*Integrated Research on Disaster Risk (IRDR)*, 2011], which draws on anthropology, sociology, and history to understand the evolution of disaster risk over the long term.

All of these "forensic" investigation methods produce important yet different insights into the field, but all of them have their own particular area of focus. To better understand and build resilience in the long term, there is a need for a disaster forensics methodology that takes a broader focus across all of these areas. PERC fills this gap. PERC is not designed to provide a great amount of detail in any one focus area that is obtained by these other methodologies. However, by touching the full range of issues, such as information regarding response and recovery, engineering performance of structures, the social and historical underpinnings of the event, and then integrating the information across all areas, we identify issues and their linkages that are often missed. For example, one theme in a PERC study might explore not only how a structure performed but how people interacted with that structure to modify their risk landscape. Furthermore, conducting this type of analysis allows the findings to be framed in very different ways from those traditionally used, clearly linking physical performance to people to policy in ways that draw new insights.

PERC is grounded in a framework that focuses on how disaster risk has evolved over time, and how, in the face of that risk, people, systems, and institutions have built and could build resilience. The frameworks that PERC builds upon are described in more detail in Section 10.2, while the operational aspects how a PERC study is conducted and how it creates value are described in Section 10.3. For those interested to learn more about or apply the PERC methodology, the authors have

Table 10.3 Disaster Forensic Methodologies Summary.

Forensic methodology	Focus	Time frame	Audience
Center for Disaster Management and Risk Reduction Technology (CEDIM) Forensic Disaster Analysis	• Hazard specifications • Assessment of direct damages • Ongoing response needs	Immediately after the event, in real time	Governments and related bodies
Damage and Loss Assessments (DaLA)	• Response needs • Assessment of physical damage	Immediately after the event, response phase	National-level response and recovery authorities
Post-Disaster and Needs Assessment (PDNA)	• Valuation of physical damage and economic loss • Human recovery needs	Ongoing throughout the response and recovery phase	Governments and related bodies
Earthquake Engineering Field Investigation Team (EEFIT)	• Assessment of structural performance during earthquake	Two to three weeks following the earthquake	Engineering audience
Forensic Investigations of Disasters (FORIN)	• Root causes of disasters, potentially going back decades or centuries	Months or years following the event	Disaster researchers
Post-Event Review Capability (PERC)	• Cross-sectorial understanding of what has happened, what worked well, and what did not at (flood) event level. • What were the underlying drivers that create or hinder risk and/or resilience?	Typically three to six months after the event	Public DRR practitioners, actors in flood resilience, decision makers

published a PERC manual [*Venkateswaran et al.*, 2015[3]]. Broadly, a PERC analysis consists of two parts: the first part is a narrative, exploring how a particular risk is understood and managed leading up to, during, and in the response and recovery phases of the event, and explores the vulnerability to that hazard; and the second part identifies opportunities for intervention and action that could strengthen the future resilience of the affected people and critical services, and how these interventions could reduce the risk posed by similar, future hazard events. The necessary data are collected using both qualitative and quantitative methods (desk research, literature review, and interviews from key actors). PERC uses a system-wide approach to review disasters, analyzing across scales (community, regional, national, transnational) and sectors (agriculture, education, heavy industries) in all phases of the disaster risk management cycle. PERC studies can be conducted in both developed and developing country contexts, and in both urban and rural areas.

Research has established the need to build resilience to disasters in infrastructure, services, agents' capacities, and livelihood systems and how they are interlinked if disaster risk is to be proactively reduced and the remaining risk judiciously managed [e.g., *Simonovic and Peck*, 2013; *Keating et al.*, 2014; *UNISDR*, 2015]. PERC is designed specifically to draw out practical, actionable lessons for promoting disaster risk reduction (DRR) to address underlying risks and enhancing resilience in the future, in addition to more oft-identified lessons around the importance of coping and recovery mechanisms. The audience for PERC reports, the individual actionable recommendations developed during PERC, and the underlying opportunities for learning processes are actors in the disaster management and risk management fields, as well as broader development, infrastructure, and land use planners. These actors include intervention organizations, bodies mandated to promote prevention and risk reduction, and those working in the DRR and climate change adaptation (CCA) space, where resilience is increasingly recognized as an important key component for growth and development.

It must be noted that PERC does not establish the immediate government needs post-disaster, nor does it work on behalf of governments or related bodies. It does not provide specific recommendations on financial or physical needs post-event. Similarly, PERC is not designed to document specific hazard specifications or technical analysis of physical performance of structures. Furthermore, PERC does not design nor recommend specific interventions or provide a framework for recovery. Rather, it identifies critical gaps and opportunities, particularly actionable opportunities, to reduce risk around which disaster practitioners, authorities, and advocates can design interventions that are grounded in the local context. PERC is not providing flood-related protection consultation, such as on the design of specific (structural) flood protection or flood risk reduction solutions. PERC is independent research to understand what happened and why, independent from insurance coverage

---

[3]*https://www.zurich.com/en/corporate-responsibility/flood-resilience/learning-from-post-flood-events*

Table 10.4 PERC Studies Conducted so far.

Flood event	Countries	Publication Year
Central European floods	Austria, Germany, Switzerland, Czech Republic	2013, 2014 - retrospective report
Surge following storm Xaver	United Kingdom	2014
Emmental valley flood	Switzerland	2014
Balkan floods	Bosnia and Herzegovina, Serbia, Croatia	2015
Karnali river floods	Nepal	2015
Sidi Ifni and Guelmim floods	Morocco	2015

and products, political views, and other vested interests and therefore from predetermined outcomes.

To date the PERC methodology has been applied exclusively to major flood events. The focus has been on floods because they cause the greatest disaster-related losses and damages, and because the number of people exposed to floods each year is increasing at a higher rate than population growth globally [*Miller et al.*, 2008; *UNISDR*, 2011; *Zurich Insurance Group*, 2013a]. Between 2013, when the concept was first implemented, and today, six PERC studies have been conducted (Table 10.4), and at the time of writing, two more were in production (the US South Carolina floods of October 2015 and the UK Cumbria floods of December 2015). From these studies, we find there are a number of similarities in what was learned across these diverse settings, which we outline below.

Despite the flood focus so far, the PERC methodology is very flexible so that it can be adapted to address other hazards, scopes, scales, and time frames if needed. For example, a mini-PERC (similar to the Balkan floods [*Zurich Insurance Group*, 2015b] and the Emmental floods [*Zurich Insurance Group*, 2014a] might be used to try and look at smaller scales or answer specific questions, for example when an event was very confined or caused very specific disaster aspects, losses, or damages that are worth investigating with a reduced time frame or investment. A multi-event, historical PERC analysis could be conducted to look at a series of similar historic disaster events to identify places where learning is or is not occurring over time. We believe it is a particular strength of the methodology that it is generalizable to other hazards, be they geophysical (earthquakes, landslides), atmospheric (hurricanes, winter storms), or human/complex (such as terrorism, pandemics). Although the specifics of the information, the way interviewees are selected, and the types of questions asked of those interviewees will vary based on the selected context and the identified priority goals, the basic PERC approach would remain the same across all of these applications.

The learning aspect of PERC is not only about information exchange; it is also about fostering a culture of iterative learning that helps create and strengthen networks, allows different stakeholders to work together, builds knowledge and capacity among people and groups,

and fosters engagement that can eventually create transformative change. This is needed to avoid continued buildup of more risk and to reduce loss and suffering during future events, at all scales from local to global.

10.2. THE POST-EVENT REVIEW CAPABILITY FRAMEWORK AND ANALYSIS

The PERC framework is not a linear process; rather, it is an iterative meta-structure that helps identify and understand the different components that create a complex system and how these different components interact to generate outcomes. The PERC analysis is structured around understanding the physical conditions that define the hazard and lead to the creation of risk, the pre-existing human, social, economic, and environmental conditions that cause and exacerbate vulnerability to hazards, and what happened during and after the disaster (Table 10.5).

To understand the physical, social, political, and economic dynamics of the disaster, and what happened in terms of disaster response and recovery, we analyze our data through the CRF and 5C-4R resilience lenses presented above.

Given that resilience is about people, their needs, and the cultural and legal norms that enable their ability to thrive in face of potential or manifested shocks, the three major components of resilience that are assessed as part of a PERC study follow:

(A) Systems – This is the "what" component of resilience. It refers to a combination of ecosystems and infrastructure systems (natural and physical capital within the 5C-4R framework), and the services they provide. Ecosystems provide basic needs (water, air, food) as well as some more advanced needs such as coastal defense, and water absorption capacity. These ecosystem services are mediated (either positively or negatively) by physical infrastructure and services (transport, water distribution, drainage, power, communications) that are central features of human settlements. The 4R-component of the 5C-4R framework analyzes how they contributed to or distracted from resilience in the disaster event, assessing the presence or absence of robustness, redundancy, resourcefulness, and rapidity within these systems.

Table 10.5 PERC Analysis Meta-Structure.

Section	Description
Physical Context	Illustrates the physical conditions (i.e., geographical, geophysical, hydrological, meteorological conditions) and the evolution of those conditions that led to the disaster event. This also involves exploration of the history of similar hazard events in the area to establish trends in event frequency and magnitude, and puts into context the relevant physical pre-conditions leading up to the disaster event. Note: In order to understand the buildup of new risk and the increase or decrease of vulnerabilities, a distinction must be made between the intensity and frequency analysis of the (natural) hazard event itself (i.e., the exceedance probability of a certain rainfall or river flood stage) and the ensuing severity of the disaster event, because this is a non-linear relationship.
Socioeconomic Disaster Landscape	Establishes the social, political, and economic conditions that cause and exacerbate the vulnerability of people to the hazard and the barriers that prevent people from adapting to such hazards. As part of this analysis and assessment work, an agent landscape map is built identifying the key players involved in disaster risk management and their roles, decision-making, and communication structures and interactions within their and across sectors.
What Happened	This section provides a detailed, compelling narrative of what happened during the disaster response and recovery phases. These are the factual descriptions of observations during the PERC research.
Key Insights	Collates the lessons learned, including critical gaps and strengths. Key questions covered might be: Were core systems flexible and redundant? Were agents able to draw on their capitals to be resourceful and responsive? Did legal and social norms enable equitable, efficient, and effective response and recovery? Have people and organizations learned from past disasters, and are processes and an enabling environment in place to enable people and organizations to learn from this disaster? What are the prevalent and/or systemic issues inhibiting disaster resilience and the disaster management system? These trends and patterns should be grounded in examples from the "What Happened" section, reflect on how the socioeconomic landscape influenced outcomes, and should be justifiable. This is the interpretation section of a post-event review capability (PERC) report.
Recommendations	Reflects and follows the key insights to identify actionable opportunities in a short, concise, and easy-to-understand form. For example, recommendations such as "the governance system needs to completely change" would be inappropriate here because this is something that is unlikely to happen. Instead, more effective recommendations are mindful of existing, deep-set constraints. Although the recommendations are designed to be actionable, they do not design specific interventions. Rather, the recommendations are built around wider trends and critical gaps identified in the "Key Insights" section. This section is designed as a standalone section and can be understood without reading the full report. They are phrased so they can contribute to the buildup of a "mineable knowledge database" where itemized findings (and other elements) can be stored and retrieved through a search algorithm. This element is absolutely critical if a forensic disaster analysis methodology such as PERC is to go beyond creating individual and unconnected reports that will be shelved and lose their value after the initial interest in the particular event itself has passed. It will also be critical to reach scale, and one of the strengths of the PERC approach is to apply the same methodological filter to different events in widely varying contexts.

(B) Agents – This is the "who" component of resilience. It refers to people and their organizations (human and social capital within the 5C-4R framework), whether as individuals, households, communities, private and public sector organizations, or companies, and their capacity to respond to and shape the world around them. Agents have different sets of assets, entitlements, and power (human, social, and financial capitals). An agent's ability to access systems, and thus their vulnerability and resilience, is differentiated on this basis. The needs, preferences, resources, and capacities of agents can be analyzed using the five capitals as described in the 5C-4R framework.

(C) Institutions – This is the "how" component of resilience. It refers to the rules, norms, beliefs, or conventions that shape or guide human relations and interactions, access to and control over resources, goods and services, assets, information, and influence (social and financial capitals within the 5C-4R framework). Although institutions shape agents, equally agents are able to shape institutions, thus opening the possibility of change.

The three components (agents, systems, and institutions) are not isolated silos; rather, they are dynamic and constantly interacting with each other. The interactions of agents and systems that lead to vulnerability and resilience outcomes are mediated by institutions. When conducting the PERC research, it is to be ensured that the aspects of our 5C-4R resilience framework are adequately covered by posing the right questions and triangulating responses with this lens in mind. (See Section 10.3 for how the PERC methodology is applied.)

Part of deconstructing the physical, human, social, economic, and natural dimensions (the 5Cs) of the disaster and understanding what happened requires an analysis of the DRM cycle as it manifested in the context of the event. The DRM cycle includes the following elements:

1. Risk reduction – This part, before the event, is about how actors tried to minimize disaster risk. It includes prospective and corrective risk reduction. A critical component to achieve risk reduction and preparedness is the awareness about and perception of risk. Risk is a fuzzy, subjective concept. Education, prior experience, culture, and many other aspects play a role in how risk is perceived and how strategies to take pre-event action (or not) are developed and implemented. We know that in practice, it is often especially hard to financially justify spending on prevention when the future return is uncertain (both in terms of when that return occurs and how large it would be) [van Aalst et al., 2013], but we also know that pre-event risk reduction and preparedness is often very cost-effective in comparison to post-event recovery [Foresight, 2012]. Prospective risk reduction comprises actions taken to avoid the buildup of more risk.[4] Corrective risk reduction comprises actions taken to reduce already existing risk.[5] Both prospective and corrective risk reduction tend to focus more on long-term processes and infrastructural change.

2. Preparedness – Crisis preparedness includes "preparedness for response" and community or localized awareness and action to help in face of the event. Preparedness is closely linked to the overall risk awareness to a particular (flood) hazard and about the long-term and short-term assessment of how an event might manifest itself. Particularly important for flooding is the availability and the credibility of forecasts and how this leads to early action achieving a high level of preparedness and readiness.

3. Response – This part is about the actions taken during and immediately after a disaster to contain or mitigate disaster impacts. This can include evacuation, search and rescue, emergency relief distribution, and so on.

4. Recovery – Recovery is concerned with the actions taken after the response phase, when the hazard event has passed, either in the short or long term. The purpose of this phase is to help people cope with or recover from disaster impacts, reconstruct damaged physical systems (i.e., infrastructure), and restore the associated services. This is also the part where learning should occur, so it is important to find out if and how relevant actors are analyzing the disaster event, what processes are in place to learn from the event, and how this learning can help adapt DRM and resilience building.

10.3. HOW A POST-EVENT REVIEW CAPABILITY IS CONDUCTED

Based on the PERC reports that have been concluded thus far, a PERC study would be conducted after the disaster response phase is over and during the recovery phase, but not so late that the momentum created by the disaster is lost and/or the next phase of the DRM cycle has already begun. In the case of floods in subtropical countries, for example, a PERC study should be conducted before the next monsoon season begins. If a PERC review is conducted too soon after a disaster (i.e., in the response phase), it will be difficult to adequately evaluate what happened and what recovery is going to look like. Pertinent information may not yet be available, and the PERC research might hinder responders in their priorities at that time. Impacted people and key DRM actors need time to overcome the initial shock and process what has happened. However, if the research is conducted too late, memory will fade, and peak information may be lost.

The studies conducted to date have taken three to six months, from the initial planning to the publication of a final report: Central European floods [Zurich Insurance Group, 2013b]; Floods in Boulder. A Study of Resilience [MacClune et al., 2014]; Storm Surge following Xaver [Zurich Insurance Group, 2014b]; Karnali River floods Nepal [Zurich Insurance Group, 2015c]; Morocco floods [Zurich Insurance Group, 2015d]. However, this timeline is

---

[4]UNISDR (2009) defines Prospective Disaster Risk Management as "Management activities that address and seek to avoid the development of new or increased disaster risks. Comment: This concept focuses on addressing risks that may develop in future if risk reduction policies are not put in place, rather than on the risks that are already present and which can be managed and reduced now. See also Corrective disaster risk management."

[5]UNISDR (2009) defines Corrective Disaster Risk Management as: "Management activities that address and seek to correct or reduce disaster risks which are already present. Comment: This concept aims to distinguish between the risks that are already present, and which need to be managed and reduced now, and the prospective risks that may develop in future if risk reduction policies are not put in place. See also Prospective risk management."

dependent on size and scope of the study and on the local situation. More or less detailed PERC analyses, or those conducted for larger or smaller events, might require different time frames.

When conducting a PERC study, before starting any fieldwork, a literature review and desk research is necessary to understand the nature and the context of the event. This research typically might cover current newspaper articles, peer-review articles, working papers, and reports about the disaster event or previous/similar events, the prevailing risk context, the physical landscape, vulnerability context, and the institutional landscape. Such background research provides the analyst with the necessary context, and helps direct the focus of the fieldwork prior to going to the field. It also aids in identifying key players in the disaster landscape and potential interviewees. The outcomes of this initial desk research are used to decide whether or not it is valuable and justifiable to continue with a PERC study.

It is important that the team of analysts establishes an initial understanding of the physical conditions on the ground during the background research phase. This is then expanded through a detailed analysis of physical and socioeconomic drivers of the disaster event. A narrative of the event is pieced together, detailing why the disaster occurred and how it unfolded. In the case of the flood PERC studies done to date, this has included a hydro-meteorological analysis of the event including the type and the scale of the flood. This analysis is then compared to previous events and the historic data catalogue available. In particular, the methodology emphasizes understanding whether this disaster event followed an expected pattern or if it was unanticipated and a very different type of event. If possible, the recurrence interval or annual exceedance probability of the hazard event is estimated to provide a sense of the relative frequency of the hazard occurring and to put this in relation to the consequences in the disaster. If there is evidence that these disaster events are occurring more frequently or with greater magnitude than in the past, this is identified and explored. Particular attention is paid to whether the perceived or corroborated increase in the magnitude or severity of the hazard events is physical (e.g., heavier rainfalls) or whether consequences are more severe due to increases in exposure and/or vulnerability (e.g., more people living in the floodplain, infrastructure changing flood water flow paths, etc.) or a combination of these factors.

When judging the severity of the event, the PERC approach assesses whether the event was of a scale that was planned for or whether it was beyond the planned (design) scale. For example, in many parts of the world, infrastructure is designed to handle the 5% annual chance to 1% annual chance event (in other words, the 20- to 100-year return period floods, respectively). Exploring how the event compared relative to the local planning standards and to the designed protective infrastructure, and to the preparations and capacities of the pertinent intervening organizations and impacted communities, is critical for outlining key lessons for the future. Some of the statements found most often during PERC interviews are those expressing "the severity of the consequences were unexpected" or that there were "surprises of the event extent and the needs for intervention." If a component of the event was due to physical structure failure, an exploration into that failure is also to be included. Understanding the severity of the event also includes an evaluation of how the impacts, that is, physical, social, and economic, of this disaster compare to the impacts of previous, similar disasters locally, regionally, and/or nationally, as well as within the corresponding watershed (in the case of flood) or impacted area of other hazardous events.

Conducting a PERC with a local partner (i.e., government agency, safety-net organization, non-governmental organization [NGO], community group) who is embedded and knowledgeable about disasters and the local context (including local languages and cultural norms, background knowledge, and access to and trust of key actors) is crucial. When working with the local partner, it is key they have both the thematic understanding as well as the know-how of the local context and the connections needed to collect useful, reliable, and accurate information. However, it is also important to work with partners that have a range of perspectives, without problematic vested interests in the outcome of the PERC, and who are open-minded and independent to any outcomes (i.e., not hostile to findings that did not prove a pre-existing hypothesis, that are unexpected, or are locally undesired). Perspectives from a range of stakeholders must be obtained to achieve the necessary credible and locally supported insights that could generate tangible actions and improvements in building resilience.

A key part of PERC is the fieldwork, with the exception of a retrospective or mini-PERC. Visiting the affected areas and speaking with those involved in the disaster provides analysts with a level of context, information, and understanding that would be otherwise difficult to obtain. It is during fieldwork that the most questions are asked and answered. The PERC manual [*Venkateswaran et al.*, 2015] provides detailed guidance on conducting fieldwork, including engagement with key partners, identifying interviewees, conducting interviews, and managing sensitive situations. In the field, the main methods are personal observation and interviews. The loose format of a semi-structured interview allows the interviewer to deviate from the plan to explore pertinent topics with the interviewee if they arise.

Initial interviewees are identified via desk research and by partner-organization contacts. A "snowball sampling" methodology is used for conducting further interviews. Emphasis is placed on ensuring engagement with a broad range of stakeholders, from different sectors and levels of action (i.e., household to national), and with different vested interests, to avoid getting a skewed or pre-arranged picture of the event. The interviewees provide the information needed to structure the agent landscape map, the narrative of what happened before, during, and after the disaster in question, and the socioeconomic and sociopolitical conditions that have led to vulnerability and the buildup of risk. Although the pool of potential interviewees will be context specific, we identify several broad groups that are likely to be needed for a full PERC analysis:

• Pertinent people and organizations in DRR, preparedness, response, and recovery processes across scales (including local, district, provincial, national, and transnational if applicable). This could include emergency response personnel, key humanitarian aid agencies, public, private, and non-profit groups working on preparedness, government officials, engineers building key disaster protection systems, groups active in recovery, and loan providers among others.

• Decision-makers and planners who make decisions that affect risk, such as planning authorities, municipal authorities or local government, community representative groups, and local and international NGOs working in the affected areas as well as flood or other hazard protection agencies, engineering companies, and the like.

• Those who are responsible for providing key services (i.e., electricity, water treatment, solid waste management, transportation, communications).

• Communities, households, and businesses that were impacted by the disaster (and possibly those who weren't if there is reason to believe lack of impacts were due to preparedness or mitigation actions that would provide a valuable story).

• Local/national academics or experts who have insight into any aspect of why the event unfolded as it did. This could include the vulnerability context, historical and current land-use, enforcement, political context, the physical science, and so on.

Interviews, such as with directly impacted communities and households, directly and indirectly impacted businesses (i.e., businesses impacted by loss of customer base) and so on, are also useful information sources for a PERC analysis. These often more informal interviews can provide a wealth of information as they take place within the context of where people live and experience their daily lives, and can serve to answer immediate questions at a particular location.

The data collected through the literature review, desk research, and interviews are analyzed using the aforementioned resilience lenses to understand how systems, agents, and institutions interact to generate vulnerability and resilience outcomes. For example, physical infrastructure, such as embankments, alone does not build or inhibit resilience. What is key is how people interact with embankments and perceive what protection they can and cannot provide. Around the issue of embankments, we have used PERC to explore how actors plan and agree on them, finance the construction, and set the protective design, and how actors have maintained and upgraded an embankment in the face of increasing hazard potentials and/or increasing risk. Most critically, PERC is designed to understand the prevailing set of both formal and informal norms and rules, and how and why they may impact other systems that people depend on. Key insights have included whether an embankment is attracting development toward it, whether relevant land use policies governing development near the embankment exist, and whether they are enforced. PERC also delves into who lives outside the embankment and who lives inside, and how the embankment changes people's behavior. In a PERC review, uncovering, assessing, and understanding these interactions are critical to grasp the local context of how resilience is built or hindered.

Looking at data critically is a key component of a PERC analysis. Analysts are encouraged to understand the political and cultural landscapes within which their interviewees and key informants are working and be sensitive to this. Contradictions or gaps in a recount of the event provide useful information about the issues within an organization or components of the disaster management system.

These applied analyses help identify useful lessons learned for action and policy that are focused on and tailored to the intended audience. These lessons learned highlight existing capacities and resilience that can be leveraged for future events and existing constraints to and/or opportunities for building resilience. Recommendations generated using PERC aim to identify the following where and how:

1. Infrastructure and ecosystems can be strengthened to reduce their vulnerability in the face of disasters and to reduce the risk of cascading failures.

2. Capacities of agents can be built to anticipate and develop adaptive responses and to access and maintain core systems. This incorporates the importance of human, social, and natural capital for both DRM and well-being.

3. Effective responses to system vulnerability and the ability of agents to take action to prevent and manage disasters are constrained by institutional factors.

4. Blockages for learning are present and how they can be overcome to ensure learning processes from past events are effective and used to improve resilience before the next event.

5. Certain aspects of the CRF and 5C-4R resilience frameworks are strong or weak and how they interact with each other.

In summary, these recommendations seek to help build resilience in the face of future uncertainty and within a changing risk landscape while taking into account the social, political, and economic drivers of vulnerability and ask the overall question of why a hazard event became a disaster.

10.4. CONSOLIDATED FINDINGS FROM POST-EVENT REVIEW CAPABILITY STUDIES CONDUCTED BETWEEN 2013 AND 2015

PERC studies thus far have identified entry points for effective action, providing organizations with a basis over which to design specific interventions that are within their capacities. The insights generated by these studies reflect the physical, human, social, economic, and natural realities of the study contexts, while also evaluating the tradeoffs between, for example, approaches that focus on physical systems and those that address community capacities or institutional issues. As PERC studies have been conducted and published, the natural tendency has been to look at emerging trends and differences between different contexts. Despite the flexibility of the PERC approach, which encourages context-specific analysis without predetermining outcomes, upon consolidating the findings (or lessons learned and subsequent recommendations) from the PERC analyses conducted to date, we find that across different contexts there are profound similarities in the points of failure, success stories, and capacities during floods. This is not to say that these lessons are applicable in every context (or have been lessons in every single PERC) but that they may be applicable across a wide range of contexts.

These emerging trends include the following:

1. **Disasters need to be managed and analyzed at their true event level, which often means crossing state, national, or other jurisdictional and administrative boundaries.** Disaster risk management and resilience building activities often stop at political, i.e., regional or national boundaries. However, events that become disasters are often bound more by geography and topography than by political boundaries. This is especially true for floods. We find that although an integrated and iterative disaster risk management cycle is well founded in theory (e.g., *Planat*, 2013; *Intergovernmental Panel on Climate Change*, 2012), it is rarely operationalized. Decisions are still taken at the state or national level, hindering efficient and effective cross-jurisdictional coordination, such as at the watershed scale.

2. **Flood planning needs to be participatory and integrated with land use, development, and other planning efforts.** (This lesson has been found for floods, although we anticipate it would apply to other hazards.) Disaster planning often exists within a silo, with a focus on reducing flood risk without entirely considering how those risk reduction decisions may differentially affect people on the ground and/or affect other sectors (i.e., through cascading failures), and how the actions of other sectors may impact flood risk. In addition, in many places DRM and infrastructure planning decisions are made at the national level, with little participation from the local levels, making it difficult to account for the spectrum of local risks, perceptions, reactions, needs, and constraints that exist. Such decision-making processes need to become more naturally participatory (i.e., not forced participation) and include voices from across scales and sectors, and be done in the context of existing development and future developmental change so that planning remains relevant under a wide range of conditions and contexts.

3. **Critical systems and protection structures need to be built so that they do not exacerbate risks or so they can at least fail safely.** Components in critical systems (i.e., food, water, shelter, power, communications, and transportation) may fail, particularly during disasters. These failures often affect not only the system in which the failure has occurred but also other systems that depend on the service(s) provided by the failed system. These are cascading failures. Cascading failures can be avoided if systems have built-in redundancies (i.e., back-up systems) and are built to fail safely (i.e., not suddenly catastrophically) while preserving as much functionality as possible. Protection structures such as levees can also fail. If they do fail, they need to do so in a safe, predictable way. Part of this requires minimizing or avoiding the "levee effect" [*White*, 1945; *Tobin*, 1995]. This is when people are attracted to build up more assets around protection structures and when settlements increase because of the (mis) perception that such structures absolutely mitigate risk and provide absolute safety. Those responsible for the construction of protection infrastructure often do not account for the (sometimes unintentional) behavioral incentives they will create to settle behind these structures, but they need to be made more aware of this "wrong incentive." If and when these structures do fail, their failure tends to be catastrophic. This is why it is necessary to have integrated DRM systems that account for how people will interact with protection structures and fully enforce regulations that reduce the additional buildup of risk that these interactions could result in (i.e., regulations that prevent development around levees or in the floodplain as a whole) and thus accomplish what we defined as prospective risk reduction.

4. **People, across scales and sectors, need access to knowledge on risk and risk reduction.** At the community level, this means gathering information on localized risk

(i.e., through mapping and participatory vulnerability and risk assessments), and making this information publicly available within a broader community discussion about risk. This information should be presented in multiple forms appropriate to the community in question, one of which includes accurate, up-to-date, and unbiased hazard maps. This information should also be accessible, by which we mean data should be easy to access physically and should be translated and presented in a way that every layperson can understand, which even (or especially) in the countries most advanced in hazard mapping is often not the case. Risk information is only one important aspect, however, it is nonetheless an essential element that helps people make informed decisions in the context of the risk they are exposed to. At a sectoral level, people benefit from access to appropriate information on the spectrum of risk reduction and protection measures and their cost-effectiveness. This information and awareness raising needs to go hand-in-hand with the appropriate incentives for implementing such measures.

5. **The tendency for emergent groups to form during times of stress needs to be recognized.** People often self-organize into groups to fill gaps in the formal disaster management system. For example, volunteer groups often form where state-mandated recovery processes do not exist, including by helping disaster-impacted households recover their belongings and rebuild their homes or by navigating formal recovery processes. Other groups may form to distribute emergency relief in the immediate aftermath of a disaster prior to the arrival of or in the absence of humanitarian organizations. Although these groups form during all disasters and provide a variety of important services, there is rarely a platform (via institutions or informal/formal processes) over which they can coordinate with formal disaster management groups. If formal and emergent groups had a platform over which to work together, there would be greater efficiency and fewer replications in and disruptions to disaster management activities.

6. **The marginalized, most vulnerable groups need to be included in the response and recovery process.** These groups are frequently excluded from disaster management processes (e.g., through exclusion from bank loans or federal funds for recovery or from relief distribution), significantly impacting their ability to cope with and recover from disasters and perpetuating and exacerbating inequity. Excluding marginalized groups from recovery not only further exacerbates their vulnerability, it also has long-term impacts on socioeconomic recovery and cohesion for all groups. In many cases, marginalized groups neglected in recovery initiatives are central to recovery efforts in their role in construction and repair industries. Highlighting and exploring the differential distribution of disaster impacts and access to recovery support is required so that the perspective of marginalized groups is heard and more equitable policies can be designed to better manage future events.

7. **The majority of these lessons can be fulfilled only if relationships are built, improved, and maintained.** Relationships need to be improved within communities, between organizations, and between government and private and/or non-governmental organizations. These improved relationships will improve efficiency and reduce friction losses and enhance access to resources, services, information, and learning from the event for both the future and to build the trust necessary to ensure efficiency in all parts of the disaster management system.

By consolidating such lessons, we are building the evidence to conclude that disasters anywhere on the globe can provide important, broadly applicable lessons learned for where and how resilience can be built. In this respect, PERC is not just an isolated event review that leads to a report. Rather, it is a part of a much wider process to understand how resilience (or lack thereof) is experienced and find actionable opportunities to fundamentally increase resilience globally. The creation of a consistent report per the PERC methodology manual ensures that the studies conducted thus far and in the future are equally rigorous, holistic, and grounded in both academic thinking and practical experience.

To ensure that these lessons have a wide impact, it is important that the key insights and provided recommendations are shared actively through various channels (e.g., a printed report disseminated to key stakeholders in the affected area and country, a freely available web download, key insights shared on social media), and presented in various formats throughout the months and years after the event (e.g., in face-to-face meetings to the relevant local agents, in topical meetings and conferences nationally and internationally). Most importantly, the lessons learned need to be easily available, at a general and not only at an event-specific level, freely accessible (online, at no cost) in a searchable format with various keywords or topical queries so that people globally can harvest the lessons as they see fit and apply them to their own contexts.

10.5. CONCLUSION

The central aim of a single PERC analysis is to provide a comprehensive picture of what happened and why, and what opportunities exist to build disaster resilience. PERC looks at disasters from a systems-wide lens, synthesizing lessons learned across sectors and scales, without anticipated outcomes. Therefore, PERC is far more than just a structured event-report and analysis; it also provides a value chain from the moment fieldwork is established by providing affected people and key actors of the event a voice all the way to long-term learning. It is not the goal of PERC to design specific interventions that deal with the minutiae; rather, it is designed to point out wider trends and systemic gaps that on-the-ground disaster management practitioners might design interventions for and to address. In this process, it also promotes the notion of learning without assigning blame, empowering actors with the idea that disaster risk is something that can be mediated through action. By documenting this phenomenon in each context, the concerns of marginalized groups are given a voice, and specific recommendations can be made to design more equitable policies.

The central goal of the collective PERC endeavor is to analyze multiple events in a consistent yet flexible manner. The consistent framework and process allows for comparative analysis and global learning as this chapter aimed to demonstrate, and this will be enhanced as further PERC studies are conducted. Beyond the stakeholders of a specific event, the PERC study and resultant reports are contributing to a much-needed body of work on addressing the proliferation of disaster risk and associated damages. Ultimately, the goal of the PERC methodology is to inform and encourage a learning and resilience-building process that prevents hazards from becoming disasters in the first place. Another goal is to support an effective recovery process when disasters do occur, while considering people and their needs, and the cultural and legal norms that enable their ability to thrive.

In summary, by using a systematic methodology like PERC to provide a "forensic" analysis of disaster events, we are identifying generalizable lessons aimed at understanding and increasing resilience and reducing vulnerability across social, political, economic, and geographic contexts.

REFERENCES

Asian Development Bank (ADB) (2013), *Investing in resilience: Ensuring a disaster-resistant future*, Asian Development Bank, Manila.

CEDIM (2015), *Forensic Disaster Analysis*, Center for Disaster Management and Risk Reduction Technology, Karlsruhe Institute of Technology, https://www.cedim.de/english/2131.php.

Damage, Loss and Needs Assessment (2010), Guidance Notes. The International Bank for Reconstruction and Development/ The World Bank.

DFID (1999), *Sustainable Livelihoods Guidance Sheets*, Department of International Development, United Kingdom, http://www.eldis.org/vfile/upload/1/document/0901/section2.pdf.

EEFIT (n.d.), *The Earthquake Engineering Field Investigation Team*, The Institution of Structural Engineers, https://www.istructe.org/resources-centre/technical-topic-areas/eefit.

Foresight (2012), *Reducing Risks of Future Disasters: Priorities for Decision* Makers, Final Project Report, The Government Office for Science, London.

Integrated Research on Disaster Risk (2011), Forensic Investigations of Disasters: The FORIN Project (IRDR FORIN Publication No. 1). Beijing: Integrated Research on Disaster Risk.

Intergovernmental Panel on Climate Change (2012), Managing the Risks of Extreme Events and Disasters to Advance Climate Change Adaptation. A Special Report of Working Groups I and II of the Intergovernmental Panel on Climate Change. C. B. Field, V. Barros, T. F. Stocker, D. Qin, D. J. Dokken, K. L. Ebi, M. D. Mastrandrea, K. J. Mach, G.-K. Plattner, S. K. Allen, M. Tignor, and P. M. Midgley, Eds.. Cambridge University Press, Cambridge, UK, and New York, NY, USA, 582 pp.

Keating, A., K. Campbell, R. Mechler, E. Michel-Kerjan, J. Mochizuki, H. Kunreuther, Bayer, J., S. Hanger, I. McCallum, L. See, K. Williges, A. Atreya, W. Botzen, B. Collier, J. Czajkowski, S. Hochrainer, and C. Egan (2014), Operationalizing Resilience Against Natural Disaster Risk: Opportunities, Barriers and A Way Forward, Zurich Flood Resilience Alliance, http://www.iiasa.ac.at/web/home/research/researchPrograms/RiskPolicyandVulnerability/Resilience-lowres_2.pdf.

MacClune, K., C. Allan, K. Venkateswaran, and L. Sabbag (2014), Floods in Boulder: A Study of Resilience. Boulder, CO: ISET-International.

MCEER (2007), Engineering resilience solutions from earthquake engineering to extreme events. Multidisciplinary Center for Earthquake Engineering Research, USA.

Miller, S., R. Muir-Wood, and A. Boissonnade (2008), "An Exploration of Trends in Normalized Weather Related Catastrophe Losses" in Climate Extremes and Society, eds. H. F. Diaz and R. J. Murnane. Cambridge, UK, Cambridge University Press, 225–247.

National Research Council (NRC) (2012), *Disaster Resilience: A National Imperative*. The National Academies Press, Washington, D.C..

Planat (2013), Sicherheitsniveau für Naturgefahren: Strategie Naturgefahren Schweiz. Nationale Plattform für Naturgefahren Planat, 18pp., August.

Post-Disaster Needs Assessments (2013), Volume A. Guidelines. Global Facility for Disaster Risk Reduction (GFDRR).

Simonovic, S. P. and A. Peck (2013), Dynamic Resilience to Climate Change Caused Natural Disasters in Coastal Megacities Quantification Framework. British Journal of Environment & Climate Change, *3*, 378–401.

Tobin, G. A. (1995), The Levee Love Affair: A Stormy Relationship. Water Resource Bulletin. *31*, 359–367.

Tyler, S. and M. Moench (2012), A framework for urban climate resilience. *Climate and Development*, 4(4), 311–326. doi: 10.1080/17565529.2012.745389.

United Nations Office for Disaster Risk Reduction (UNISDR) (2009), Terminology http://www.unisdr.org/we/inform/terminology.

United Nations Office for Disaster Risk Reduction (UNISDR) (2011), Global Assessment Report on Disaster Risk Reduction, United Nations International Strategy for Disaster Reduction, Geneva.

United Nations Office for Disaster Risk Reduction (UNISDR) (2015), Global Assessment Report on Disaster Risk Reduction, United Nations International Strategy for Disaster Reduction, Geneva.

van Aalst, M., J. Kellett, F. Pichon, and T. Mitchell (2015), 'Incentives in Disaster Risk Management and Humanitarian Response,' Background note for World Development Report 2014, at http://siteresources.worldbank.org/EXTNWDR2013/Resources/8258024-1352909193861/8936935-1356011448215/8986901-1380568255405/WDR14_bn_Incentives_in_disaster_risk_management_vanAalst.pdf, 2013 (retrieved 28 March).

Venkateswaran, K., K. MacClune, A. Keating, and M. Szoenyi (2015), *Learning from Disasters to Build Resilience: A Simple Guide to Conducting a Post Event Review*. Boulder, CO: ISET-International & Zurich Insurance Group.

White, G (1945), *Human Adjustment to Floods*. Department of Geography Research Paper no. 29, The University of Chicago, Chicago.

Worldbank (2010), Damage, Loss and Needs Assessment. Guidance Notes. Volume 1. Global Facility for Disaster Risk Reduction GFDRR.

Zurich Insurance Group (2013a), Enhancing community resilience: a way forward. Zurich Insurance Group.

Zurich Insurance Group (2013b), European floods: using lessons learned to reduce risks. Zurich Insurance Group.

Zurich Insurance Group, (2014a), Emmental, Switzerland floods of July 2014. On a hot, sunny day, a flood alert! Zurich Insurance Group.

Zurich Insurance Group (2014b), After the storm: how the UK's flood defences performed during the storm surge following Xaver. Zurich Insurance Group, September.

Zurich Insurance Group (2014c), Central European floods 2013: a retrospective. Zurich Insurance Group, May.

Zurich Insurance Group (2015a), Turning knowledge into action: processes and tools for increasing flood resilience. Zurich Insurance Group, September.

Zurich Insurance Group (2015b), Balkan floods of May 2014: challenges facing flood resilience in a former war zone. Zurich Insurance Group.

Zurich Insurance Group (2015c), Urgent case for recovery: what we can learn from the August 2014 Karnali River floods in Nepal. Zurich Insurance Group

Zurich Insurance Group (2015d), Morocco floods of 2014: what we can learn from Guelmim and Sidi Ifni. Zurich Insurance Group.

11

Defining Complete Post-Flood Scenarios to Support Risk Mitigation Strategies

Scira Menoni[1], Funda Atun[1], Daniela Molinari[2], Guido Minucci[1], and Nicola Berni[3]

ABSTRACT

The chapter starts with a definition of a complete damage scenario after an event. Then, the specific post-event assessment methodology, prepared by the collaboration of Umbria Civil Protection Authority and researchers from Politecnico di Milano, is explained. The methodology has been applied to two cases: the floods that occurred in the Umbria Region, Central Italy, in November 2012 and November 2013, which provoked significant damage to multiple sectors. The primary focus of the research is to complete a post-event scenario that highlights the different combination of hazard, exposure, and vulnerability factors and specific characteristics that emerge from those factors across space and time.

11.1. INTRODUCTION

The chapter illustrates the concept and the methodology to develop what we call a "complete event scenario" of past events to provide a descriptive and interpretative framework of the different types of damage and losses that have actually occurred. The reporting of the complete event scenario has multiple aims, ranging from providing an explanation of the damage that can be taken into consideration for a more resilient reconstruction, to offering insight into real complex events, to permitting to enhance risk modelling capabilities for the future [*De Groeve et al.*, 2013].

The methodology is then applied to two cases, the floods that occurred in the Umbria Region, Central Italy, in November 2012 and November 2013. The choice of this particular case derives from the long lasting tight collaboration that has been established between the research group of the Politecnico di Milano and the Umbria Region Civil Protection Authority, as described in Chapter 6 in this book. This collaboration proved to be a win-win situation, in which researchers could obtain the best possible data and the most complete information they rarely have access to, while the public administration departments could profit from the time and the analytical systemic expertise of researchers to obtain a report showing and explaining how scenarios of damage and losses unfolded over time in their region.

In the second section, the concept of complete post event scenario will be described. In the third section, the methodology and the data requirements will be explained. In the fourth section, the application of the concept and method in the real case of the Umbria Region floods will be illustrated. The last two sections discuss the result and the challenges ahead.

---

[1] *Department of Architecture and Urban Studies, Politecnico di Milano, Milano, Italy*

[2] *Department of Civil and Environmental Engineering, Politecnico di Milano, Milan, Italy*

[3] *Umbria Region Civil Protection Authority Foligno (PG), Italy*

Flood Damage Survey and Assessment: New Insights from Research and Practice, Geophysical Monograph 228,
First Edition. Edited by Daniela Molinari, Scira Menoni, and Francesco Ballio.
© 2017 American Geophysical Union. Published 2017 by John Wiley & Sons, Inc.

11.2. DEFINITION OF THE "COMPLETE EVENT SCENARIO" CONCEPT

The word "scenario" was first used by Khan from the Rand Corporation [*Kahn*, 1960], a consultancy firm working for the US Army, to label the task he was given, which was to provide a realistic forecast of a likely future after a nuclear war. *Kahn* [1960] depicted the scenario with the aim of drawing attention to the many negative and dramatic aspects of a nuclear war in order to discourage the militaries from initiating one. Since then, the scenario planning approach has been used by several disciplines. Scenarios are used by economists as a business planning tool [*Wilkinson and Kupers*, 2013], by environmental scientists to conduct impact assessments [*Liu et al.*, 2008], by researchers in climate change [*McCarthy et al.*, 2001] to forecast possible futures given different levels of temperature increase, and by land use planners and water managers to estimate how urbanization may develop in the future years and how the demand for water supply will be distributed and up to what levels it might raise [*Hulse et al.*, 2004]. Yet there are differences in the way the scenario approach is used by those different scientists and professionals. In terms of time frames, for example, economists are interested in short-term forecasts, ranging between the next three to the maximum of five years. Instead, researchers in environmental sciences or climate change tend to consider a much longer timeline, looking at consequences of human action on different environmental assets including climate in the next 50 or even 100years.

A scenario is a tool to provide a dynamic picture of the future environment. On one hand, it helps to predict potential situations in the future; on the other hand, it shows how single actions could change the entire outcome. *Staley* [2007, 2009] indicates that "the purpose of scenario planning is not to predict the future, but rather, to show how different forces can manipulate the future in different directions." Scenarios, in fact, are used to explore alternative futures, given different conditions of critical determinant variables on the general behavior of the studied systems, and given alternative sets of decisions and interventions that can be taken to intentionally change the path of an unwanted likely future. It is not surprising then that also in the field of climate change and natural hazards scenarios have been developed with a similar intent: to address those factors that can reduce expected damage given temperature rise and potential severity of natural calamities.

11.2.1. The Scenario Approach in the Natural Hazard Field

A widely accepted definition has been provided by the *Intergovernmental Panel on Climate Change* [2008]: "A scenario is a coherent, internally consistent and plausible description of a possible future state of the world. It is not a forecast; rather, each scenario is one alternative image of how the future can unfold."

In the field of natural hazards, scenario modeling has been used as an alternative to probabilistic risk assessment, by addressing the impact a deterministic hazard input (be it an earthquake, a flood, or a landslide) may have in a given area. In some instances, scenarios limit their consideration to one or two variables identifying key damage aspects, for example, the number of affected people and the number and location of collapsed and severely damaged buildings after an earthquake of predefined magnitude and epicenter. In those applications, a quantitative estimate of the expected damage can be provided, because the number of considered variables is limited. In what we call a "complete event scenario," the interdependency among the various components of affected systems and among systems, as well as interface between societies and built and natural environments are accounted for [*Kropp et al.*, 2011, in *Menoni and Margottini*, 2011]. This requires development of scenarios that are semi-quantitative, as the key numbers defining the magnitude and the severity of described events are accompanied by a more qualitative description of how the interaction among systems and components shape a specific post-disaster situation. Such qualitative description highlights the dysfunction of critical facilities, the delays in return to normalcy, the second and higher order damage [*Rose*, 2004] such as those due to the closure of roads and to interruption of businesses. In theory, most descriptions could also be provided in quantitative terms, however, very seldom such effects are reported if not in a very anecdotic way. Examples of this kind of mixed qualitative and quantitative scenarios can be found in the seismic field in particular. For example, one may consider the exercise carried out by *Geohazard International* [1994] in the Quito area or the Japan International Cooperation Agency's (JICA) study in the Istanbul metropolitan area (IMM) [*IMM and JICA*, 2002].

11.2.2. Linking the Past to the Future

In general, the scenario approach is taken to describe what the future may look like given a strong stress in the environment similar to that produced by a natural hazard. Even though scenarios are projected into the future, knowledge about how past events evolved is necessary to produce them. The past is the only source of knowledge regarding what may occur not only as a direct effect of the external stress but also as induced and indirect consequences, due to the interconnection and interaction between complex systems. *Galderisi and Ceudech* [2010] stress the importance of combining the characteristics of past physical hazardous events and the present vulnerability and exposure of the affected areas, such as the characteristics of buildings (age, structure, and maintenance), morphology of urban pattern, population density, and the accessibility and permeability of urban areas in order to explain the damage patterns.

Even though the importance of back analysis is recognized as essential [*Ringland*, 1999; *Rescher*, 1991], few attempts have been made to comprehensively analyze past events, describing the damage as resulting from the interaction among given physical phenomena and complex exposed systems. See in particular *EQE International* [2002] developed for the 1995 Kobe earthquake; *US Geological Survey (USGS)* [1996]; *Galderisi and Ceudech* [2005], describing the impacts of the 1980 earthquake in Southern Italy; *Mori et al.* [2011]; and *Holguin-Veras* [2012] on the Tohoku, Japan earthquake in 2011.

In the case of floods, comprehensive former post-flood reports have been produced to fine tune analysis of the losses and impacts on multiple sectors to identify key lessons and weaknesses to be addressed by national policies. This is the case of the Pitt Report after the 2007 Severn flood in the United Kingdom, and for the various "return of experience" reports that have been produced in France after a severe storm and flood events [*Agence de l'Eau Artois-Picardie*, 2000; *Direction Generale Cerema*, 2014]. Those reporting efforts, though, are still carried out as single spot initiatives and are not conceived yet as a standardized tool to represent the multidimensionality of damage and losses.

In Figure 11.1, the similarities between scenarios of the past and scenarios of the future are shown, providing the conceptual basis for the approach that is proposed in this chapter. In the picture, the two scenarios are separated by the actual time of the event occurrence, shaping the way the damage and eventually cost-benefit analysis can be carried out and correctly used. Before the event occurrence, the damage is estimated on the basis of information regarding the hazards existing in the area of concern, the exposed assets, and the vulnerability of both assets and communities. Cost benefit analysis is then used to compare alternative risk mitigation strategies aimed at reducing one or all components of the risk function. After the event occurrence, instead, the damage is a "known" factor, shapes the scenario that has actually occurred and which can be analyzed so as to identify how and to what extent the risk factors of hazard, exposure and vulnerability had actually concurred to produce the observed levels of damage. In this case, cost-benefit analysis is aimed at comparing different recovery and reconstruction options, with the aim of improving the pre-disaster situation and achieving a more resilient condition.

In more general terms, post-event scenarios are useful for identifying and eliciting lessons learned to be applied in risk management. A systematic reporting of post-disaster scenario could contribute in this regard, to create a larger

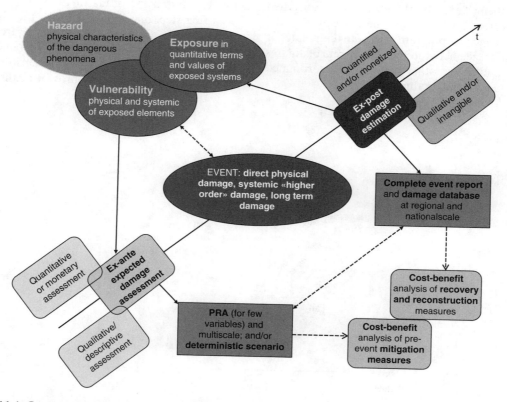

Figure 11.1 Connection between pre-event and post-event scenarios.

evidence base for lessons learned in the past and provide the knowledge base for identifying key recurrent aspects that can be generalized to other contexts and situations. This was certainly the intention of the NEDIES project run by the European Joint Research Centre (JRC) for some years, providing a large collection of events described in detail and lessons learned in a wide range of disasters, including in particular floods that have occurred in Europe between 1984 and 2001 are discussed [*NEDIES*, 2002].

Disaster data and information are required nowadays by a number of international organizations, from the United Nations (UN) for measuring the advancement toward the goals set by the Sendai Framework for Action, to the European Union (EU) to support the implementation of the Flood Directive.

11.3. KEY COMPONENTS OF A COMPLETE POST-FLOOD EVENT SCENARIO IN THE PROPOSED METHODOLOGY

To accomplish the objective of developing a comprehensive post-flood damage scenario, some limits of current practices of damage data collection and analysis need to be overcome. This is not a new issue, as *Hoyt and Langbein* [1955] were discussing the reasons for the unreliability of US flood damage data already in 1955. According to them, "each of the authorities concerned with flood protection examines microscopically the sources of damage its program seeks to eliminate or reduce. The difference between the several sets of figures are confusing, and argue rather strongly for coordinated and impartial fact finding in this field."

The Hoyt and Langbein book has been quoted much more recently by *Pielke* [2000], suggesting that the situation has not changed much in the last 50 years or so.

In Figure 11.2, the key conceptual aspects that need to be considered to perform the complete event scenarios of past events are represented. Such aspects relate to how different sectors are impacted by the event, how damage unfolds over time giving rise to different types of damages, how the latter are spatially distributed, and what the explicative variables are that need to be considered to characterize different types of damages across temporal and spatial scales [*Menoni et al.*, 2007]. We may find elements of such multifaceted framework for scenario development also in recent guidelines and methodologies.

For example, the Post Disaster Needs Assessment (PDNA) [see *Wergerdt and Mark*, 2010] methodology, resulting from a coordinated effort among the World Bank, the UN, and the European Commission, was an important source of inspiration for us. The methodology, especially in one of its components, the Damage and Loss Assessment (DaLA), addresses the issue of developing a cross-sector analysis of the damage, quantifying and describing losses suffered by residential houses, economic activities, environmental assets, and lifelines.

In a similar way to the PDNA methodology, we also have identified key sectors that we wish to address in order to evaluate the damage a society has suffered in its activities, capital, and infrastructures.

As a first step, one should describe the physical event, in our case, the flood that affected the area, providing analyses, graphs, and maps both at the regional and the local scale. Then the sectors and systems that have been considered for evaluation in our own method have been

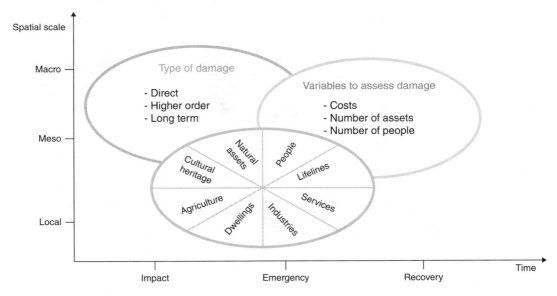

Figure 11.2 Methodological framework to build the complete event scenario.

synthetically grouped by (as synthesized in the rose ellipse in Figure 11.2.) affected people, damaged lifelines, public services and facilities, agriculture, industrial and commercial activities, residential houses, cultural heritage, and environmental assets.

The direct physical damage to those systems is provoked by the flood and more specifically by sediment, contamination, and debris carried by the flooding water as well as by the water itself. Equally important to grasp the overall impact of an event is to consider second or higher order effects [*Rose*, 2004], by which we mean disruption of services and dysfunctions in different sectors that are consequent to the physical damage that has occurred in systems and in systems' components. According to *Van der Veen and Logtmejer* [2005], such second order damages are due to the systemic vulnerability to the first order physical losses and not directly by the physical stress (the flood). While the physical damage has to be assessed locally, second and higher order damages emerge at larger scales, that is, on a regional, national or even global scale. It is important therefore to be able to connect the second and higher order damage to the triggering physical event, in order for the analysis to be consistent. *Pielke* [2000] refers in particular to challenges related to "contingency," the term he uses to address the problems in identifying impacts beyond the more evident direct ones, and "attribution," which he uses to refer to the difficulties in identifying the "real" causes of the damage that may be social (creation of vulnerability) rather than natural. In the proposed methodological approach particular care is payed to connecting the event to the various impacts the physical stress has induced in a complex built and natural environment, and to attributing the damage to individual event contingency by addressing the dynamic evolution of disasters in time and space.

First and higher order damages unfold also over time. The time scale is an essential component in the dynamic development of any disaster scenario. In the case of floods, for example, physical consequences due to humidity and mold may appear weeks and months after the peak event; indirect effects on economic activities and services may be recorded months and sometimes even years after the event occurrence. This is the reason why the *World Meteorological Organization (WMO)* [2013], in its guidelines, recommends conducting flood loss assessments at different times, right after the occurrence and three to six months later. The iterative effort is necessary on the one hand to gather more reliable and stabilized information, but also, on the other, to get information that does not pertain to the first emergency, such as the time when activities were fully restarted in a firm or a public facility. The WMO guidelines indicate how the reporting effort may support decision-making at different stages of the disaster. The initial report has to provide rapid damage assessments to relief forces to coordinate the required level of assistance. A second stage report is useful to provide evidence for collecting insurance claims or requests of compensation. A third stage and more complete report can be used to appraise the effectiveness of flood defenses and compare mitigation options.

The relevance of the spatial scale should not be neglected either. In fact, damage and losses become manifest and relevant at different spatial scales. At the local scale, most physical damage to objects and systems' components can be traced; however, some damages to critical infrastructures and lifelines often affect much larger levels that may be regional, national, or international, depending on the relative importance of the affected item and system. Also, the interconnection among damages to different objects and systems develops across spatial scales in a complex way, particularly when urban and metropolitan environments are concerned.

Summarizing, in the proposed methodological approach, the post-event scenario is a sort of articulated and advanced back analysis that does not cover only one or two explicative variables but aims at intercepting also the qualitative aspects of a disaster and the relation among damage to different sectors, among different types of damage, and between the risk components. Building on the experience that has been developed in the French "return of experience" reports, our own methodological framework prescribes a sort of standardized index that can be considered in any reconstruction of a complete event scenario. The standardized schematic index is represented in Table 11.1.

The various sections will be developed in depth each on its own or jointly, depending on the type and extent of damage to assets and systems pertaining to each of the considered sectors. In some cases, it will make sense to join the discussion regarding, for example, lifelines and public facilities if only minor damage has been suffered by both or by one of the two sectors. Similarly, some sections can be split in two, as an example in the case of damage to both the agricultural and the industrial/commercial sectors having been significantly high, making it preferable to treat them separately.

11.3.1. Understanding the Data and Their Characteristics

To standardize the reporting methodology, data need to be pre-processed and structured in such a way that queries for multiple purposes will be possible in a consistent way. However, as discussed in literature and in the working group organized by the JRC on loss reporting and assessment [*De Groeve et al.*, 2014], there are a number of problems in the way post-disaster data are collected and processed. In our proposal and experience in the Umbria region, we are actually trying to overcome such problems and provide a methodology that will be feasible, adaptable

Table 11.1 Standardized Index for Post-Event Damage Reporting.

STANDARDIZED INDEX	
Description of the physical triggering event	in case of relevance of cascading events at the ***regional*** and the ***local scale***
Affected people (victims, displaced/evacuated)	***by space*** (municipality), ***by time*** (duration), and ***by variables*** such as costs of assisting the population (civil protection).
Damage to **lifelines**	***by type*** (transport, water, energy), ***by space*** (local physical damage; regional: disruption, maps and total direct losses and indirect effects), ***by time*** (duration of disruption, temporary recovery, permanent repair) and ***by type of damage*** (physical to assets and higher order to systems), ***by variables*** (costs of repair and amelioration)
public facilities	***by type*** (school, hospital, theatre), ***by space*** (local physical damage; regional: areas deprived from services; cost of repair), ***by time*** (duration of disruption, temporary relocation, for permanent repair) and ***type of damage*** (physical to assets and higher order to service), ***by variables*** (costs of repair and amelioration)
economic activities	***by affected subsector*** (agriculture, industrial, and commercial activities), ***by space*** (local: individual firms and farms affected; reported damage in terms of repair costs and lost revenue; regional: most affected economic sectors, map of clustered damage); ***by time*** (duration of disruption, time needed to fully restart) and ***by type of damage*** (physical to assets and higher order to sub-sector), ***by variables*** (cost of recovery, cost of repairs, cost of lost production, cost for unemployed subsidies)
residential buildings	***by space*** (municipalities and local physical damage by forms; regional: most affected municipalities for residential buildings), ***by variables*** (cost of repair; intangible loss of memorabilia)
cultural heritage	***by type of use*** (public facilities, residential buildings, public areas); ***by space*** (individual buildings, historic centers); ***by variables*** (associated cost of repair and loss of cultural value)
environ-mental goods	***by space*** (local: individual areas damaged; regional: contamination, large recreational, park areas); ***by time*** (longer term physical damage, cost of reclaim, decontamination, etc.) and ***by type of damage*** (direct loss of species, vegetation, and fauna; higher damage to ecosystems); ***by variables*** (loss of biodiversity; hectares of lost vegetation)
Synthetic overview of the event	***comparative analysis*** of physical damage, systemic damage, to ***multiple sectors across time and space***. Identification of the most critical and specific aspects of the particular event
Evaluation of the effectiveness of pre-event risk mitigation measures	***by space*** (local and regional plans, hazard maps, etc.) and ***by time*** (contingency plans, sectoral plans like the flood risk management plan etc.)

to different administrative contexts and necessities, and in the meantime, more effective in getting the type of results we want (see Chapter 6). Such results imply the possibility to use the data for several purposes, including accounting, identifying post-disaster needs, carrying out forensic analysis on the direct and indirect causes of damage, including root causes [see *Oliver Smith et al.*, 2016], and providing elements to improve future risk assessment and scenario modeling capacity.

Two major elements are at stake when developing post-event complete damage scenario. On one hand, the organizational requirement to appoint a data manager or a data coordinator [*JRC Technical Group*, 2015], that will be in charge of collecting the data from different authori-ties and different agencies in charge of tasks including damage assessment in the aftermath of an event. On the other hand, the data coordinator should work in close contact with (or be the same person as) the event scenario analyst so that the latter will have a deep understanding and knowledge of the data to be used in the scenario development. Data need to be deeply known regarding the unit of measures used to collect them, for the spatial scale at which they were collected, and to the agency that provided them. Also, the original purpose or mandate for which the data were collected must be known to fully appraise what the data actually represent before they are used in the scenario reporting. A very fine tuned data characterization needs to be performed that goes beyond what is generally considered as data reliability and quality check. Because under the same "label," different types of damages and losses may be reported, data need to be analyzed according to several criteria as follows:

• The original scale and level of detail at which it was surveyed. It is clearly very different to get information about damaged buildings from a satellite data or from a survey on the ground. The first may cover much larger areas, but be less detailed; the second may suffer from the dishomogeneity due to the large number of individual surveyors.

• The administration that is responsible for declaring the damage and to restore it. The mission of the administration affects the way the data are collected and the damage described.

• The damage may be provided in physical terms, both with quantitative data (for example, number of affected buildings, surface inundated for each building, etc.) or qualitative, describing what elements have been damaged and providing an address, a geographic description of the affected areas.

• Losses are generally considered to be costs necessary to restore the damage or corresponding to the lost item. Here, a very important issue arises: physical (quantitative and/or qualitative) and monetary damage assessments do not necessarily match. Although the damage description is an objective representation of what can be seen and surveyed on the ground, associated costs may correspond to rather different values. Costs may refer to the cost of repair and/or replacement when the latter is inevitable, and they generally correspond to the amount an insurance company would cover under a certain insurance policy. However, a careful examination of the data sets of the Umbria Region Civil Protection Authority, shows at least two other cost categories emerged, even though not explicitly referred to as such. Some costs include also an amelioration component, particularly for infrastructures and mitigation defenses. It is certainly unwise to reconstruct an item bringing the situation back to the pre-event condition if significant vulnerabilities have been clearly recognized as the main damage caused. Instead, it makes sense to improve the pre-event situation by adding to the pure reconstruction and the cost of prevention and mitigation. In other cases, costs that are submitted to the compensating authorities are actually new defenses or mitigation measures that were waiting for financing and take "the window of opportunity" provided by the event to be implemented. While in the previous case the argument of resilient reconstruction and recovery can be brought forward, in this latter case, it is clear that an anomaly is created, because money for reconstruction is actually used for mitigation, while the latter should have been done before the disaster. For the Italian situation, see the relevant study by *Cellerino* [2004]. However, this is

often the case, as the needs that can be identified after a disaster do not totally derive from the direct damage provoked by the event but may result from inadequacies that existed before.

• Finally, there is the entire set of costs that are related to the second and higher order damage that refers to the consequences the emergency brought by critical facilities' and lifelines' malfunction. However, this type of cost is only seldom provided in official databases and compensation requests documents. Some components of such second and higher order damage may be covered by insurance, for example, business interruption or unemployment subsidies in case of jobs permanently or temporarily lost because of an emergency.

11.4. REPORTING THE COMPLETE DAMAGE SCENARIO AND APPLICATION TO THE CASE STUDY AREA

As discussed in the introduction section, a post-event complete damage scenario constitutes a qualitative and quantitative description of the damage and losses that occurred as a consequence of a given hazard that affected a territory characterized by certain levels of exposure, type of exposed systems and their intrinsic physical and systemic vulnerability. What the scenario does is offer a comprehensive overview of what happened, in exactly what areas, at what time, and how damage in one system was related to the damage in another. Finally, the scenario also allows for explaining the damage in terms of its main driver, be it hazard, exposure, vulnerability, but also other contingent conditions that may regard the presence of population in that period of the year, the hour when the event occurred, the season, and so on.

What is certainly necessary therefore is to be able to use the information and the data that have been described in the previous section to reconstruct the dynamic of the disaster over time, at different spatial scales, and to match spatially and temporally the different risk components (hazard, exposure, and vulnerability). This is the type of match that is not allowed at the moment by available data sets worldwide as described elsewhere in the book (see Chapter 1 and Chapter 2).

In the following paragraphs, therefore, the methodological framework described in section 11.2.2 will be applied to the case study of the Umbria Region in Central Italy.

11.4.1. Characterization of the Case Study Area

The Umbria region is located in central Italy (Figure 11.3) and covers 8,456 km^2 with a population of 883,000 inhabitants [National Statistical Office, 2011]. Most of its territory is made of the Appenines Mountains

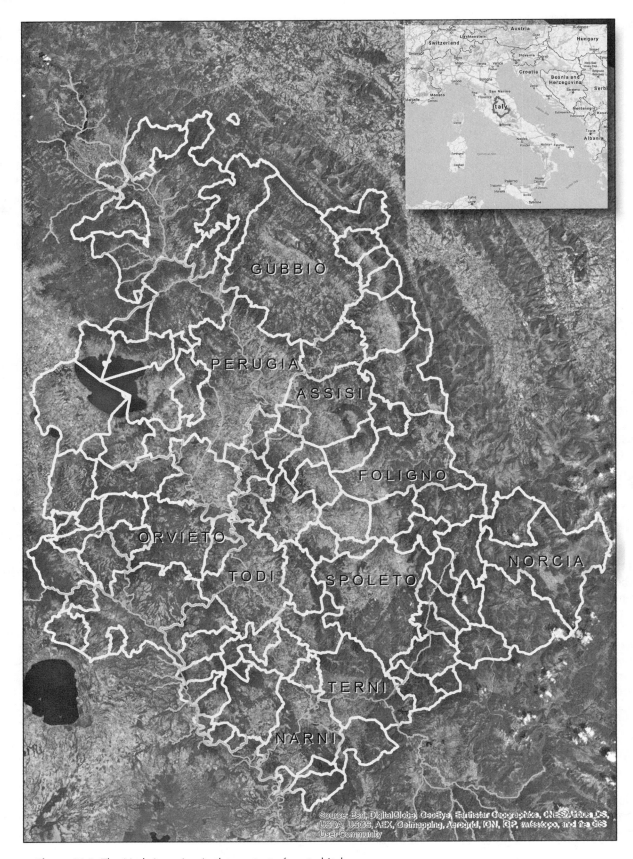

Figure 11.3 The Umbria region in the context of central Italy.

and hillside of which 46% of its territory is covered by forests and 7% of which are preservation areas or protected parks. Urbanization has taken place along the two main valleys: the Tiber Valley stretching from north to south for almost 100 km from the borders with Tuscany to the Todi Municipality in the south at the borders with the Lazio region, and the Umbria valley in the direction orthwest to southeast, between Perugia and Spoleto for 40 kms with a larger width ranging between 5 to 10 kms. Both valleys are highly infrastructured with highways, main roads, and railways. The valleys by being the only plain available space conditioned the present linear urban development. The Tiber River basin is the third largest national river by length and the second largest by area (12,700 km²) and is characterized in its highest part by a complex topography with elevations comprised between 50 and 2500 meters above sea level.

Available historic data confirm the hazardousness of the area, characterized by a variety of phenomena, including debris flow, flash floods, and riverine floods in reclaimed land and suspended levees. In the last century more than 100 of these events occurred, with peaks in the winter time in the months of November and February. The Tiber provoked several indundations, but significant damage has also been caused by its tributaries and by minor channels and creeks. In 2010, two major floods occurred in Umbria. The overall economic impact, to both private and public assets, amounts to 1 billion euros damage for the last 10 years only.

In the following sections, the methodology for developing a post-flood damage scenario is applied to the two floods and mixed floods and debris flows events that occurred in November 2012 and 2013. Figure 11.4 represents using different colors for the municipalities that have reported damage in 2012, 2013, or both.

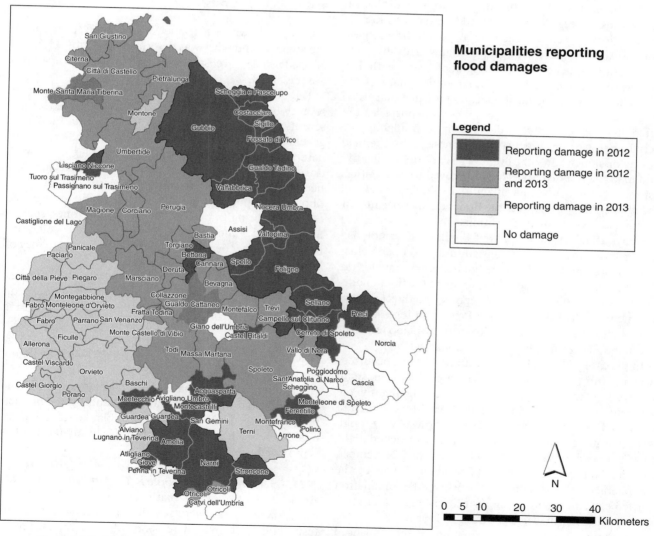

Figure 11.4 Affected municipalities in the 2012 and 2013 floods.

11.4.2. Reporting the Physical Event Scenario

The back analysis of the physical event that triggered the sequence of damage and losses, both physical and functional, starts with a description of the hazardous event evolution over time until the most acute phase was reached up to its end. In the case of a flood, trivially this means to be able to reconstruct peak discharges and water depth at different locations, that are relevant at a regional level (at the catchment scale, in correspondence with relevant artifacts like dams, viaducts, etc.) and at a local scale (in correspondence to bridges, manholes, etc.).

11.4.2.1. Reconstructing the Physical Event Scenario in the Umbria Case Study Area. The description of the physical event scenario is produced out by the Regional Civil Protection Authority within 48 to 72 hours after the event. The report generally combines data deriving from meteorological forecasts and observations, hydrological monitoring, forecasts and observed hydrographs in relevant sections, which was the case of the Umbria Region downstream of the existing large dams. However, the information is generally at a regional scale, with few downscaling at the local level. In particular, maps of the inundated zones at the local scale, that are necessary also to guide in the surveys for compensation purposes are often missing, because they depend on the availability of aerial photos or on direct surveys conducted in the field during or shortly after the event. The capacity to identify and map the flooded areas locally proved to be a critical aspect because of the unavailability of satellite data and the impossibility to fly over the flooded zones because of prolonged bad weather conditions. In the meantime, a few days after the flood, signs of the flood were not visible anymore, particularly in agricultural and open fields.

The maps and the data in the flooded zones used for the two reports were obtained by constraining the flood modeling to available photos, citizens' reports, and direct surveys that were conducted by the Civil Protection together with the volunteers, including the Politecnico di Milano researchers, in the affected dwellings and economic activities.

In the case of the 2012 event, different floods occurred, ranging from flash floods to riverine floods in small valleys due to the levee failures, affecting mainly the municipalities of Todi, Città della Pieve, Marsciano, and Orvieto. In the city of Orvieto, the flood was very rapid. A peak water depth of almost 3 meters was reached in the lowest zones at around 8:00 or 9:00 a.m. on 12 November, but then water receded rather fast about 3 p.m. In the peripheral zone of Città della Pieve, the inundation started early in the morning on 12 November and the 2 meters water depth lasted for three to four days.

An important result of this mapping is the evidenced lack of reliability of official hazard maps held by the Tiber River Basin Authority in the city of Orvieto. The unreliability of the maps was known to experts for a long time, however, it was only after the flood that the mistake was corrected in official maps that are the basis for land use planning and granting construction permits.

In the November 2013 event, the situation was more mixed and complex with respect to the previous year. In fact, a mix of hazards occurred, ranging from floods, landslides, and debris flows with more or less sediment content. Furthermore, while the 2012 inundations occurred over large areas in the affected municipalities, the 2013 flood was a typical "multi-site" event, with relatively minor single events, but all together affecting a rather large area along the Apeninnes and at the Eastern border with the Marche region.

11.4.3. Damage to People (Including Health Issues), Civil Protection Intervention, and Associated Costs

Coherently with the framework adopted for modeling the scenario and what was described in section 11.2, first affected people and costs associated to population assistance are presented.

The death toll and the number of the injured and the severity of injuries are clear indications of the extent and severity of the flood and hydrogeological events. Floods are generally associated with fewer victims than other types of natural hazards, typically earthquakes that are much more dreadful in this respect. Information acquired insofar, particularly in developed countries, suggests that during a flood people put their life at risk while trying to save their car and/or being trapped in the car in low zones, such as underground tunnels, low passages, or on collapsing bridges. Another deadly case is related to the occupancy of basements when the flood peak occurs.

Regarding assistance to the population when evacuation or longer term relocation is required, there is an increased number of manuals and guidelines addressing the issue of necessary material and psychological support, including the best sheltering solutions that can be provided at an affordable price, part of which is generally paid through public solidarity funds.

This type of information is collected by both the Regional Civil Protection Authority and by the municipal civil protection. A cost related to blankets, food, beds, or aid to rent a house or spend a couple of nights in hotels is provided.

11.4.3.1. Damage to People in the Umbria Case Study Area. In the two events that occurred in Umbria, fortunately there were no victims, unlike in neighboring regions (especially Tuscany) where people died in circumstances such as those described above.

Costs associated to population assistance and first intervention amount to 150,000 euros in 2012 and to 30,000 euros in 2013. In the former, 175 isolated buildings and 10 villages were evacuated, totaling about 95 families with 300 people. The muncipaplities where the larger number of people were evacuated were Marsciano and Todi. To coordinate the evacuation activities and also the shelters that were required, 15 operational centers, 14 municipal emergency control rooms, and the central emergency coordination center in Perugia were activated. In the 2013 event, 11 families were evacuated, mainly in the municipalities of Gualdo Tadino and Scheggia Pascelupo, 40 volunteering organizations were mobilized, activating more than 200 people.

11.4.4. Damage to Lifelines and Facilities

Damage to lifelines and critical facilities has to be considered as an important priority during the emergency because lifelines and facilities like hospitals, municipalities, emergency control centers, and fire stations are strategic for managing the crisis and rescuing the population. The recovery of facilities such as schools, theaters, conference centers, and libraries is fundamental for signaling the return to normal life conditions. It is also important to consider the interaction between lifelines and critical facilities. Access to the latter may be hampered by the flood or by landslides, and electricity and water outages may cause some facilities such as hospitals to be less operational than needed.

At the regional scale, disruptions that provoked general outages and lack of services ranked as having a regional or even national relevance must be identified. Clearly, those are the first to be repaired. At the local scale, a multitude of damages may be reported, making hurdles for the rescuers and delaying return to normalcy.

Different stakeholders are responsible for managing lifelines. Some are public entities (in particular for water and sewerage), others are private (telecommunication), and others may be a mix of public and private organizations (roads, energy, gas). Each entity collects data for its internal needs, be it repairing or keeping track of the effectuated intervention.

11.4.4.1. Damage to Lifelines and Public Services in the Umbria Case Study Area.
In the case of the 2012 and 2013 floods in Umbria, a variety of sources have been addressed. Reports and data were provided in paper forms or through interviews by lifelines managing companies. Requests of aid in the first hours of the emergency were collected by the regional emergency control center. Public authorities at the municipal level collected information in cases when they had to intervene directly to facilitate urgent interventions and when managing companies requested a public contribution for the sustained repair costs.

Data and information had to be pre-processed in order to obtain a comprehensive overview of the damage suffered by lifelines and critical facilities and that affected lifelines' and services' interdependency.

11.4.4.2. Damage to Lifelines and Services in the Umbria Case Study Area.
In the 2012 flood, electrical, water, and sewerage systems were affected, with a total expenditure of around 1.1 million euros as can be seen in Table 11.2. As for the electrical system, the latter was affected, particularly in Orvieto, where transformation cabins and medium voltage lines were submerged. It took almost a year to fully recover the electrical system in the area, because damaged components also had to be relocated from the floodplain zone (given also the redrawing of the hazard map as described above). In the meantime, the electrical company had to guarantee service by generators first and then by temporary repairs.

As for the water systems, depuration plants, conduits, and pumping systems capacity were overwhelmed by the peak water volumes experienced during the flood.

In the 2013 event, damage suffered by electrical and water systems totaled 2.2 million euros. An electrical cabin remained inaccessible until the large volumes of debris that covered the road were removed. A conduit of a hydroelectric power plant had to be reconstructed due to another debris flow. Water and sewerage systems were also affected but far less than in the previous year's event as can be seen by comparing Table 11.2 and Table 11.3.

The most affected lifelines in both events were certainly roads. Only in the 2012 event entrances to the main highway connecting southern to northern Italy in the western side of the peninsula in Orvieto and Città della Pieve experienced minor problems. In both events, the entire regional network that is made of local to provincial roads connecting a large number of villages, medium cities, and small cities was severely affected, particularly by debris flows and landslides. Some accessibility problems to critical facilities were reported. In the 2012 event, the two parts of the city of Orvieto were disconnected because the main bridge was closed early in the morning. Schools as well as the hospital that is located on the left bank of the Paglia River could not be reached by inhabitants on the other bank. Emergencies had to be dealt with by helicopters. In the 2013 event, the main access to the hospital of Gualdo Tadino (in the Branca locality) was obstructed by the inundating water, and the only access left was the old narrow road across the mountains at the border with the Marche region.

Damage to schools, municipal buildings, fire departments, and hospitals was reported by municipal authorities as having physical damage and associated costs.

As for public facilities, 2012 proved to be more problematic, with medium size cities and their modern industrial areas affected. In Orvieto both the fire

Table 11.2 Affected Lifelines in the 2012 Flood.

	Affected lifelines in the 2012 event	
Affected sector	Affected system	Total cost
Lifelines	Road system	€ 10.615.262,78
	Parking lots	€ 60.000,00
	Railways (regional network)	€ 1.131.650,00
	Electric power system	€ 479.594,00
	Public lighting	€ 34.620,59
	Sewerage	€ 460.530,40
	Acqueducts	€ 179.057,20
Structural defenses	Levees, dams, landslide consolidation, slope stabilization	€ 20.815.000,00
Total		€ 33.775.714,97

Table 11.3 Affected Lifelines in the 2013 Flood.

	Affected lifelines in the 2013 event	
Affected sector	Affected system	Total cost
Lifelines	Road system	€ 17.540.591,40
	Electric power system	€ 230.000,00
	Sewerage	€ 250.000,00
	Aqueducts	€ 1.576.980,00
Structural defenses	Levees, dams, river banks landslide consolidation, slope stabilization	€ 29.044.380,76
Total		€ 48.641.952,16

stations and local police stations were flooded, with significant damage to the trucks and cars, and the same buildings hosting respectively firemen, policemen, and their offices. In 2013, public facilities such as schools, a historic theater, and a municipal building suffered damage. Such damage, however, was not due to the flood but to the very intense precipitation that affected old, not sufficiently maintained roofs in the municipality of Città di Castello.

11.4.5. Damage to Economic Activities: Agriculture, Industrial, and Commercial

Damage to economic activities is not easy to reconstruct, as it is clearly not limited to the structural failure of a building but implies losses and disruption of several assets and stocks that are necessary for the firm production or logistics. Damage curves that have been developed insofar for the industrial and agricultural sectors significantly underestimate the total damage due to a multiplicity of failures that may occur in different components of a production and commercial chain. Furthermore, they also neglect the very large variability that is intrinsic in economic activities. In our own experience and considering that Umbria is not among the richest Italian regions nor the most diverse in terms of

manufacturing, we surveyed a rather wide range of different damages and losses, depending on the intrinsic characteristics of affected activities.

As for agriculture, *Brémond* [2013] published a useful review of the existing literature regarding damage to the agricultural sector, showing that under the label "damage to the agricultural sector" different types of affected items have to be considered. Those include lost crops and yield reduction, damage to the perennial plant material (like vineyard or orchard), damage to the soil structure (due for example to salinization or contamination), damage to machinery, stocks of products, and buildings used for agricultural activities. Each of those items actually requires a set of functions, because different indicators of vulnerability need to be considered.

In the following section, we will address damage to the agricultural, industrial, and commercial sectors, respectively.

11.4.5.1. Damage to Agriculture in the Umbria Case Study Area. As a consequence of the EU legislation, in Italy the agricultural sector is treated in a different way with respect to all the other sectors in that a declaration of state of exceptional meteorological events must be issued by the Ministry of Agriculture. Such declaration covers under the same label phenomena that are rather diverse in severity and geographic coverage, such as floods, extreme winds, hail, or storms. Furthermore, this declaration only loosely relates to the declaration of the state of emergency decided by the Presidency of Ministries Council under the technical reporting prepared by the National Civil Protection Department [*Mysiak et al.*, 2013] for a specific event. To be clear, the same notion and definition of event diverge between the two ministries, so it is not always easy to find the exact correspondence between the damage related to the specific emergency under study and damage suffered by the agricultural sector in a given period of time. Furthermore, also the format of damage reporting seems to be less structured than that required nowadays by the National Civil Protection Department. In the case of the 2012 flood, the only available information refers to the total declared loss amount of 12 million euros and the list of affected properties identified by their cadastral code within the most affected municipalities (Figure 11.4). For the 2013 event that provoked a much less severe damage of 3.8 million euros, more detailed data could be obtained through declarations made to municipal authorities. Structures, internal fields infrastructures, crops, stocked materials, and machinery were all damaged either by flooding water or by landslides. In one case, an entire property was covered by the debris carried by the mountain flood; in another case, fields were contaminated by sewerage waters leaking from damaged pipes.

11.4.5.2. Damage to Industries in the Umbria Case Study Area.

As already mentioned, damage to the industrial sector diverge significantly from one place to another due to the intrinsic characteristics of the flooded firms and activities.

Two sources of information were provided: from self reporting and direct surveys, using the forms appositely developed and described in Chapter 6. In the latter case, the description of the physical damage permitted to detail the type of affected assets, distinguishing between stocked material, finished products, and machinery. Apart from the time needed for cleaning up, the duration of business interruption was recorded as well as information regarding unemployment subsidies for temporarily or permanently jobless people. In the 2012 event, data regarding costs sustained by firms because of the flood were managed by the Regional Department for Economic Development. This department carried out a very attentive evaluation of reported damage, assessed eligibility for refunding to allocate the amount that was granted by the central government to partially compensate damage sustained by the private sector. Within the latter, a high priority was granted to economic activities with respect to damage to residential buildings that were compensated for later and to a lesser amount.

Given the two data sources that we could rely on in the Umbria 2012 and 2013 event, we represented damage to industries at different scales and with different levels of detail.

Figure 11.5 shows the municipalities reporting damage to industrial facilities in the 2012 flood. The city of Orvieto was clearly the most affected; the entire industrial area was flooded and as many as 127 economic activities were affected.

Given the detail that was required to obtain compensation, it is also possible to assess the relative contribution of structures, internal equipment and machinery, and stock of input material and finished products, that was actually equal in the 2012 event.

Direct surveys provided us with a better perception of what the damage actually entailed and of the real possibility for firms to repair and recover machinery in each visited site. In total we visited 14 firms after the 2012 event, 11 in Orvieto, and 3 in Città della Pieve. During surveys additional and more detailed information regarding the vulnerability of firms and the different types of losses (direct, second order) they suffered was collected. We learned that firms did not have any plan in case of emergency and did not expect a major flood, even though they had experienced measurable centimeters of water in the most rainy years in the past. Ironically, most owners reported that they were keeping old and less important document as well as material they used only very rarely in the highest parts of the building.

None had insurance coverage, and none was able to put in place any mitigation measure because of the timing of the flood (early on a Monday morning) and lack of precedent alert.

An important component of damage to the industrial sector that has to be carefully considered is the large quantity of waste and especially polluting and contaminating waste that can be "produced" by a flood. In fact, even in areas where no hazardous installation was present (according to the definition provided by the Seveso Directive in Europe), the municipality of Orvieto spent as much as 900,000 euros for safe disposal of such waste.

In the 2012 event, the higher order damage due to the duration of the business interruption also was investigated. A year after the event, the same firms were visited again, and they all reported to have restarted full activity four months after the disaster. Some reported that they had to ask for subsidies for their temporarily unemployed workers, and some others said that they had to reduce the number of workers. Firms able to restart had to anticipate payment for damaged machinery and electrical equipment hoping for a subsequent at least partial refunding granted by the state.

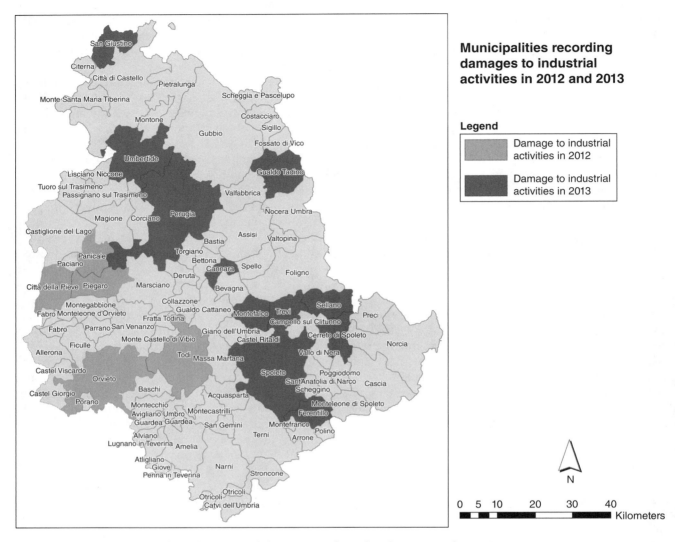

Figure 11.5 Municipalities that reported damage to industrial and commercial activities.

In the 2013 event, mainly commercial activities were affected with few exceptions. One exception is interesting to mention because it highlights the very large variability of activities and the consequent difficulty in summarizing everything in a damage function where only two variables are related (water depth and losses expressed in costs). An area where mineral water sources are located of a renowned firm was very severely affected by a debris flow that also covered the access path to the five individual water sources. One of the conduits from the water source to the bottling plant had to be closed for precautionary reasons. While the bottling plant, located in another part of the city of Gualdo Tadino, suffered no damage, the area where the sources are located is still covered by debris and the mitigation works for securing both the access to the sources and the entire area amount to around 2 million euros. A controversy arose between the

municipality and the firm regarding who should pay for it, as the area is public subject to concession to the firm that pays a fee for its exploitation.

11.4.5.3. Damage to the Commercial Sector in the Umbria Case Study Area. Under this label, we may include retail and professional services ranging from private medical offices, to lawyers and other types of experts and consultants. Damage to this type of facility is often smaller in the case of small bureaus and surfaces, totaling an average damage of 20,000 euros each, due to the loss of computers, furniture, and electrical equipment. An important damage component is related to stock in the case of shops and retailers and to documents in the case of private offices. In the two floods in the Umbria region, damage to the commercial sector totaled 3.5 million euros in the 2012 flood and almost 700,000 euros in the

2013 flood. A key variable to keep in mind is the relative extent of the surface. In the case of a large supermarket that is damaged, losses can easily reach up to 200,000 euros given the large amount of perishable stocked goods.

Table 11.4 shows the costs that were considered eligible for refunding for the commercial sector in 2012, and Table 11.5 represents the self-declared costs in the flood of 2013.

11.4.6. Damage to Residential Dwellings

Damage to residential buildings is perhaps the most investigated in literature, because most damge functions have been developed for this sector. In our methodology, we propose to use survey forms to collect information regarding not only the damage that has been suffered by structures, equipment and furniture but also vulnerability factors, such as material, typology, number of floors, and presence of a basement as discussed in Chapter 6.

11.4.6.1. Damage to the Residential Sector in the Umbria Case Study Area. In both events, two sources of information were available: self reporting by citizens made to the municipal authority, and direct surveys conducted by volunteers and ourselves together with officials of the regional civil protection using forms that are described in Chapter 6. In the first set of data, only the monetary cost to repair windows, plasters, doors, equip-

ment, and furniture is provided; in the second, a much more detailed description of both the vulnerability of buildings and dwellings and the description of the physical damage is reported. Furthermore, some information regarding the time and resources needed for cleaning up was collected as well.

In the 2012 event, damage to dwellings reported to the regional authority received compensation totaling 2.9 million euros, and for the 2013 event, over 6 million euros were requested. It must be highlighted that the highest damage reported in the 2013 event is due partly to structural problems suffered by roofs because of rain infiltration (that cannot be related to the flood) and partly to a number of landslides that provoked severe structural damage including partial, and in one case, total building collapse.

11.4.7. Damage to Cultural Heritage (Connection with Residences, Public Facilities), Open Recreational Spaces, Environmental Systems

Damage to the cultural heritage is generally reported by municipalities as physical damage and associated costs of repair. Damage to the environment is reported either by municipalities or in the case of parks and preservation areas, by agencies and organizations that manage them. Damage to the environment is reported as a physical damage description and associated costs.

Table 11.4 Commercial Activities Eligible Losses by Sub-Sector, 2012 Flood.

Type of activity	Municipality	Total declared damage
hotel and restaurants	Orvieto	€ 705.628,02
gas pumps, garages, and car dealers	most in Orvieto, one in Città della Pieve, and one in Mars ciano	€ 937.247,00
engineering, computer, and graphic design offices	Orvieto and Marsciano	€ 474.709,40
food, clothing, furniture, and other shops	most in Orvieto, three in Mars ciano, one in Città della Pieve	€ 1.187.670,10
artisans	Città della Pieve	€ 148.932,05
Total		**€ 3.454.186,57**

Table 11.5 Commercial Activities, Self-Declared Losses by Sub-Sector, 2013 Flood.

Type of activity	Municipality	Total declared damage
sports center	Cannara	€ 230.000,00
hotel and restaurant	Cannara and Sellano	€ 158.879,49
large construction material retail	Cerreto di Spoleto	€ 17.000,00
denstist, medical clinic lawyer		130527,84
iron crafting shop	Umbertide	€ 13.500,00
hardware, flowers, bicycles, and other shops	Gualdo Tadino	€ 71.550,00
photos and graphic editiing	San Giustino	€ 40.700,00
Total		€ 662.157,33

Under the label cultural heritage, different types of artifacts can be found as it may refer to archeological sites, entire historic centers, or individual monuments. The type of building can vary significantly from churches to theaters, to villas, palaces, etc. In fact, the cultural heritage label means that the asset is listed in the national or international ranking of sites to be preserved. Private residential and public facilities such as municipal buildings, for example, may well be listed as cultural heritage. The reason for grouping together cultural heritage and environmental systems stems from the intrinsic difficulty to find an adequate monetary value that will represent the loss in case of partial or total disruption. This has to be taken into consideration when a complete event scenario is developed. Damage to cultural heritage means often damage to the identity of the settled community, the loss of paintings, original plasters, and parts of the structure that make the asset so valuable and attractive for both inhabitants and tourists.

The costs that are reported by local governments related to cultural heritage (and also to environmental assets) mainly refer to the repair and the preservation of what is left to avoid further deterioration but do not embed also the evaluation of intangible components of the loss itself.

11.4.7.1. Damage to Cultural and Environmental Assets in the Umbria Case Study Area.
In the case of the reported events in Umbria, fortunately damage to cultural heritage has been limited and mainly due to the intense precipitation that affected roofs and infiltrated into the lower floors. Mostly urban walls were affected by the 2012 flood, as well as a theater and historic palaces where the municipal offices and a school are respectively located were affected in Città di Castello in the 2013 event.

As for the damage to the environment, pollution and contamination can be distinguished from damage to, for example, parks and preservation areas. In the latter case, it is very interesting to understand to what extent a flood that is a natural event, that is usually considered to be a hazard only if urbanized areas are exposed, may actually disrupt environmental assets. A reference to the cases that were witnessed in the 2012 event may be useful in this regard. Apart from the contamination industrial waste provoked in the Paglia River, the case of the preservation oasis of Alviano along the Tiber River can be mentioned because it reported a total damage of about 150,000 euros. The flooding water destroyed the light structures that are used when visiting the park and observing the species. An important component of the damage was due to the waste content in the water. Large garbage (such as electrical equipment, cars, sediments) that was carried by the currents was deposited in several parts of the park and had to be removed manually and taken to disposal

sites. Polluted water provoked the death of fishes and microorganisms that reside in the river and constitute the food of some rare species of birds that previously found the oasis to be an exceptional habitat. The oasis manager suggested that two to three years are needed to understand if the new ecological equilibrium reached by vegetation and the new species that reconquered the waters constitutes just a substitution or a loss of biodiversity. Similarly, it was still too early at the time when we developed the scenario to understand if the rare bird species will come back or not.

11.4.8. A Synthetic Overview of the Scenario

At the end of this detailed analysis of the damage scenario sector by sector at different spatial scales and also considering how damage may unfold along the time scale, a more synthetic and systemic overview of the damage suffered by systems also as a consequence of their interconnection is provided.

At the regional level, one gets to know what the total numbers and facts are related to damage due to the specific event at stake or to the chain of interconnected phenomena. At this stage, it is possible to understand what the relative weight of each sector is on the total amount of damage (see Figure 11.6).

This is very important for a number of reasons. First, because such analysis allows characterization of each event with respect to its specific characteristics in terms of type of phenomena, geomorphological features of the affected areas, nature, extent, and value of exposed assets and their vulnerability. Secondly, this type of analysis is relevant in order to identify a typology of different combinations of hazard, exposure, and vulnerability components that determine different patterns of damage and damage distribution across sectors. For example, in the 2012 event, the industrial sector was among the most affected, because industrial areas were flooded, especially in the city of Orvieto. Damage to the industrial sector was far less relevant in the 2013 case, also due to the fact that the event occurred in mountain areas that are generally less developed. Instead, a much larger rate of damage is sustained by the structural risk mitigation measures and by lifelines, particularly roads.

Another resulting synthetic information relates to the spatial and time distribution of the damage at a regional level, identifying, for example, what the municipalities that were most affected are, ordering also by type of main and secondary events that had produced that level of damage. This information is particularly relevant if compared to the mitigation measures that existed before the event occurrence. It can be understood for example, if land use and emergency planning were consistently taking into account the known level of hazards, if hazards

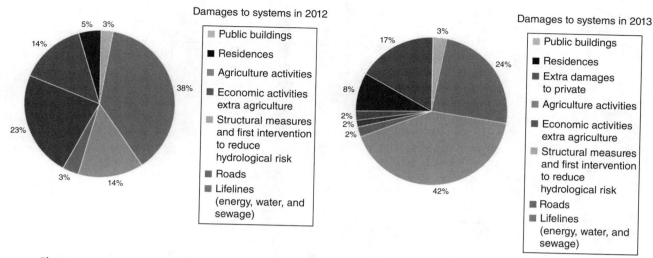

Figure 11.6 Distribution of damage across sectors in the 2012 and 2013 floods.

had been assessed correctly, if structural and non-structural mitigation measures were effective or not and in the latter case why they failed given the scenario that occurred.

11.5. DISCUSSING THE UTILITY OF THIS WORK

We believe the idea and methodology that are proposed here to develop post-event damage scenario assessment is useful for three key purposes.

First, it provides a sounder basis on which damage and loss data are not only collected but also organized and framed in a way to provide a clear and, as much as possible, comprehensive overview of the event and its impact in the affected area. Such information is valuable for large scale data sets, such as those at national and international levels, required by different legislations (national and European, for example) but also by international agreements. (The need to better assess post-disaster data is put forward by the Sendai Framework for Action in a rather strong way.) The point in the development of complete event scenarios is to get a clear picture of what had happened and of the most relevant damage in the context of the affected region. Given this type of comprehensive understanding, it is possible to elicit data to feed the indicators that the various legislations and agreements require. Data may be either representative of the most relevant loss or be derived from a careful aggregation of finer and more detailed data recorded at local and regional scales. Presently, there is not such care in selecting the data that will enter into the database so that at the end no meaningful comparison can be made between countries or among events occurring in the same country at different times.

In fact, the comprehensive scenario allows identification of data that are of general relevance from data that

have limited capacity to represent the overall magnitude and significance of an event and the losses it provoked. Somehow, improving global databases starts from the ground level, that is, from the way damage data are collected in the field. By systematizing an improved collection methodology, it will be possible to compare events across regions and countries to provide more reliable guidance for national and international policies.

Second, a complete post-event scenario allows a so-called "forensic investigation" by presenting all factors that have contributed to the disaster in terms of hazard, exposure, vulnerability, response capacity, etc. [*Oliver-Smith et al.*, 2016; *De Groeve et al.*, 2013]. Such an explanation of the event should be considered when assessing the priorities and the modalities of recovery and reconstruction to achieve a more resilient condition for the future. Currently losses are assessed incrementally according to the funding that becomes available through national and international grants, but there is little if any strategy that drives recovery and reconstruction toward more resilient outcomes. Locally mitigation measures are addressed, but with little consideration regarding what type of measure would be preferable (structural, non-structural, a mix of the latter), and without an overall assessment that connects the local to higher scales of analysis and decision. In the Umbria experience, conducting this comprehensive damage assessment has provided the public administration with relevant arguments to prioritize among interventions and to initiate an assessment process of the effectiveness of both structural and non-structural mitigation measures that had been put in place in the last decades.

Third, a certainly useful exercise is to compare the post-event damage scenario, based on data collected in

Table 11.6 Comparison Between the Pre-Event Damage Assessment in the Three Most Affected.

| Municipality | Pre-event damage assessment | | Post-event damage estimation | | |
Sector	Industrial and commercial	Residential	Industrial and commercial: declared	Industrial and commercial: eligible	Residential
Cittá della Pieve	€ 6.152.775,00	€ 958.373,00	not available	€ 258.384,75	
Orvieto, using available Corine Land Cover	€ 3.106.488,00	-	-	€ 4.335.576.95	
Orvieto corrected Corine Land Cover	€ 28.691.776,00	€ 1.799.389,00	€ 28.064.922,85		
Marsciano	€ 2.484.938,00	€ 884.408,00	€ 1.286.012,77	€ 368.893,97	
Total	€ 37.329.489,00 considering industrial land use for Orvieto	€ 3.642.170,00	€ 29.350.935,62	€ 4.962.855,67	€ 2.900.000,00

the field after a real event, with the pre-event damage scenario that could have been obtained based on hazard, exposure, and vulnerability assessments that were available before the event. This has been done for the 2012 flood, and some results were achieved. As already mentioned, the reliability of the official hazard map was questioned based on the assessment and the river basin authority was forced to change them. Second, for the land use map that was used, the Corine land cover provided by the EU Commission, proved to be wrong by attributing the industrial area of Orvieto to mainly "residential" use. Correcting the input data on the hazard and the exposure map and applying the damage functions that are available in literature, the results were not so far from the real occurrence of the 2012 flood (see Table 11.6).

As discussed in literature [*Oreskes et al.*, 1994; *Oreskes*, 2000], scenarios cannot be verified or falsified, so results that are matching the observed damage cannot corroborate in our case the damage function model. However, the discrepancy by order of magnitude between the forecasted and the verified certainly highlights some problems. The post-event damage scenario cannot say the definitive word on the correctness of risk assessment, but it may significantly improve the evidence base on which such assessments are carried out. It can also be suggested that such evidence base relevance is not restricted to the area where the event has actually occurred, but it can be extended to similar cases in other regions or countries. The example of the Umbria floods may find correspondence in areas that are similar as far as geographic and population distribution are concerned, such as alpine areas, including countries such as Switzerland or Austria and more generally in mountain regions, characterized by relatively large exploitation and urbanization of valleys and problematic accessibility to the more remote settlements and villages.

11.6. CHALLENGES

It is important to also mention what the current challenges are with the methodology we face and that need to be addressed by future research and practice.

The main challenge is to provide a manageable overview and interpretation of the complexity implied by the need to describe how a territory has been affected by a disaster. As we have learned by practice, such complexity is similar in efforts to forecast future scenarios and to reconstruct post-event damage scenarios. The types of "ignorance" that *Rescher* [2009] mentions in his insightful book regarding the unavailability of the "future" and the "undetectable past" are both hindering the efforts of depicting the "full picture" of the complex situation expected or experienced.

It is important to note that the applications that are presented refer to relatively small areas regarding the strengths and the weaknesses of the work that has been carried out in the Umbria region. As for the strengths, a small case gave us the opportunity to develop and fine tune the methodology while testing it in an area where the complexity mentioned above is mangeable. The weakness is intrinsic in the overall extent of surveys that have been conducted and in the lack of striking evidences of failures due to the fact that no critical infrastructure of national relevance is located in Umbria, a region that is somehow isolated from the heaviest traffic and the main lifelines systems. Yet, regarding the scale of the Umbria region, the events clearly had significant social and economic impact.

We are aware that for very large events, some adaptations would be required, starting from the need to carry out sampling assessment on the sectors that have been more widely affected, be it residential houses or industries. The PDNA methodology proposes to overcome the problem by overlapping a spatial grid on the affected country (as in the example, carried out in Myanmar after the Nargis cyclone in 2008) and conducting selected surveys within each mesh area of the grid. Other methods may be considered, for example, deforming a rigid and regular grid to adapt to the administrative borders and to areas of differential concentration of population and assets. However, this has still to be tested and therefore is not fully covered at this stage by the proposed methodology.

Second, one has to ask what the minimal event magnitude is beyond which such an effort of data collection and analysis is not justified by the required costs in terms of time, resources, and personnel dedicated to this work. Even though a number of smaller events all together may provoke large expenses for a region or a country, the idea of reconstructing a post-event scenario is somehow fading away. This is because in smaller events the effects on multiple sectors and the ripple effects due to higher order damage tend to range from much less visible to non-existent.

A third order of considerations relates to the treatment of data necessary to conduct the complete event scenario. To develop the latter as described in section 11.3, data need to be reorganized and analyzed according to the main criteria and parameters that are useful to assess where damage occurred, how damage was distributed at a regional and local scale across time, and what the main and secondary causes of damage are. This type of analysis requires an information technology system that will allow multi-modality queries and uses with the data acquired in the same survey campaign or collected from various agencies. As discussed in Chapter 15, the design of such a complex information system is on its way but still requires further development and testing. The final aim is to shift the time and work burden from data input and organization to the analytical and assessment phases that require a much more careful intervention of the scenario analyst to interpret and explain the data and the results of the data analysis given in maps, graphs, and tables. Furthermore, such effort implies a larger collaboration among the different stakeholders who need to be facilitated in the phase of data collection and better engaged in the analytical phase, when their specific expertise is required to develop the full damage scenario reporting.

The person or the unit that is developing and writing the post-event scenario needs to be the one that acts as data coordinator, as correctly put by the *EU Technical Group* [2015]. As mentioned in the latter report, the data coordinator "must ensure the application of a coherent methodology and foster the sharing of good practices. [...] The coordination body is a person or a group of persons in an administration who is responsible for collecting and assembling data coming not only from different sources (field survey, satellite data, other data sets) but also for proactively asking them from other administrations." In our view, the data coordinator should coincide with the "data curator, responsible for processing the collected data." This is necessary because only the unit or person who is actually developing the scenario knows what the required data are and can identify informational gaps.

The unwillingness of agencies and especially of private partners such as lifelines managing companies to participate in the data provision effort cannot be neglected. Experiences that have been already conducted in recent times to also federate private partners into the large effort of getting data for risk assessment and management purposes must be permitted to overcome traditional barriers. We may refer here to the significant example of the ONRN platform (http://www.onrn.fr/) and to the data and maps sharing platform developed for Italia Sicura (SafeItaly, see http://italiasicura.governo.it/site/home/dissesto.html), the new structure under the prime minister for structural defenses enforcement at the national scale, and the platform that was created for managing the safety and security plan for the EXPO exhibition in Milan in 2015 (see http://www.e015.regione.lombardia.it/PE015/).

Among the private stakeholders that may be both interested in the results of the scenario development and act as data providers, insurers certainly hold a primary position. In Italy though, the insurance coverage against natural calamities has still a very limited penetration, plus it is not available for households. While the situation remains as it is, it will be difficult to enroll insurance companies in this data collection and interpretational effort. The already mentioned ONRN platform is a remarkable example also in this respect, because it provides tangible evidence of the fact that insurance data can be used for knowledge management purposes thereby avoiding publishing sensitive data.

11.7. CONCLUSION

In this chapter, we have first defined what we mean by "complete damage scenario" of past events, then a methodology for developing the latter has been proposed and detailed in its various components and steps. An application of the methodology has been described for the Umbria region floods that occurred in two consequent years, given the tight collaboration existing between the Regional Civil Protection Authority and the researchers of the Politecnico di Milano [*Ballio et al.*, 2015].

The results showed that significant differences exist between two events, even though they have occurred in the same region and in similar geographical contexts. This reinforced the idea that the effort required to draw a complete post-flood damage scenario is valuable in that it highlights the different combination of hazard, exposure, and vulnerability factors and specific characteristics that emerge of those factors across space and time. In fact, it is suggested that if such an effort is extended to other cases, a typology of different combinations of risk factors may be identified and constitute a basis for a sort of standardized abacus of contexts and conditions that may lead to certain damage patterns. In the meantime, such comprehensive quantitative and qualitative reporting may permit a much better elicitation of key indicators and values to be aggregated to become part of the enhanced losses databases at the national and international levels.

Finally, some considerations are addressed regarding the limits of the current state of the methodology and future research and practical application needs.

REFERENCES

Ballio, F., D. Molinari, G. Minucci, M. Mazuran, C. Arias Munoz, S. Menoni, F. Atun, D. Ardagna, N. Berni, and C. Pandolfo (2015), The RISPOSTA procedure for the collection, storage and analysis of high quality, consistent and reliable damage data in the aftermath of floods. J Flood Risk Management, doi:10.1111/jfr3.12216.

Cellerino, R. (2004), L'Italia delle alluvioni, Franco Angeli, Milano.

De Groeve, T., K. Poljansek, and D. Ehrlich (2013), Recording Disasters Losses: Recommendation for a European Approach, JRC Scientific and Policy Report, available at: http://publications.jrc.ec.europa.eu/repository/bitstream/111111111/29296/1/lbna26111enn.pdf

De Groeve, T., K. Poljansek, D. Ehrlich, and C. Corbane (2014), Disaster loss data recording in EU member states. A comprehensive overview of current practice in EU Member States, JRC Scientific and Policy Report.

Direction Territoriale Méditerranée du Cerema (2014), Retour d'expérience sur les inondations du département du Var les 18 et 19 janvier 2014 Volet 2 - «Conséquences et examen des dommages». [Accessed: 14 May 2015, available at: http://observatoire-regional-risques-paca.fr/sites/default/files/rapport_rex83_2014_dommages_sept14_0.pdf].

Euopean Union expert working group on disaster damage and loss data (2015), Guidance for recording and sharing disaster damage and loss data. Towards the development of operational indicators to translate the Sendai Framework into action, JRC Scientific and Policy Report.

EQE International (2002), The January 17, 1995 Kobe Earthquake [online], available at: http://www.eqe.com/publications/kobe/kobe.htm.Galderisi, A., and A. Ceudech (2005), Il terremoto del 23 Novembre 1980 a Napoli: la ricostruzione del danno funzionale nella prima settimana post-evento. In S. Lagomarsino, P. Ugolino (eds.), Riscio sismico, territorio e centri storici, Franco Angeli, Milano.

Galderisi, A., and A. Ceudech (2010), The "seismic behavior" of urban complex system, in S. Menoni (ed.) Risks Challenging Publics, Scientists and Governments, Taylor and Francis Group, London.

Geohazard International (1994), The Quito Earthquake Risk Management Project: an overview.

Holguin-Veras, J., E. Taniguchi, F. Ferreira, M. Jaller, and R. G. Thompson (2012), The Tohoku Disasters: Preliminary Findings Concerning the Post Disaster Humanitarian Logistic Response, Annual meeting of the Transportation Research Board (TRB), Washington, D.C. Available at: http://docs.trb.org/prp/12-1162.pdf.

Hoyt, W. G., and W. B. Langbein (1955), Floods. Princeton, NJ: Princeton University Press.

Hulse, D. W., A. Branscomb, and S. G. Payne (2004), Envisioning alternatives: using citizen guidance to map future land and water use, Ecol. Appl. 14 (2), 325–341.

Intergovernmental Panel on Climate Change (2008), [Accessed: 15 January 2016, Available at. http://www.ipcc-data.org/ddc_definitions.html]

Istanbul Metropolitan Municipality (IMM) and Japan International Cooperation Agency (JICA) (2002), The study on a Disaster Prevention/Mitigation Basic Plan in Istanbul Including Seismic Micronization in the Republic of Turkey. Final Report. Japan International Cooperation Agency, December.

Kahn, H. (1960), On Thermonuclear War. Princeton University Press. ISBN 0-313-20060-2.

Kropp, J., G. Walker, S. Menoni, M. Kallache, H. Deeming, A. De Roo, F. Atun, and S. Kundak (2011), Risk Futures in Europe, in Menoni, S., and C. Margottini (Eds.), Inside Risk: A Strategy for Sustainable Risk Mitigation. Springer, Milan, London.

Liu, Y., M. Mahmoud, H. Hartmann, S. Stewart, and T. Wagener (2008), Formal Scenario Development for Environmental Impact Assessment Studies. U.S. Environmental Protection Agency Papers.University of Nebraska.

McCarthy, J. J., O. F. Canziani, N. A. Leary, D. J. Dokken, and K. S. White (2001), Climate Change 2001: Impacts, Adaptation, and Vulnerability. Contribution of Working Group II to the third assessment report of the Intergovernmental Panel on Climate Change, Cambridge University Press, Cambridge.

Menoni, S., F. Pergalani, M. P. Boni, and V. Petrini (2007), Lifeline Earthquake Vulnerability Assessment, in I. Linkov et al. (Eds.), Managing Critical Infrastructure Risks, Springer.

Menoni, S., and C. Margottini, Eds. (2011), Inside Risk: A Strategy for Sustainable Risk Mitigation. Springer, Milan, London.

Ministère de l'Écologie et du Développement Durable (2005), Réduire la vulnérabilité des réseaux urbains aux inondations. Direction de la Prévention des pollutions et des risques - Sous-direction de la Prévention des risques majeurs, Paris. [Accessed: 14 May 2015, available at: http://www.ecologie.gouv.fr - http://www.prim.net].

Mori, N., T. Takahashi, T. Yasuda, and H. Yanagisawa (2011), Survey of 2011 Tohoku Earthquake Tsunami Inundation and Run-up, Geophysical Research Letters (38) L00G14.

Mysiak J., F. Testella, M. Bonaiuto, G. Chorus, S. De Dominicis, U. Ganucci Cancellieri, K. Firus, and P. Grifoni (2013), Flood risk management in Italy: challenges and opportunities for the implementation of the EU Floods Directive (2007/60/EC), Nat. Hazards Earth Syst. Sci. 13: 2883–2890.

National Statistical Office (ISTAT) Census (2011), [Accessed: 25 July 2016, Available at: http://www.istat.it/it/censimento-popolazione/censimento-popolazione-2011].

NEDIES Project, A. Colombo, and A. L. Vetere Arellano (2002), Lessons learnt from flood disasters, European Commission DG Joint Research Centre.

Oliver-Smith A., I. Alcántara-Ayala, I. Burton, and A. M. Lavell (2016), Forensic Investigations of Disasters (FORIN): a conceptual framework and guide to research (IRDR FORIN Publication No. 2). Beijing: Integrated Research on Disaster Risk. 56 pp.

Oreskes N., K. Shrader-Frechette, and K. Belitz (1994), Verification, validation, and confirmation of numerical models in the earth sciences, in Science, 4 February, 263.

Oreskes, N. (2000), Why predict? Historical perspectives on prediction in earth sciences, in D. Sarewitz, R. A. Pielke Jr., and R. Byerly (Eds.), Prediction. Science, decision making and the future of nature, Island press, Washington D.C.

Pielke, R.A. (2000), Flood impacts on society. Damaging floods as a framework for assessment, in D. Parker (Ed), Floods, vol. 1. Routledge.

Pitt, M. (2008), The Pitt review: learning lessons from the 2007 floods [online]. [Accessed: 05 May 2015, available at: http://archive.cabinetoffice.gov.uk/pittreview/thepittreview/final_report.html].

Rescher, N. (1991), Baffling phenomena and other studies in the philosophy of knowledge and valuation, Rowman and Littlefield Pub.

Rescher, N. (2009), Ignorance (On the wider miplications of Deficient Knowledge), University of Pittsburg Press.

Ringland, G. (1999), Scenario planning, Managing for the future, John Wiley & Sons Ltd., 1999.

Rose, A. (2004), Economic Principles, Issues, and Research Priorities of Natural Hazard Loss Estimation, in Y. Okuyama and S. Chang (Eds.), Modeling of Spatial Economic Impacts of Natural Hazards, Heidelberg, Springer.

Shell Scenarios (2013), 40 Years of Shell Scenarios available at: http://s05.static-shell.com/content/dam/shell-new/local/corporate/corporate/downloads/pdf/shell-scenarios-40years book080213.pdf.

Staley, D. (2007), History and future: Using historical thinking to imagine the future, Lexington Books, Lanham, MD.

Staley, D. (2009), Imagining possible futures with a scenario space. Parsons Journal for Information Mapping, 1(4), 1–8.

US Geological Survey (USGS) (1996), USGS Response to an Urban Earthquake. Northridge'94, Open-file report 96-263, Denver.

Van der Veen A., and C. Logtmeijer (2010), Economic hotspots: visualizing vulnerability to flooding, in "Natural Hazards," vol. 36.

Wilkinson, A., and R. Kupers (2013), Managing Uncertainty, living in the future, Harvard Business Review, May.

World Meteorological Organization, Global Water Partnership (2007, revised 2013), WMO methodology Integrated Flood Management Tools Series. Conducting Flood Loss Assessments. Associated program on Flood Management. Issue 2. Available at: http://www.apfm.info/publications/tools/APFM_Tool_02.pdf.

Wergerdt J., and S. S. Mark (2010), Post-Nargis Needs assessment and monitoring. ASEAN's Pioneering Response, Final report, Asean Secretariat.

12

Rebuild and Improve Queensland: Continuous Improvement After the 2010–2011 Floods in Australia

Brendan Moon

ABSTRACT

An unprecedented season of natural disasters in Queensland, Australia led to the establishment of the Queensland Reconstruction Authority (QRA) in 2011.

The QRA administers a reconstruction program currently valued at more than $13 billion, funded by the Australian and Queensland governments.

Since its establishment QRA has implemented a range of systems and processes to manage and monitor reconstruction activities. The reconstruction process in Queensland is now more efficient in time and cost, and supports communities to recover as quickly as possible in the aftermath of a disaster.

With a focus on continuous improvement and maximising spatial technology, Queensland is developing world best-practice techniques in the areas of disaster management, recovery, and minimizing the impact of future events.

12.1. INTRODUCTION

Australia is exposed to a broad range of natural disasters including storms, cyclones, floods, bushfires, and earthquakes. From 1967 to 2012, Australia experienced, on average, at least four major natural disasters per year where insurance losses exceeded $10 million [*Deloitte Access Economics*, 2013]. The State of Queensland (the State), the most northeasterly state of Australia, has experienced the highest natural disaster costs in recent years as a result of cyclones and flooding rains. Between 2011 and 2015, there were more than 25 such events, with 2011 in particular seeing devastating flooding across much of the State.

Following flooding and cyclones in late 2010 and early 2011, the Queensland Reconstruction Authority (QRA) was established under the Queensland Reconstruction Act 2011. The QRA manages and coordinates the Queensland

Government's program of infrastructure renewal and recovery, helping communities effectively and efficiently recover from the impact of natural disasters. Initially established for a two-year period, the QRA's tenure was extended several times and in June 2015 became a permanent part of the Queensland government. The QRA's role is to work with local government authorities and State departments and agencies to deliver value for money, best practice expenditure, and acquittal of public reconstruction funds while reconnecting, rebuilding, and improving Queensland communities following natural disasters.

The QRA administers the State's Natural Disaster Relief and Recovery Arrangements (NDRRA) program, currently valued at more than $13 billion, funded by the Australian and Queensland governments. The Queensland NDRRA program is the most significant disaster reconstruction program in Australia's history.

This chapter will explore how the establishment of the QRA enabled the State of Queensland to implement

Queensland Reconstruction Authority, Brisbane, Australia

Flood Damage Survey and Assessment: New Insights from Research and Practice, Geophysical Monograph 228,
First Edition. Edited by Daniela Molinari, Scira Menoni, and Francesco Ballio.

173

robust systems and processes to manage, monitor, and ensure communities recovered as quickly as possible in the aftermath of a disaster while delivering optimal value-for-money reconstruction activities. It will discuss and illustrate how these activities have helped to guide and make the reconstruction process in Queensland more efficient in both time and cost, through a range of systems and processes, with a heavy reliance on the use of spatial technology. Further, it will demonstrate how Queensland is developing and implementing world best practice techniques in the areas of disaster management, recovery, and minimizing the impact of future events.

January 2011 Brisbane

January 2011 Mud Army Brisbane

February 2011 Cardwell

February 2011 Tully

January 2013 Bundaberg

January 2013 Gladstone

12.2. AUSTRALIA AND QUEENSLAND NATURAL DISASTER SITUATIONS (2011-2015)

Natural disasters are an inherent part of the Australian landscape, with floods, bushfires, and tropical cyclones occurring regularly. It is estimated they cause, on average, more than AUS$1.14 billion in damage annually to homes, businesses, and infrastructure, and cause serious disruption to communities. Recent natural disasters across Australia have had a devastating and ongoing effect on Australia's economic and social well-being.

According to the *Australian Government* [2014], "Since 2009, natural disasters have claimed more than 200 lives and directly affected hundreds of thousands of Australians. Deloitte Access Economics estimates that the total economic cost of disasters for 2012 alone exceeded $6 billion. It predicts total economic costs will double by 2030, and will 'rise to an average of $23 billion per year by 2050', even without any consideration of the potential impact of climate change."

It is important to highlight that insured losses represent only a portion of the total economic costs of natural disasters. Total economic costs (incorporating broader social losses related to uninsured property and infrastructure, emergency response, and tangible costs such as death, injury, relocation, and stress) are far greater.

12.3. NATURAL DISASTER EVENTS IN QUEENSLAND

Queensland is Australia's most exposed State to natural disaster risk and has incurred the nation's highest natural disaster costs in recent years. The impacts of these natural disasters on communities, infrastructure, and economy are enormous. Figure 12.1 demonstrates the increasing cost to Queensland under NDRRA alone. Of the NDRRA program, almost 90% of damage costs since 2002 occurred after 2009 when the State moved out of several years of *El Nino*-induced (drought) conditions into 'wetter' *La Nina* (and more problematic from an extreme climate event perspective) conditions.

Queensland, due to its location, experiences rapid onset natural disaster events, and in the past decade, parts of the State have been impacted by repeat events, including the summer of disasters in 2010 to 2011 and severe Tropical Cyclone Oswald in 2013 (see Queensland government photos). Such natural disasters have claimed over 40 lives and cost more than $13 billion in reconstruction expenditure under NDRRA.

The World Bank estimates that the losses associated with the Queensland flooding of 2010–2011 amounted to more than $15 billion. It was one of Australia's largest natural disasters and one of the major international disasters of the past decade:

All 73 Local Government Areas (LGAs) or Councils in Queensland were activated for natural disaster funding. Queensland experienced both slow-onset and deep inundation events as well as flash floods in various low-lying parts and valleys of Queensland. The floods inflicted significant damages and losses to private properties and businesses, and a vast amount of public infrastructure.

"Ballpark" estimates indicate cumulative damages and losses from the floods and cyclones in the 2010/2011 period reached at least AUD$15.7 billion resulting in a consequent lowering of Queensland growth estimates from 3 percent to 1.25 percent. These damages include:

- *damages to more than 9,100km of state road network and approximately 4,700 km of the rail network;*
- *power disruptions to approximately 480,000 homes and businesses;*
- *97,000 insurance claims in respect of damages to private assets, of which 50–60 percent are for privately*
- *owned residential properties;*
- *damages or disruptions to 54 coal mines, 11 ports, 139 national parks and 411 schools;*
- *estimated losses of $875 million to primary industries, primarily the sugar, fruit and vegetable sub-sectors [The World Bank, 2011].*

In January 2013, ex-Tropical Cyclone Oswald hit the coast of Queensland with catastrophic effects. This event brought damaging winds including mini-tornados, extreme rainfall, and flooding across the State. The flooding caused destruction to Queensland Local Government Authorities, with more than 70% activated under NDRRA, many of which were still recovering from the unprecedented loss and damage inflicted by the destructive natural disasters of 2010–2011. Regions that were particularly devastated by the 2011 floods were not spared, with communities again being impacted by floodwaters (Figure 12.2).

The Australian government has recognized the significant financial and economic impacts disasters have had on all levels of government and their expected increase in future years as well as the importance of a measured approach to recovery and reconstruction. This led the Australian government to call for a Productivity Commission Inquiry into the current national natural disaster funding arrangements. The purpose of this inquiry was to 'analyse the full scope (incorporating the quantum, coherence, effectiveness and sustainability) of current Commonwealth, State and Territory expenditure on natural disaster mitigation, resilience and recovery.'

The final report following the inquiry, released in May 2015, highlighted the liability natural disasters had on governments, with assets being repeatedly damaged by successive natural disasters and repeatedly rebuilt in the same location to the same standard:

• *Governments generally overinvest in post-disaster reconstruction, and underinvest in mitigation that would limit the impact of natural disasters in the first place. As such, natural disaster costs have become a growing, unfunded liability for governments, especially the Australian Government.*

• *Government investment in mitigation tends to be outweighed by post-disaster expenditure. For example, Australian Government mitigation spending was only 3 per cent of what it spent post-disaster in recent years [Australian Government – Productivity Commission, 2014].*

Funding for natural disaster events in Queensland

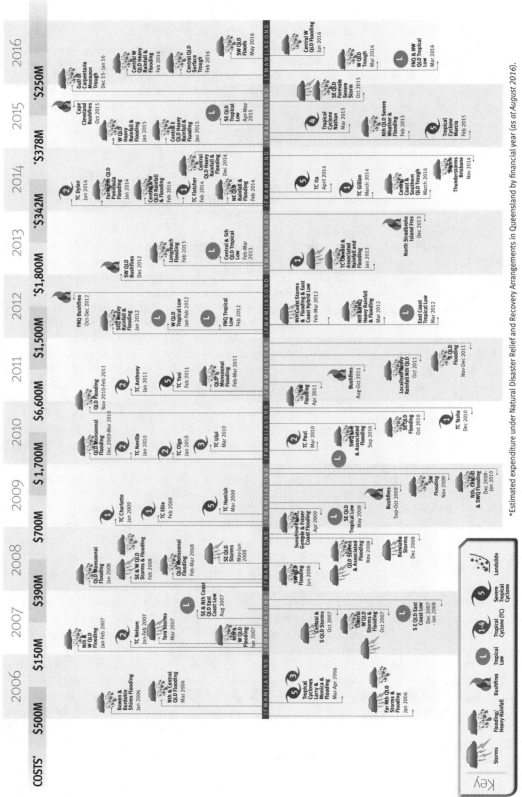

Figure 12.1 Natural disaster events in Queensland 2006–2015 and expenditures under NDRRA.

*Estimated expenditure under Natural Disaster Relief and Recovery Arrangements in Queensland by financial year (as at August 2016).

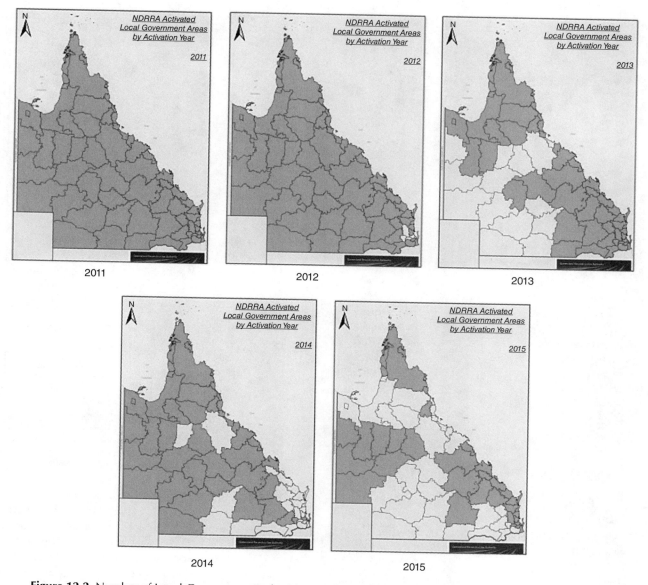

Figure 12.2 Number of Local Government Authorities activated under NDRRA in Queensland 2011–2015.

12.3.1. Cyclones and Flooding Rains

In Australia, the Bureau of Meteorology refers to large-scale storms as tropical cyclones, similar in size and impact to hurricanes in the United States of America and typhoons in Asia. A cyclone is an extreme storm event characterized by inward spiraling winds. The winds (often approaching 300 km per hour) flow clockwise in the Southern Hemisphere and counterclockwise in the Northern Hemisphere.

The warm waters of the Pacific Ocean, particularly in the Coral Sea, provide ideal conditions for the formation of cyclones between the end of November and early May. In the average cyclone season, 10 tropical cyclones develop over Australian waters, with 6 crossing the coast, mostly over northeast Queensland (between about Mossman and Maryborough) and northwest Australia (between Exmouth and Broome) (Figure 12.3).

In addition to the highly destructive winds associated with a cyclone, the accompanying storm surge (an increase in the water level above regular tidal ranges caused by wind shear and reduced atmospheric pressures associated with the cyclone) often causes major damage.

Between November 2010 and April 2011, Queensland suffered extensive flooding following periods of extremely heavy rainfall caused by storm cells including Cyclone Tasha, Cyclone Anthony, and Severe Tropical Cyclone Yasi, and subsequent monsoonal rainfall. As a result,

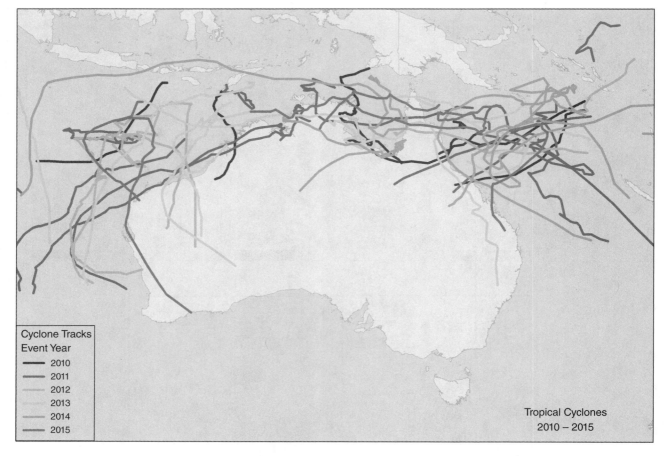

Figure 12.3 Queensland cyclone activity 2010–2015. Data source: Australian Government, Bureau of Meteorology.

100% of the local government authorities across the State were activated for disaster funding, more than 70 towns were evacuated, over 200,000 people were affected, and there were numerous fatalities [*Queensland Reconstruction Authority*, 2011].

Between 2011 and 2015, Queensland experienced 9 cyclones and 16 major flood events. Such major natural disaster events will continue to occur in the future, and may increase in severity and frequency under a climate change paradigm. The State's vast area, covering more than 1,850,000 km², adds significant challenges to its preparedness and recovery following natural disasters.

12.4. RESPONSE–DAMAGE ASSESSMENT IN QUEENSLAND

Damage and Loss Assessment and Post Disaster Needs Analysis techniques have long been recognized as key tools to scoping recovery and reconstruction programs in the post-disaster context. A quantitative as well as qualitative assessment process is acknowledged by the World Bank and the United Nations as providing the optimal approach to identifying the most affected social, economic, and environmental sectors post-disaster [*The World Bank*, 2011].

In 2011 and before, response agencies had a relatively small team of trained operators based in the State's capital of Brisbane that undertook Rapid Damage Assessments (RDA). These assessments were the primary means whereby the State government could accurately determine the immediate impact of a disaster on residential and business structures.

In the months following these disasters, it was impossible to accurately determine the level of recovery, such as the number of residential homes that remained uninhabited, or homes and businesses that were not fit for use due to the damage sustained and were not yet repaired. Furthermore, there was no coordinated approach to sharing information between government agencies, water utilities, waste collection services, and insurance companies, and not all impacted areas were assessed. Initial attempts to gauge housing reconstruction were paper-based, resulting in difficulties relating to unanswered questions, missing information, and difficulty deciphering handwritten material, which limited the usability of the collected data.

Early in its establishment, the QRA recognized that its approach to damage assessment must inform the recovery strategy and identify priority social and infrastructure reconstruction tasks. This included the need for

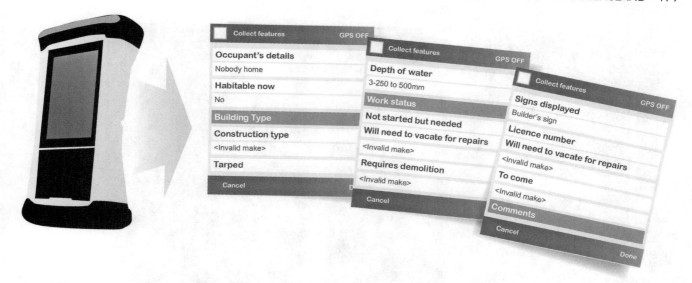

Figure 12.4 Typical questions used in DARMSys™.

rapid and detailed assessment of early recovery needs to enable swift mobilization of community and financial support mechanisms.

The QRA developed an Australian-first system that enabled the gathering of early and accurate information to allow speedy damage assessments. This system is the Damage Assessment and Reconstruction Monitoring System (DARMSys™) (Figure 12.4).

The DARMSys™ was based on the RDA system that was used by Queensland Fire and Emergency Services (QFES) at the time in 2011. The QRA enhanced the system, taking into account mobile hardware and mobile broadband Internet available in 2011. The QRA also consulted with other key agencies including the Department of Housing and Public Works and the Building Services Authority to develop key data input requirements that would be the basis of the innovative DARMSys™, designed for reconstruction monitoring. The QRA trademarked the DARMSys™ as it was unique to reconstruction monitoring.

The DARMSys™ process consists of a three-phased approach for all hazards: initial response, damage assessment, and reconstruction monitoring. These phases are outlined in more detail below.

12.4.1. Phase One, Initial Response

The initial response phase is conducted within the first 72 hours of an event. The timing of the initial response commences after the impacted area is safe to access. The primary function of responding agencies in this period is to ensure the safety of residents and response personnel.

The collection of damage assessment data is a secondary priority.

Initial assessments are undertaken in the disaster area, and the data are used to identify the broad levels of damage, loss of life and injuries, and the impact on critical infrastructure, services, businesses, and housing (Figure 12.5).

Planning for this phase begins before an event and focuses on preparation (assessing risk) and identifying potential immediate life-threatening situations.

This phase is considered to be complete once the disaster area has been inspected, no further life threatening situations exist, the number of deceased, injured, and missing persons is established, and the nature and extent of damage to critical infrastructure are identified.

12.4.2. Phase Two, Damage Assessment

The damage assessment phase aims to provide detailed and comprehensive information to support disaster declaration and response and recovery decisions. This assessment is more focused and detailed than the rapid damage assessment conducted in the initial response phase and includes quantifying damaged public assets, businesses, and private properties.

The focus of this phase is to capture data associated with damage to housing, businesses, infrastructure, and property. Field staff use a handheld device to capture detailed damage data that is uploaded to a central storage point coordinated and managed by QRA for analysis and presentation to recovery and reconstruction managers at state, district, and local levels.

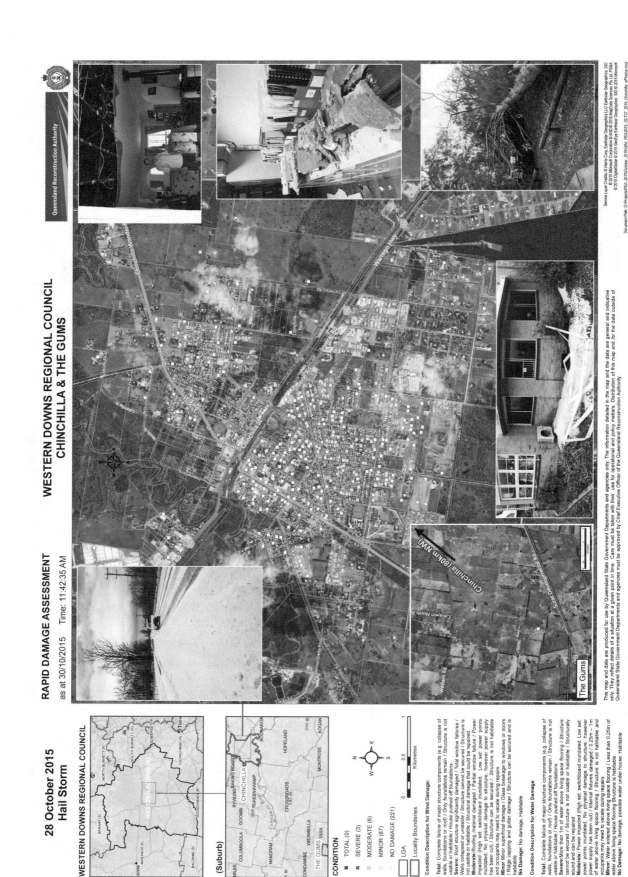

Figure 12.5 Example of Rapid Damage Assessments – Chinchilla, Queensland, October 2015. Microsoft product screenshot(s) reprinted with permission from Microsoft Corporation.

Information is also disseminated to key recovery and reconstruction organizations and supporting authorities to improve understanding of the needs and priorities of impacted communities. These data are crucial to decision-makers responsible for managing and responding to community needs both during and following natural disasters.

This phase is considered complete when the entire disaster area has been identified and the following occurs:

• public assets, businesses, and private properties have been inspected

• the number of impacted private houses, businesses, and government buildings and the levels of damage have been quantified

• the extent of damage has been mapped and damage assessment data and information has been disseminated

• reconstruction planning has commenced

12.4.3. Phase Three, Reconstruction Monitoring

The reconstruction monitoring phase commences once the restoration of or initiation of processes to restore services such as transportation, sanitation, power, and communications has been achieved. Reconstruction monitoring is an ongoing process designed to determine the progress of recovery and reconstruction efforts. This phase provides regular updates on the progress of reconstruction and builds upon the accumulated information and data from phases 1 and 2.

Managed by QRA and in conjunction with other agencies, reconstruction monitoring involves quarterly audits of progress. Phase 3 data are analyzed in the context of that collected in Phase 2 and measures the speed and completeness of the reconstruction effort. Information and data collected in this phase are disseminated to key recovery and reconstruction organizations and supporting agencies to assist in the recovery effort.

This phase is considered to be complete when the disaster area has been re-inspected sufficient times to determine that the reconstruction effort is complete.

12.4.4. DARMSys™ in Action

The initial pilot DARMSys™ reconstruction audit began in May 2011 as a three-month follow-up of the damage caused by Cyclone Yasi. This field activity was undertaken in partnership with QFES, which provided personnel with experience at house-to-house assessments. Additionally, the Queensland Building Services Authority provided expertise in assessing the magnitude of impact and consequential repairs (Figure 12.6).

Following the May 2011 pilot, the QRA acquired additional hardware and software and developed processes for its implementation. This included funding additional RDA equipment to allow QFES to greatly increase its RDA capability and to have RDA equipment pre-deployed around the State.

In July 2011, using the newly developed system, the QRA worked with the Queensland Building Services Authority and assessed 11,500 properties in the Brisbane and Ipswich areas covering all non-industrial properties within the defined flood line. This involved 26 assessment teams being allocated target addresses to inspect each day, with daily re-evaluation of results and re-tasking where necessary. The coordination and planning of this activity was a significant factor in its success.

Further follow-up assessments were undertaken by the QRA in October 2011, February 2012, and May 2012 to track reconstruction, with each assessment targeting properties identified as not having been repaired in the previous inspection. Similarly, properties impacted by Cyclone Yasi were re-assessed in September and November 2011 and in March and June 2012.

By undertaking structured and planned assessment of damage every three months, the QRA accurately reflected the impact and progress of recovery of disasters on residential and business structures. Figure 12.7 illustrates the local government authority of Brisbane (of which Rocklea is a suburb) during the Brisbane floods.

Because the State was impacted by further disasters in the summer of 2011–2012, DARMSys™ three-monthly reconstruction activities were undertaken for affected communities. Beginning in May 2011 for 18 months, 35,000 door-to-door DARMSys™ assessments were carried out.

DARMSys™ provided useful information in isolation. However, combining information from other sources such as census information from the Australian Bureau of Statistics, weather information from the Bureau of Meteorology, and housing and land zoning information from local and state governments further increased its value to key recovery agencies. For example, in response to the 2010–2011 disasters, the Queensland Premier's Disaster Relief Appeal (PDRA) provided financial and material assistance to impacted residents. By combining DARMSys™ assessment data with specific financial, insurance, and social information from the individual PDRA applications, QRA was able to assist other agencies apportion and distribute the PDRA funds as quickly as possible to those most in need.

By 2013, it was evident that there were significant advantages to having an existing, established organization focused on reconstruction following natural disasters. Due to its established structure, resources, and systems, the QRA had the ability to respond quickly and effectively, swiftly deploying staff to the most affected communities following disaster events including Tropical Cyclone Marcia in 2015 (Figure 12.8).

DARMSys™ Case Study: Cyclone damaged residence, Cardwell – Cassowary Coast Regional Council

In 2012, a specialist DARMSys™ team undertook an audit of homes that were demolished or needed to be demolished due to the level of damage as a result of Cyclone Yasi.

A previous audit had identified that an occupant in Cardwell was living in less than habitable conditions where their home had suffered structural damage from the cyclone. In response, QRA provided information to the relevant government authorities to assist prioritising community response activities.

When the DARMSys™ team revisited the occupant, they had subsequently received financial assistance through a housing assistance program and received building management advice.

The occupant was re-housed in a new cyclone-compliant home and was very grateful for the role of the QRA in assisting them to secure safe living conditions.

Before – damaged residence

After – the elderly resident back in their home

Figure 12.6 DARMSys™ case study – Cardwell, Queensland.

This resulted in expeditious rapid damage assessments of commercial, residential, and public infrastructure in affected areas, enabling State government assistance to be targeted where needed. It also meant that affected local government authorities received immediate assistance to inspect damage and to help them keep their reconstruction programs moving.

DARMSys™ has been acknowledged by the World Bank as having played an instrumental role in enabling Queensland to recover quickly from the natural disasters of recent years. Today, social media and wide area vehicle-based assessments provide the initial indication of natural disaster impact and assist in the planning and targeting of the response by defining specific areas where house-to-house rapid damage assessments are required. New technologies are continuing to be explored in future disaster management assessments.

12.5. RECOVERY-THE QUEENSLAND EXPERIENCE

12.5.1. Natural Disaster Funding Arrangements

Prior to the establishment of QRA, the Queensland Government used a range of procedural and administrative arrangements to oversee its reconstruction and recovery program. The former Department of Local Government and Planning (DLGP) and Emergency Management Queensland (EMQ) managed these activities. Emergency Management Australia (EMA) provided general oversight of such actions at the national level.

The NDRRA, established in 1974, is a cost-sharing approach between the states and territories and the Australian government to help manage relief and recovery costs in the aftermath of large natural disasters. The arrangements provide grant and loan assistance to

Rocklea January 2011 flood line

Rocklea July 2011 – 835 assessments

Rocklea May 2012 – 36 remain damaged

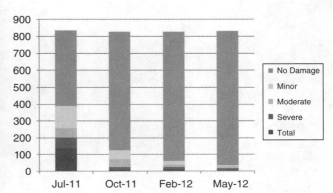

Rocklea – properties impacted

Figure 12.7 Data collection showing damage severity and progress of reconstruction. Microsoft product screenshot(s) reprinted with permission from Microsoft Corporation.

disaster-affected communities, small businesses, not-for-profit organizations, primary producers, and state and local government authorities following defined disasters. The intent of the NDRRA program is to assist in the recovery of communities severely affected socially and financially following the impacts of a disaster event. The Commonwealth Government's NDRRA Determination sets out the financial arrangements, which provide up to 75% Commonwealth financial assistance to states in the form of partial reimbursement of actual expenditure related to natural disaster events.

Although the NDRRA assists impacted communities in a variety of ways, including the provision of grants and subsidies for individuals and business owners, by far the largest portion of funding provides for the restoration of essential public infrastructure such as roads and bridges. In general terms, these essential public assets must be restored to pre-disaster standard to meet the requirements of the Commonwealth's Determination.

Between 2002 and 2007, Queensland's total cost of reconstruction under the NDRRA program was approximately $740 million. The increase in scale, scope, and complexity of natural disasters, which occurred in the period following 2010–2011, means the QRA is now required to administer more than $13 billion in reconstruction works (primarily for events since 2010–2011 to 2015). This required an alternative delivery model that took into consideration the complexity of managing a program of works with more than 2,500 individual funding applications and in excess of 10,000 reconstruction work sites across Queensland.

QRA implemented a range of measures to address these challenges with improvements to infrastructure assessments, flood mapping, and value-for-money and reporting processes to provide increased confidence in the expenditure of public funds. These activities are outlined in more detail below.

February 2015 – 270 damaged

May 2015 – 155 damaged

August 2015 – 82 damaged

November 2015 – 48 damaged

Figure 12.8 Example DARMSys™ data in Yeppoon following Severe Tropical Cyclone Marcia 2015. Microsoft product screenshot(s) reprinted with permission from Microsoft Corporation.

12.5.2. Adapting DARMSys™ into Infrastructure Assessments

The capability and success of DARMSys™ was leveraged and adapted to provide a platform for the collection of data on damaged infrastructure, International Defense Acquisition Resource Management (IDARM). IDARM enables local government authorities to commence data collection immediately after an event, which has proven to expedite the grant funding application and approval process, enabling recovery to commence rapidly. IDARM provides a solution for local government authorities to easily log damage to their essential public assets such as roads and bridges (Figure 12.9 and Figure 12.10).

A combination of cross-government collaboration and state-of-the-art technology is used in the field to assist local government authorities to capture infrastructure damage data in the immediate aftermath of a disaster event. Assessors collect real-time data using a rugged GPS-enabled handheld device, which is uploaded by Wi-Fi directly to a web-based portal. This is used to "ingest"

field data and photographs and allows applicants to edit the collected information and supplement it with additional information or evidence, including photographs and engineering reports.

Technical staff and assessment officers from QRA are able to work in the field with local government personnel to assist them with identifying priority projects and focus on developing and submitting funding applications as soon as possible. The web-based portal enables applicants to appropriately verify their submissions, reducing paperwork and making the process of applying for NDRRA funds simpler and faster.

Reports mapping the location and showing the damaged condition can also be produced and updated after reconstruction works are complete, validating the restored assets and assisting with the NDRRA submission acquittal process. The QRA has deployed officers post-disaster since 2013 to assist with infrastructure damage assessments, assisting local government authorities to conduct more than 16,000 assessments and capture more than 130,000 damage photos.

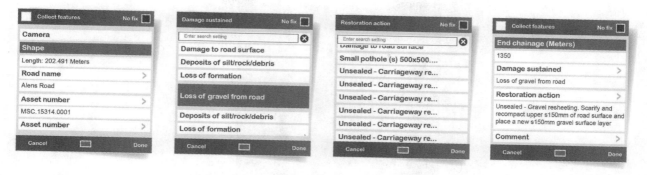

Figure 12.9 Infrastructure damage capture – feature breakdown screens.

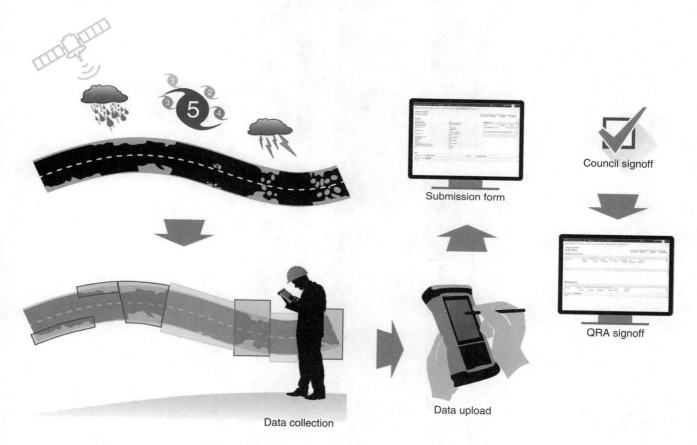

Data collection

Data upload

Submission form

Council signoff

QRA signoff

Figure 12.10 IDARM Streamlined NDRRA submission workflow.

12.5.3. Delivering Value for Money

In identifying how best to deliver this complex portfolio of NDRRA works, the QRA developed and applied a value-for-money strategy that underpinned the coordination and monitoring of state-wide reconstruction efforts. The strategy was based on three key principles: transparency, efficiency, and effectiveness. It was rigorous and flexible and balanced the need to rapidly mobilize applicants to restore essential economic and community infrastructure with the objective of delivering outcomes that offered value for public expenditure.

The QRA identified five key phases within the life span of a reconstruction program activity (Figure 12.11) that resulted in optimal cost-benefit outcomes and that could be demonstrated to all key stakeholders.

The QRA's risk-based approach meant focusing its efforts on those activities or applicants that faced challenges

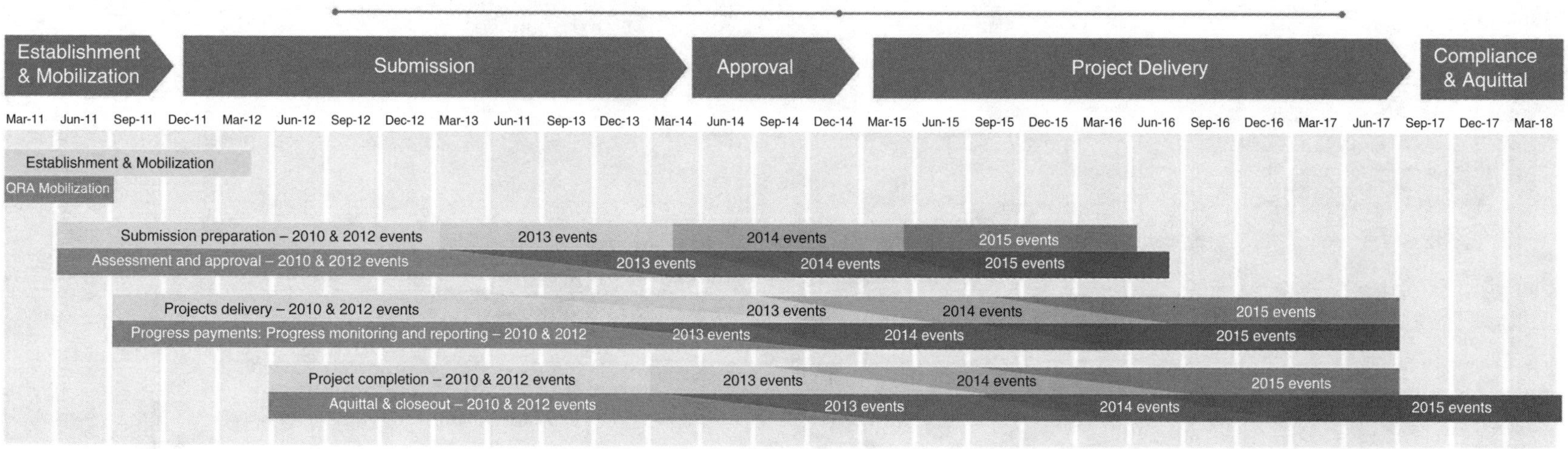

Figure 12.11 Reconstruction program phases 2011–2015.

with key aspects of program delivery. These challenges include the following:

- Project management skills
- Experience
- Constrained local sources of supply
- Whole-of-state issues, such as the potential competition for resources and skills that drives up cost
- Possible diseconomies of scale arising from the challenges of coordinating activities across a vast geographical area

The QRA's early establishment of its value-for-money strategy has provided immediate and significant cost and time savings and set the conditions that enable reconstruction works to be completed within specified timeframes and often below forecast cost estimates. Additionally, the knowledge and expertise developed through the process has provided the Queensland government with enduring benefits that will support the successful planning and delivery of future natural disaster funding programs.

By focusing on a comprehensive value-for-money framework, QRA delivered cost-effective reconstruction programs that were easily tracked, ensuring transparency of activities. This approach also balanced local priorities with State-level considerations.

12.5.4. Development of the Grants Management and Reporting System

Although the QRA improved its damage assessment and monitoring process, the successful reimbursement of grants funding relied heavily on the expediency of submission processes, assessment, and acquittal of funding applications. QRA identified that government authorities relied heavily on paper-based funding applications and seized the opportunity to streamline and redefine the funding application system, recognizing the opportunity for automation, performance, and efficiency improvements.

A Grants Management and Reporting System (GMRS) was developed by QRA, allowing local government authorities to upload funding applications, including photographic evidence and supporting documentation, directly to the QRA. The QRA staff can then assess the application, and all documentation and decisions are tracked using the system, which can also provide reporting formats to assist with acquittal.

Based on a Customer Relationship Management system model, GMRS provides a seamless system for the application, assessment, auditing, and archiving of grant applications (Figure 12.12 and Figure 12.13).

QRA's web portal allows rapid update of damage assessment locations, engineering treatment types, and benchmarked rates to inform the assessment of submissions. QRA's ability to manage the cost of reconstruction works is supported by the use of external scope and cost references. These references ensure the QRA understands the differing costs of reconstruction works and ensures value for money is achieved.

In contrast to the experiences of 2010–2011, the QRA now hosts a sophisticated grants management system that provides for the automation of briefing documentation, a single repository for data storage, digital capture with immediate retrieval and access to documentation, and a unified structure to manage funding application phases. Stakeholders are also able to submit documents and evidence, such as photographs via Dropbox.com, web portals, and email.

The success of GMRS has resulted in approaches from other Queensland government agencies seeking either to replace their current information and communications technology systems or to replace laborious paper-based processes with the QRA's grants management system.

12.5.5. Flood Mapping

In response to the floods of 2010–2011 and the Queensland Floods Commission of Inquiry, the QRA collaborated with the then Department of Local Government and Planning, including Building Codes Queensland (BCQ), the then Department of Environment

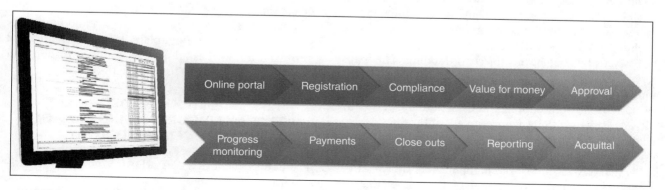

Figure 12.12 Business process management life cycle.

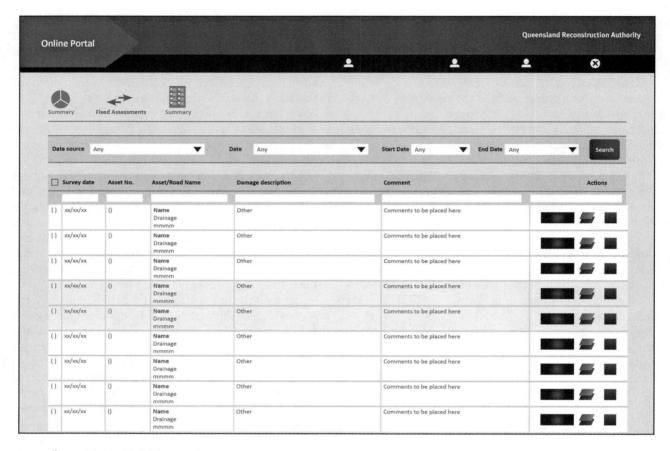

Figure 12.13 GMRS integration.

and Resource Management (DERM), and the then Department of Community Safety (DCS) to undertake the Queensland Flood Mapping Program (QFMP) (Figure 12.14). The program ensured that Queensland learned from natural disasters and assists local government authorities to deliver a body of work supporting greater resilience and understanding of Queensland floodplains. The flood mapping program cost approximately $2 million.

The QRA maps give an indication of the likely extent of Queensland's floodplains. They are not intended to represent a specific flood event, identify exact flood heights, or indicate with certainty that any particular property will or has been subject to flooding. The maps were developed based on high-level, statewide information including 10-meter contours, historical flood records, vegetation and soil mapping, and satellite imagery. The QRA consulted with local government authorities during the development of the maps and gave them the opportunity to provide input based on local experience. During the January 2013 floods, the mapping was shown to be remarkably accurate across many affected regions of Queensland.

The QFMP is the largest floodplain mapping exercise in the State's history. The maps contained in the toolkit, *Planning for stronger, more resilient floodplains*, are drawn from evidence of past flooding, including soils, topography, and satellite imagery. This approach was adopted because while the whole of Queensland was affected in 2010–2011, we also know that there have been larger localized floods in the past. What the maps show are areas where inundation had previously occurred or where it may occur.

Queensland is now the only jurisdiction in Australia with a state-wide catchment-based understanding of its floodplains (including floodplain maps) to help local government authorities adopt and apply consistent and specific planning controls to manage flood risk.

12.6. MITIGATION AND BETTERMENT PROGRAMS INCORPORATING DAMAGE ASSESSMENT METHODOLOGY

The development of QRA's IDARM and GMRS systems have ensured a smooth process from initial response to recovery and acquittal of works under NDRRA.

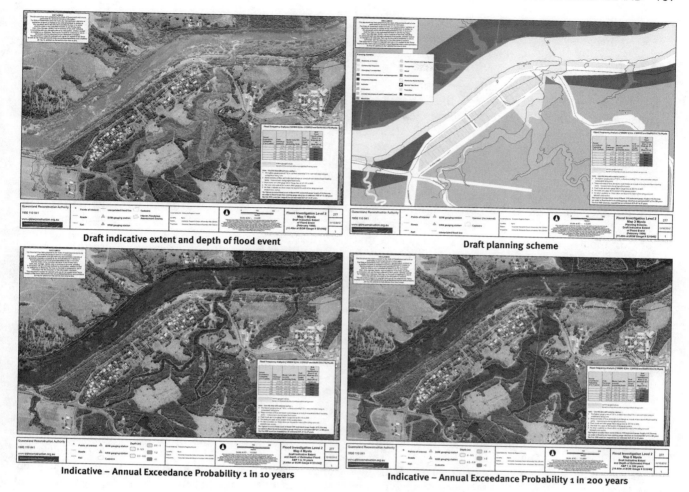

Figure 12.14 Indicative flood mapping – Tablelands Regional Council. Microsoft product screenshot(s) reprinted with permission from Microsoft Corporation.

These systems, coupled with the State's flood mapping tools, enables the state and local governments to be prepared for potential natural disasters and to make considered and informed decisions regarding the implementation of mitigation and resilience works into recovery programs.

The United Nations Development Programme's (UNDP) Hyogo Framework for Action 2005–2015 (HFA) and its successor, the Sendai Framework for Disaster Risk Reduction 2015–2030, are the internationally accepted frameworks for building resilience. Since adoption of the HFA and the release of the Sendai Framework (the successor instrument to the HFA), there has been an international acknowledgement that efforts to reduce disaster risks must be systematically integrated into government policies, plans and programs.

According to the *UNDP* [2014], "At any point in time, more than a third of all countries are recovering from disasters, and for many communities recovery is an ever-present concern. This is therefore a critical moment not only for reassessing risk and pushing forward on risk reduction reforms and investment, but also for building comprehensive resilience to disaster. Ill-informed recovery often worsens the underlying conditions of risk and can lead to future events having even worse effects."

Experience in Australia and around the world demonstrates an upward trend over time in the regularity of natural disasters and the economic burden they cause.

"The impacts and costs of extreme weather events can be expected to increase in the future with population growth and the expanding urbanisation of coastlines and mountain districts near our cities" [*Australian Government – Productivity Commission*, 2014]. Given all of the above, it is relevant to highlight the implementation and successes of a Queensland government initiative, the "Betterment Fund" (jointly funded by the Australian and Queensland governments) that adopted the approach of future risk reduction in the wake of Tropical Cyclone

Oswald in 2013. This event caused approximately $2 billion of damage to many public assets that had been repeatedly impacted and restored following earlier disasters in 2011 and 2012.

Traditionally under the Australian government's rules for NDRRA, funding was restricted to the restoration of essential public assets to their pre-disaster standard, leading to repeated damage of vulnerable assets. Although there has been provision for betterment funding under the NDRRA model since 2007, the process has been cumbersome, with only one project approved.

Under the Queensland Betterment Fund, an exceptional circumstances allocation of $80 million was approved for eligible local government authorities to restore or replace damaged assets to a standard that would be more disaster resilient, reducing risk to the community and reconstruction costs from future events. The Betterment Fund framework also streamlined the process of eligibility, funding application submission, assessment criteria for funding, and distribution of betterment funds. The 2013 Queensland Betterment Fund delivered more than 230 projects with a total cost of more than $160 million, which included almost $80 million in betterment funding. Local government authority contributions and natural disaster reconstruction and recovery arrangements funding supported the program.

A key test for the betterment program is whether it leaves infrastructure and communities less vulnerable to the natural hazards of Queensland's climate; that is, does the change allow communities to recover faster from a natural disaster and allow communities to return to normal quicker?

Repairs and restoration of damaged infrastructure and public assets is the most significant component of the cost of natural disaster events. A study of 5,500 mitigation grants approved by the United States Federal Emergency Management Agency (FEMA) between 1993 and 2003 report an overall cost-benefit ratio of 4:1 for mitigation investments [*Rose et al.*, 2007]. Queensland's experience supports this finding. Queensland betterment projects approved following the 2013 event were completed by June 2015 over a period of almost two years. Of those completed in 2014 and early 2015, 71 were in the impact zone of eight separate natural disaster events, most notably Category 5 Tropical Cyclone Marcia, with 12 of these projects impacted by two subsequent events. Of these, 92% were either undamaged or remained functional with minimal damage following the event(s).

In total, betterment projects have sustained less than $160,000 in damage costs as a result of natural disasters since their completion, saving $22 million for a $16 million investment (Figure 12.15). That is, building back to pre-disaster standards for these projects would have cost more than $22 million in restoration funding while enhancing the resilience of these assets through betterment required an additional investment of $16 million. Thus, an additional investment of $16 million in these cases has led to an avoided cost of some $22 million, in year two after construction alone.

If this benefit is extrapolated to multiple events, the avoided costs (or benefits) accumulate with each event and the benefit-cost ratio for the investment progressively improves over time, an attribute of very few financial transactions or investments.

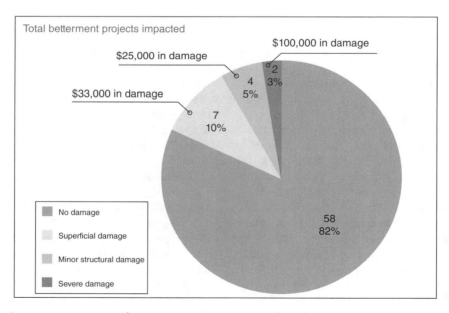

Figure 12.15 Betterment project performance post Severe Tropical Cyclone Marcia.

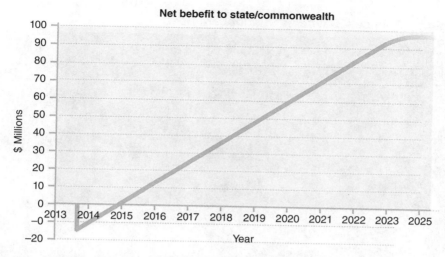

Figure 12.16 Betterment Project financial benefit forward projections.

An analysis of recent event years suggests that these projects will be impacted every two years on average. Using past events to extrapolate, this gives Queenslanders an overall cost-benefit ratio of almost 7:1 on these 71 projects, saving seven times the amount invested over a 10-year period (Figure 12.16).

Local government authorities around Queensland have been empowered to assess, plan, and implement disaster recovery at the grass-roots level. This enables the facilitation of betterment works to begin as soon as possible following a natural disaster, mitigating the impact on their local communities. Examples of resilience that have been delivered using the betterment framework are shown in Figure 12.17 and Figure 12.18.

Following the success of the 2013 Queensland Betterment Fund, the Australian government agreed to an additional $20 million Betterment Fund for Queensland communities impacted by Tropical Cyclone Marcia in 2015.

In addition to the Queensland Betterment Fund, other exceptional circumstances arrangements have been approved by the Australian government to help Queensland communities recover following natural disasters. One of the most significant was the $18 million "Rebuilding Grantham" grant, which saw almost the entire town moved to higher ground following its devastation in 2011 (Figure 12.19).

Through the Queensland Reconstruction Authority Act 2011, the QRA was able to establish a regulatory framework for the local government authority that ensured proposed rebuilding efforts and approval processes were fast-tracked, including the facilitation of development application approvals with relevant agencies.

More than 100 lots were "swapped," and within 11 months of the disaster, the first residents moved in. Grantham was impacted only two years later by another flooding event, which inundated the original town area.

Lockyer Valley Regional Council Mayor Steve Jones said had Grantham not been relocated, the council would have been looking at a $20 or $30 million bill for flood damage in that year. Instead, they expected damage to the Grantham area to total about $20,000.

The Australian government is currently reviewing its current national natural disaster funding arrangements. To help inform its approach, a public inquiry was undertaken by the Productivity Commission. The report, released in May 2015, recommended the Australian Government increase its mitigation funding to states to $200 million a year and be matched by the states (Table 12.1).

The Queensland government welcomes an increase in mitigation funding, however, a redirection of national funding for recovery to offset mitigation and betterment initiatives is not acceptable; it would result in a reduction of national funding. Discussions on natural disaster funding reforms in Australia are ongoing, with no clear outcome for betterment and mitigation funding.

12.7. THE FUTURE

In the short period of time that it has been in operation, the QRA has delivered, on behalf of the population of Queensland, significant improvements in the way that post-disaster reconstruction activities across the State are

North Burnett Regional Council – Gayndah-Mundubbera Road

Gayndah Mundubbera Road Damage by Year

— 2015
— 2013
▨ Betterment works zone

Burnett River

Kilometres 0   0.25   0.5

Avoided cost: $6,785,707 – one event

Project details	
Restoration	$6,785,707
Betterment	$1,308,863
Council contribution	$100,000
Avoided Cost	$6,785,707

- Damaged in 2011 and rebuilt, only to be damaged again in 2013.
- Project: relocated a two-kilometre section of road up to 11 metres up hill.
- The road was re-opened within three hours of the flood waters receding following Tropical Cyclone Marcia, with minor expenditure required in emergent works to clean up and remove debris. This compares with the road's closure for four months in 2013.

Figure 12.17 Betterment example – Gayndah-Mundubbera Road, North Burnett Regional Council. Microsoft product screenshot(s) reprinted with permission from Microsoft Corporation.

assessed, implemented, and managed. The streamlining and introduction of efficiencies with regard to how damage assessments are undertaken and the approval of funding applications processed has more than halved the time taken for reconstruction works to be approved compared to that which the QRA inherited following the natural disasters of 2010–2011.

These actions are now also being implemented in such a way, using "betterment" principles, that future reconstruction actions and costs should reduce over time. Using improved statewide flood mapping information, future development activities will be better informed to ensure that construction occurs in flood-free (and hence less disaster-prone) locations. These are all major and

highly beneficial developments for the State and have potential for application worldwide. The Queensland approach to betterment has delivered compelling proof that this approach works.

The journey taken by the QRA, however, is not finished. Several enhancements to the portfolio of tools and knowledge-based techniques at its disposal are currently in development and will be implemented by the QRA in the near future. These include improving the linkage of damage assessments with benchmarking for estimates of reconstruction costs, improving mobile damage assessments such as the use of drones and vehicle remote sensing technology, and gearing GMRS toward total system integration.

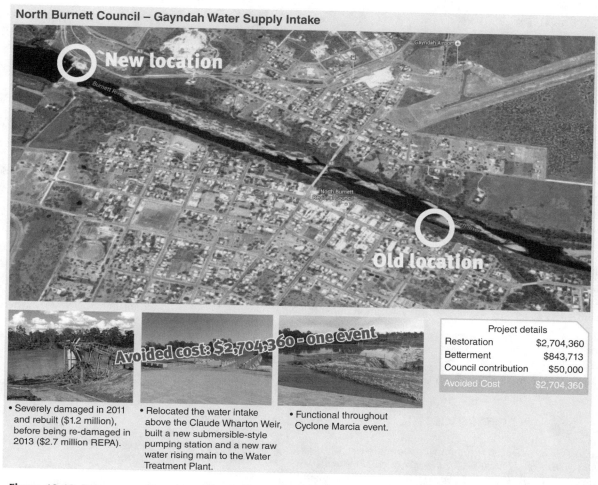

North Burnett Council – Gayndah Water Supply Intake

New location

Old location

Avoided cost: $2,704,360 – one event

Project details	
Restoration	$2,704,360
Betterment	$843,713
Council contribution	$50,000
Avoided Cost	$2,704,360

- Severely damaged in 2011 and rebuilt ($1.2 million), before being re-damaged in 2013 ($2.7 million REPA).
- Relocated the water intake above the Claude Wharton Weir, built a new submersible-style pumping station and a new raw water rising main to the Water Treatment Plant.
- Functional throughout Cyclone Marcia event.

Figure 12.18 Betterment example – Gayndah Water Supply Intake, North Burnett Regional Council. Microsoft product screenshot(s) reprinted with permission from Microsoft Corporation.

Figure 12.19 The Master Plan for Grantham involved the relocation of the community to higher ground.

Table 12.1 Summary of Reforms (Extract).

Current problem	Proposed response	Main benefits of change
Recovery funding		
Reforming recovery funding arrangements to provide stronger incentives for effective natural disaster risk management by governments		
The current reimbursement model drives behaviors and reduces incentives for state and local governments to implement the most appropriate and cost-effective options for disaster recovery.	Move toward: • assessed damages and benchmark prices for essential public assets • an united grants model for community recovery funding. **(Recommendations 3.1, 3.2, and 3.4)**	States have greater autonomy in how they allocate recovery funding to respond to local circumstances. More neutral incentives are needed to consider replacement, betterment, and abandonment of assets.
Mitigation funding		
Increasing investment in disaster mitigation, with robust governance and transparent decision-making		
Total mitigation expenditure across all levels of government is more likely to be below the optimal level than above it.	Australian government to increase mitigation funding to states to $200 million per annum: • initially distributed according to the current allocation under the NPANDR • subsequently according to a more risk-based allocation. **(Recommendation 3.5)**	Increased investment is needed in appropriate mitigation activities, which may reduce the future economic costs of natural disasters and insurance premiums where natural hazard risks to private property have been materially reduced.

[Australian Government – Productivity Commission, 2014]

REFERENCES

Australian Government – Productivity Commission (2014), *Natural Disaster Funding Arrangements – Productivity Commission Draft Report Volume 1*. Retrieved November 2015, from http://www.pc.gov.au/inquiries/completed/disaster-funding/draft/disaster-funding-draft-volume1.pdf.

Australian Government (2014), *Attorney-General's Department Submission – Productivity Commission Inquiry into Public Infrastructure*. Retrieved November 2015, from http://www.pc.gov.au/__data/assets/pdf_file/0004/133933/sub101-infrastructure.pdf.

Deloitte Access Economics (2013), *Building our nation's resilience to natural disasters*. Retrieved November 2015 from http://australianbusinessroundtable.com.au/assets/documents/White%20Paper%20Sections/DAE%20Roundtable%20Paper%20June%202013.pdf.

Queensland Reconstruction Authority (2011), *Monthly Report June 2011*. Retrieved February 2016 from http://www.qldreconstruction.org.au/u/lib/cms2/ceo-report-jun-11-full.PDF.

Rose *et al.* (2007), Benefit-Cost Analysis of FEMA Hazard Mitigation Grants. http://earthmind.org/files/risk/Nat-Haz-Review-2007-CBA-of-FEMA-Grants.pdf.

The World Bank (2011), *Queensland - Recovery and Reconstruction in the Aftermath of the 2010/11 Flood Events and Cyclone Yasi*. Retrieved November 2015 from http://qldreconstruction.org.au/u/lib/cms2/world-bank-report-full.pdf.

United Nations Development Programme (2014), *The Future of Disaster Risk Reduction: UNDP's Vision for the Successor to the Hyogo Framework for Action*. Retrieved November 2015 from http://www.undp.org/content/dam/undp/library/crisis%20prevention/UNDP_CPR_DRR_The%20Future%20of%20DRR_July2014.pdf.

13

Forensic Disaster Analysis of Flood Damage at Commercial and Industrial Firms

Martin Dolan, Nicholas Walliman, Shahrzad Amouzad, and Ray Ogden

ABSTRACT

Direct effects of flooding on business premises can range from mild to catastrophic. Where buildings are penetrated by flood water there is likely to be significant damage and commercial/industrial processes could be compromised. Particular concerns include loss of essential equipment (expedient replacement of which can frequently be impossible), loss of data, loss of operations in the period to remediation, and significantly the effect of failure to meet contracts or the needs of clients. This latter issue often resulting in long term lost business opportunities as clients engage alternative suppliers. Very often therefore the immediate cost of physical damage to business infrastructure (buildings, plant, machines and systems) can be small in relation to the broader damage sustained as a result of lost turn-over. Businesses can be made more resilient to indirect or direct flood damage (including through the use of flood defences or back-up utility systems), but economic appraisal of such measures can be difficult as there is little data or established methodology.

13.1. INTRODUCTION

The economic and social impacts of natural disasters are increasing throughout the world, particularly in developing countries. Essential economic and social developments are repeatedly experiencing setbacks due to the effects of large scale, or series of smaller scale natural disasters, such as floods, hurricanes and typhoons, and earthquakes. This is despite the increased knowledge about these events and better technology to mitigate their effects. More effort appears to have been devoted to research into the effects of flooding of domestic properties, despite the fact that businesses provide the basis for economic stability and growth [*Tierney*, 1995; *Dahlhamer and Tierney*, 1998; *Webb et al.*, 2000; *Rodriguez et al.*, 2006].

Direct effects of flooding on business premises can range from mild to catastrophic. Where buildings are penetrated by floodwater, there is likely to be significant damage, and commercial/industrial processes could be compromised. Particular concerns include loss of essential equipment (expedient replacement of which can frequently be impossible), loss of data, loss of operations in the period to remediation, and significantly, the effect of failure to meet contracts or the needs of clients. This latter issue often results in long-term lost business opportunities as clients engage alternative suppliers. Very often therefore the immediate cost of physical damage to business infrastructure (buildings, plant, machines, and systems) can be small in relation to the broader damage sustained as a result of lost turnover. Businesses can be made more resilient to indirect or direct flood damage (including through the use of flood protection measures or back-up utility systems), but economic appraisal of such measures can be difficult because

School of Architecture, Oxford Brookes University, Oxford, UK

Flood Damage Survey and Assessment: New Insights from Research and Practice, Geophysical Monograph 228,
First Edition. Edited by Daniela Molinari, Scira Menoni, and Francesco Ballio.

there is little data and established methodologies are not available.

The focus of this chapter is on identifying the vulnerabilities that lead to damage to commercial and industrial activities and on using this information to examine what is needed to improve resilience to flooding events in the business sector. In detail, observed direct and indirect damages are analyzed according to a forensic approach. Particular care is put on analyzing indirect damage, as well as damage to infrastructure, as it is often the case that business activities suffer damages because of the disruption of essential services.

This chapter will aid an understanding of the following:
• the nature of forensic analysis when applied to flood events
• the data required for forensic analysis
• examples of forensic analysis of flood event through an example of the procedure
• business vulnerability to natural disasters
• the use of forensic analysis for business continuity measures

13.2. FORENSIC ANALYSIS: QUANTITATIVE AND QUALITATIVE APPROACHES

There is a growing international interest in using forensic techniques to analyze flood and other natural disaster events more scientifically in order to reveal the complex underlying causes and effects that result in the growing disaster losses. The notion of "Forensic Disaster Investigations" has been coined by the Integrated Research on Disaster Risk (IRDR) initiative. It has been defined as an "approach to studying natural disasters that aims at uncovering the root causes of disasters through in-depth investigations that go beyond the typical reports after disasters" and that "will help build an understanding of how natural hazards do – or do not – become disasters".

According to McBean, despite the recent growth of knowledge of the causes of natural disasters and more effective management of disaster risks, there is a need for an enhanced rigor in the use of collected event data in order to gain a better understanding of the actual causes of disasters [IRDR, 2011]. This would help to underpin evidence-based decision-making and policy making in disaster risk reduction. Cooperation across a broad range of fields of expertise is required that brings in researchers and practitioners in diverse fields of study. The collection of data about flooding events and other natural disasters is the first step in learning from the events in order to mitigate those in the future.

Risks of flooding are the result of the interaction between social, technical, and natural processes. *Vojinović* [2015] argues that the holistic study of risk must acknowledge its inseparability from root causes that encompass the structure and processes that explain why disastrous impacts occur. He maintains that this type of analysis can be considered as a form of forensic investigation. He identifies four levels of drivers of vulnerability and risk: the natural, technical, and social drivers and those factors that are related to our values and beliefs. Through this process, a broad investigation of the conditions and profiles of risk and losses is possible, in order to determine how, where, and to whom losses occurred, as well as how and why others escaped those losses.

Forensic analysis can be seen as the second step in the application of loss data. The first step is loss accounting, which consists of recording impacts and measuring trends. The second, forensic analysis, consists of analyzing the causes and learning from the past. Finally, risk modeling consists of predicting future losses and calibrating and validating model results. The data required for each should be collected at local, regional, and global scales. For forensic analysis, the data need to be at a detailed level to understand the local context of the disaster, e.g., source of flooding, topography of affected area, local mitigation measures, length of warning, etc., and at a coarser scale for the wider context, e.g., regional disaster management systems, flood defence policy, infrastructure resilience, etc. [*De Groeve et al.*, 2013].

Although the selected research approaches will be determined by the nature of the events, their context and the motivating interests involved, according to the Forensic Investigations of Disasters (FORIN) template as developed by *IRDR* [2011], the suggested research methodology of forensic investigations can be divided into four steps:
• Critical cause analysis
• Meta-analysis
• Longitudinal analysis
• Scenarios of disaster

Wenzel et al. [2012] state the analysis needs to investigate a number of specific issues. These include the identification of the critical factors contributing to losses, damages, failures of infrastructure and impacts on economy, and the critical interactions between the hazard, socioeconomic, and technological systems. The nature and extent of the protective measures and their performance during the event should also be assessed. This will enable predictions to be made of patterns of possible future losses and socioeconomic implications due to future events based on parameters of hazard, historical evidence, and socioeconomic conditions that have useful implications for reconstruction efforts. It will point out what can be done within the social, cultural, and economic context taking into account the limitations and opportunities to take actions to prevent, withstand, or avoid the hazards.

As with any form of research, it is essential to clearly define the questions to which answers are required in order to guide the investigation efforts. The data collection

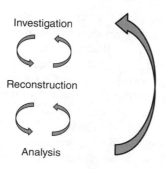

Investigation

Reconstruction

Analysis

Figure 13.1 Process of critical cause analysis.

and analysis will vary greatly between say, a study to find out the reason for varied losses in different types of business when subjected to the same flood incident, and a study of an individual business on the causes of damage and the possible risk reduction measures to be taken.

13.2.1. Critical Cause Analysis

The process of cause analysis is one that is difficult to define in terms of process since investigation, reconstruction, and analysis can all feed into each other. For example, the reconstruction of an event based on investigations can lead to the revelation of new hypotheses on what the root causes were [*IRDR*, 2011]. See Figure 13.1.

A critical cause analysis is a method of investigation that aims to identify the root causes of a disaster event or situation with the aim of avoiding or eliminating these to prevent future disasters. The approach is multi-disciplinary and should integrate social, environmental, and technical assessments in order to incorporate the complex range and interaction of factors [*Burton*, 2010]. The following tasks are involved in a critical cause analysis:

1. Identifying critical cause factors such as root causes, contributory factors, and contextual factors
2. Identifying proactive actions that could have been taken in order to reduce the risk, e.g., land use planning, environment management, etc.
3. Devising the types of resilience measures that could have been taken to limit or prevent damage due to the hazard
4. Defining critical practical maximum and minimum limits to possible resilience measures
5. Establishing monitoring requirements at critical risk locations to provide sufficient warnings of impending failure
6. Taking corrective actions that are appropriate to conditions and funding

Critical cause analysis is carried out by following three methodologies: expert judgment, statistical techniques, and analytical techniques.

13.2.2. Meta-analysis

Meta-analysis is the systematic review of the available literature carried out to identify and assess consistent findings across diverse studies and offers the ability to look for causal linkages, the strength of relationships among factors, and the effectiveness of interventions by coding observations and findings [*Burton*, 2010]. This is necessarily a transdisciplinary exercise incorporating social, technical, and environmental aspects.

Meta-analysis can be carried out at different scales, ranging from the concentration on one particular form of hazard or a single type of event to overarching international studies of themes such as public policy or climate trends. It is a useful method of systematically synthesizing the results of numerous studies in order to identify trends, common causes, and successful mitigation measures. Statistical methods are commonly used to analyze numerical data.

13.2.3. Longitudinal Analysis

Longitudinal analysis is the reconstruction of an event in order to allow repeated observations. These are detailed, place-based re-analyses of events and can be geographically comparative or comparative in situ. This methodology provides in-depth understanding of the causes and consequences of disasters and can offer insight into what mitigation strategies worked and the lessons learned [*IRDR*, 2011]. A good example of this is the study by the University of Wisconsin Green Bay Center for Organizational Studies on what happens when small businesses and not-for-profit organizations encounter natural disasters [*Alesch et al.*, 2001].

13.2.4. Scenarios of Disaster

This type of investigation aims to reconstruct and understand the interactions between the different root causes to provide a map of, for example, how a flood event was generated. The scenario is aimed at producing a realistic account of what would happen to people, infrastructure, buildings, and organizations, as well as the natural environment with a given set of disaster conditions. It serves to provide a vehicle for rapid communication about likely events and concerns among many different stakeholders, and enables feedback in order to refine ideas and clarify the problems [*Rosson and Carroll*, 2009].

Scenarios are based on a solution-first problem solving strategy, with the attendant dangers that the designers of the solutions become too fixed in their approach and adopt familiar solutions that might not be appropriate, and are reluctant to explore very different alternatives [*Cross*, 2001]. By stressing the "quick and dirty" nature of the solution, a more relaxed attitude to the solutions can be achieved, and the involvement of stakeholders emphasises the use appropriateness of the ideas and the possible need to find different solutions to raised problems. Despite the designed solutions being incomplete, it is important that they provide concrete proposals that can be evaluated, while being flexible to allow for alteration and refinement.

13.2.5. Data Required for Forensic Analysis

For forensic analysis to be possible and ultimately, useful, there is a need for a wide range of data to be made available regarding the event under scrutiny. For analyzing natural disasters, the following types of data can be regarded as the basic minimum level of detail:
- Spatial scale
- Time
- Geo-referencing
- Description of the physical damage
- Description of indirect damage
- Monetary damage (physical)
- Costs of emergency
- Costs of intervention
- Source
- Level of trust

These data provide a general picture of the disaster impacts and identify important features of data (scale, quality, whether or not the historical values are required, etc.). The main reasons for the collection of such large and varied data are addressed in Chapter 2.

In order to analyze these data to identify the main causes and drivers, data on the hazard factors (i.e., of the physical event) and the vulnerability factors (physical vulnerability, socioeconomic data, master plans, emergency plans, laws/regulations) are required. Two steps are necessary:

1. Identify links among cause/drivers and damage to each affected sector to identify most critical factors.

2. Prioritizing factors (sector by sector) to rate the factors in order of importance.

As a summary, Figure 13.2 illustrates the comprehensive range of data that are specifically related to businesses and flooding and are relevant for forensic analysis.

Despite this comprehensive demand for data, wider investigation into the political, economic, and social conditions may be needed in order to set the context within which the scenario evolved. The methods for collection of such large amounts of data refer to Chapter 6 and Chapter 7.

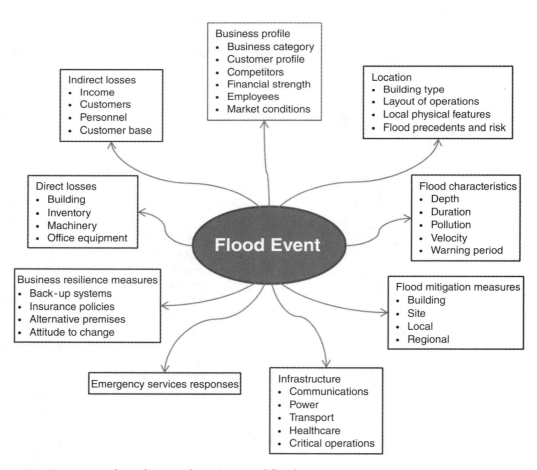

Figure 13.2 Data required in relation to businesses and flooding.

Sector	Data	Status			Source
		Acquired	In acquisition	Not available	
Base maps	DTM	x			EA
	Administrative boundaries	x			EA/GCC/DC
	Land use	x			GCC/DC/EA
	Census zones		x		GGC/DC
Physical event	Hazard zones	x			EA
	River track for floods	x			EA/IDB
	Monitoring data	x			EA/IDB
	Forecasting data	x			EA/IDB
	Affected areas	x			EA/IDB
	Hazard intensity		x		EA/IDB
	Induced landslides		x		EA
Protective measures for floods	Location/vulnerability	x			EA/LC
	Direct damage (physical)	x			EA/LC/IPP
	Direct damage (economic)	x			EA/LC
	Indirect damage (physical)		x		EA/LC
	Indirect damage (economic)		x		EA/LC
	Mitigation actions		x		EA/LC
People	Exposed people		x		GCC/LC
	Number of deaths	x			NHS/GHT/GFRS
	Number of injured	x			NHS/GHT/GFRS
	Number of affected people	x			NHS/GHT/GFRS
	Number of evacuees	x			GFRS
	Intangible damage	x			
	Mitigation actions (before/after)		x		
Infrastructure lifelines	Lifelines location/vulnerability		x		UC
	Direct damage (physical)		x		UC
	Direct damage (economic)		x		UC
	Indirect damage (physical)		x		UC
	Indirect damage (economic)		x		UC
	Mitigating actions (before/after)		x		UC

District Council (DC), Environment Agency (EA), Gloucestershire County Council (GCC), Gloucestershire Fire and Rescue Service (GFRS), Gloucestershire Hospital Trust (GHT), Internal Drainage Board (IDB), Individual Private People (IPP), Local Council (LC), National Health Service (NHS), Utility Companies (UC)

Figure 13.3 Data sources and categories (sectors) for 2007 Gloucestershire flood.

13.2.6. Sources of Data

One of the problems associated with forensic investigations is where to source the required data in order to make a useful analysis. Multiple agencies have to be consulted, including local authorities, emergency services, insurance companies, the affected businesses, research organizations, members of the general public who observed the event, etc. Figure 13.3 provides a short summary of where some of the flood event data is stored in the case of the flood in Gloucestershire, United Kingdom (UK) in 2007 with the Environment Agency (EA), Gloucestershire County Council (GCC), Internal Drainage Board (IDB), and district councils being identified as likely sources.

13.2.7. Categories of Data

As can be seen in Figure 13.3, data can be categorized under the type of information concerned. By establishing early on which categories are most relevant to achieving the most accurate analysis of the event, a more focused search for relevant data can be carried out. In terms of examining the effects of hazard events on businesses and establishing indirect damage, it is essential to include as many categories of data as possible. The particular characteristics of the affected business(es) must be investigated in order to establish the pattern of their vulnerabilities and hence the most influential aspects of the hazard. These are explored in more detail in the section on business vulnerability below. In the case of the study on the Gloucestershire floods in 2007, the categories involved (shown in Figure 13.3 under the column "Sector") were base maps, physical event, protective measures (e.g., dikes, walls, weirs) for floods, people, lifelines (roads, railways, electric lines, water supply, sewage, telecom), public items (public buildings/public spaces), strategic buildings (hospital, schools, headquarters, etc.), economic activities (commercial, industrial, agricultural), residential buildings, environment, cultural heritage, and emergency management.

13.2.8. Scales of Data

When considering what data are required for the forensic examination of a flood event related to commercial properties, the issue of scale (often also referred to as granularity) is a basic consideration. Commercial enterprises are affected by factors at a variety of scales, e.g., the direct damage to buildings and contents at the smallest scale, to disruption to infrastructure and community mechanisms at the middle scale, and local and national policy regarding flood prevention and mitigation at a national scale, and climate behavior and geo-characteristics, such as climate change on a global scale. The most influential factors for a particular event must be decided upon at the outset in order to limit the extent of necessary data gathering and can be categorized by scale as, for example, international, national, regional, local, or individual.

13.3. EXAMPLE OF FORENSIC INVESTIGATION PROCEDURE

To illustrate the steps required in forensic investigation, an example of a critical cause analysis relating to a flood event in Gloucestershire, UK in 2007 was developed and carried out as part of the European Union's Improving Damage Assessment to Enhance Cost-Benefit Analyses (IDEA) project.

It was decided that in order to establish a clear picture of the interrelations between infrastructure and business it was necessary to simultaneously establish the direct impact on infrastructure and the major contributing factors involved while also establishing the consequent impact on businesses. To this end, a forensic investigation was carried out that aimed to examine the complex dependencies of the infrastructure system.

Water supply relies on electricity supply, and normal operation of electricity relies on the road network working effectively. Once failure occurs in one element of the system, a knock-on effect can impact other elements. For example, once the Mythe Water Treatment Plant in Gloucestershire was shut down and water had to be distributed, the roads network became a crucial factor in the effectiveness of distribution.

Thus, a fragmented approach was adapted that entailed examining individual elements of the entire system separately at first with the idea of then collating all findings and establishing a more holistic view of the relationships and interdependencies of different elements. Water, electricity, sewerage, roads, and telecommunications were treated separately at first for ease of analysis before a system map was then drawn up to further visualize the operation of the infrastructure system.

Through this approach, it was possible to then analyze the impact on businesses of the failure of any one single element of the system. In the case of Gloucester, it emerged that failures, restrictions, closures, and systemic vulnerabilities meant that the network of infrastructures affected businesses in multiple ways.

Figure 13.4 shows a basic concept of how different infrastructures relate to each other. Electricity is essential

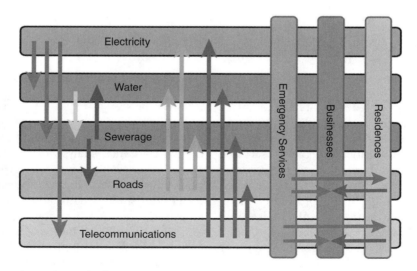

Figure 13.4 Interrelationships of infrastructure system and civil society.

for the functioning of all other infrastructures, as is telecommunications. Water and sewerage are mutually dependent. Roads are essential for the maintenance of all other infrastructure systems.

Emergency services, businesses, and residences as consumers rely on all infrastructures. Emergency services, although considered a critical infrastructure, are here illustrated separately from the others because it is considered that they are generally only essential in the case of emergencies and thus, during the event of a flood, are consumers. The diagram shows how there is also interaction between these groups via infrastructure. For example, businesses need roads and telecommunications to interact with residences or customers. The complexity of the interrelationships in this diagram show clearly how failure in any one infrastructure can result in a consequential impact in another infrastructure and may ultimately impact businesses and residences.

In this particular case, it was extremely valuable to gain insight into how particular sectors were affected in different ways by varying vulnerability factors in order to establish the most beneficial actions to take in building the resilience of businesses. This section explains how to arrive at the point where this is possible.

The forensic analysis began with the creation of a timeline of events as they unfolded, beginning with the first evidence of heavy rainfall. The entire event unfolded over quite a long period, and emergency operations and other actions took place outside of the research area and also unrelated to infrastructure and business. These events were considered to be outside the limits of this study.

It was deemed necessary to treat each individual element and sector of the infrastructure network separately by carrying out individual forensic investigations for each in order to firstly establish the direct impact and then take an overarching approach to how direct impact to each of these sectors indirectly impacted businesses. This decision was taken due to the complexity of the infrastructure network and the interrelations within the system.

From the timeline, an extensive list of the perceived contributing factors was created. This included such aspects as the suitability of emergency response plans in place before the flood and the times of flood warnings. Following this, the factors were categorized depending on whether they were considered to be related to the hazard or to physical, social, organizational, or systemic vulnerability. Some factors fell into several of these categories simultaneously and the categorization reflected this. This process is illustrated in Table 13.1.

Following this a judgment was made as to whether these factors contributed to or reduced damage, or both. This is relevant in that some factors were a combination of both. For example, the shutting down of a water treatment plant protected the machinery of the plant and thus the water company's assets and reduced long-term damage, but it had an impact on businesses and residences in the short term.

This exercise built up a strong picture of where the largest vulnerabilities were. For example, the majority of contributing factors may lie in the organizational vulnerability of a company or agency. This is to say that emergency planning, management, and decision-making would be the primary contributing factors in causing damage. Additionally, systemic issues may be prevalent in the causing of damage as the failure of a single element of an infrastructure system can result in large-scale impact.

Table 13.1 Forensic Analysis Listing Contributing Factors Assigned to Vulnerability Class.

Factor	Physical vulnerability	Social vulnerability	Organizational vulnerability	Economic vulnerability	Systematic vulnerability	Reduction of damage	Increasing of damage
Flood warnings			X				X
Many customers dependent on a single water main					X		X
Plant shut down as precautionary measure			X		X		X
Limited flexibility in system to draw water from other sources in event of failure					X		X
Misuse of bowsers resulting in loss of water	X	X	X				X
Road diversions	X		X				
Located on banks of River Severn near confluence with Avon	X						X
Increase in water usage following news of imminent flooding		X			X		X

Table 13.2 Assigning Impact to Affected Sectors.

Factor	Residences	Business	Infrastructures	Emergency services	Council	Environment Agency
Flood warnings	X	X		X		X
Many customers dependent on a single water main	X	X				
Plant shut down as precautionary measure	X	X				
Limited flexibility in system to draw water from other sources in event of failure	X	X				
Misuse of bowsers resulting in loss of water	X	X				
Road diversions	X	X	X			
Located on banks of River Severn near confluence with Avon	X	X				
Increase in water usage following news of imminent flooding	X	X				

Table 13.3 Assigning of Weight to Impact on Affected Sectors. Total is Calculated by Summing Weights Across all Sectors.

Factor	Residences	Businesses	Infrastructures	Emergency services	Council	Environment Agency	Total
Flood warnings	1	1		2		2	6
Many customers dependent on a single water main	2	2					4
Plant shut down as precautionary measure	3	3					6
Limited flexibility in system to draw water from other sources in event of failure	2	2					4
Misuse of bowsers resulting in loss of water	2	2					3
Road diversions	2	1	2				5
Located on banks of River Severn near confluence with Avon	1	1					2
Increase in water usage following news of imminent flooding	2	1					3

Following this step, an estimated weight was assigned to the direct impact of each factor on the main affected sector. In order to do this, first a judgment was made on which sectors were affected. For example, this could include businesses, residences, infrastructure companies, emergency operations, government agencies, etc. Once the affected sectors are decided on, a decision is made regarding which sector was affected by each factor. This is done by entering an "X" under the affected sector as shown in Table 13.2. In this case, it is clear that the hazard affected all sectors, and the large amount of information to deal with affects mostly emergency services and the Environment Agency.

The next step in the process was to make a judgment regarding the weight of the impact on each sector for each factor. This is done on a scale of 1 to 3 with 1 being minimal, 2 being significant, and 3 being major.

A value of −1, −2, or −3 may be assigned to factors if it is believed that these factors contributed more to a reduction of vulnerability than to an increase. The importance of a negative value is revealed in the following step when values are accumulated and thus reflect on the overall weight of contributing factors whether they reduce or contribute to damage. The value of this is clear where one particular factor may both reduce and contribute to damage for varying sectors. This is shown in Table 13.3.

In order to be able to analyze the real weight of factors of vulnerability and be able to comment on the varying degrees of relevance for different sectors, it was essential to be able to apply values to the importance of each factor. In order to do this, informed judgments were made, and there was an element of subjectivity. So the next step is a return to the previous section where the above weighted values are added together and input into the vulnerabilities columns as shown in Table 13.4. This gives a cumulative number of the impact of each vulnerability factor based on comparable parameters. This then allowed for the calculation of the total impact of each vulnerability in order to establish which is the most critical as shown in Table 13.5.

Having carried out forensic investigations for each of the previously outlined infrastructure sectors, the factors

Table 13.4 Total Weight of Factor is Substituted Back into Vulnerabilities Columns to Calculate total Significance for Each Vulnerability.

Factor	Physical vulnerability	Social vulnerability	Organizational vulnerability	Economic vulnerability	Systematic vulnerability	Reduction of damage	Increasing of damage
Flood warnings			6				6
Many customers dependent on a single water main					4		4
Plant shut down as precautionary measure			6		6		6
Limited flexibility in system to draw water from other sources in event of failure					4		4
Misuse of bowsers resulting in loss of water	3	3	3				3
Road diversions	5		5				5
Located on banks of River Severn near confluence with Avon	2						2
Increase in water usage following news of imminent flooding		3			3		3

Table 13.5 Total Weight of Each Vulnerability Calculated by Summing Accumulated Impact.

Factor	Hazard	Physical vulnerability	Social vulnerability	Organizational vulnerability	Economic vulnerability	Systematic vulnerability
Plant shutdown as a precautionary measure				−3		−3
Obliged to supply 10 liters of water per person			−4	−4		−4
Hub and Spoke approach to bowser refill				−2		
Built on artificially raised ground						
Reorganization of command structure. STW Gold Team established under control of Managing Director		−2		−2		
Site evacuated				−1		
Installation of semi-permanent flood protection after the plant had already been flooded		0		0		
Total Reduction	0	−2	−9	−28		−18
Total Increase	0	11	1	11		9
Overall	0		−8	−17		−9

that affected businesses indirectly were translated to a forensic investigation of the business sector. These factors were categorized as indirect factors, and direct factors were added that were not related to infrastructure. From this point, a similar process was followed for the infrastructure sectors. The aim of the forensic investigation for businesses was to compare and contrast the effect of direct and indirect vulnerabilities.

13.3.1. Indirect Damages

After accumulating and analyzing the impact of indirect damages to businesses caused by direct damage to infrastructure, it can be seen in Figure 13.5 that organizational vulnerabilities contributed more to damage than any other.

Organizational vulnerabilities of infrastructure refer to the decisions taken by management in the infrastructure

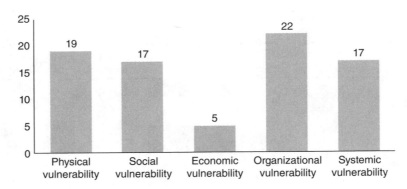

Figure 13.5 Significance of indirect vulnerabilities to business.

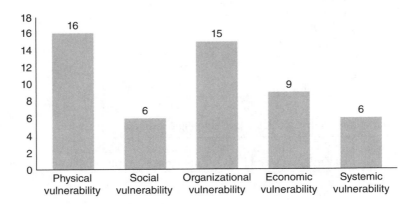

Figure 13.6 Significance of direct vulnerabilities to business.

companies that lead to increased impact on businesses. These decisions include the decision to shut the water treatment plant and the electricity supply plant, for example. In relation to roads and transport, they relate to the lack of communication to motorists about what to do in emergency situations.

Physical vulnerabilities were the second most significant for businesses. This includes the location of the water treatment plants and the power supply plants and roads that were all in physically vulnerable locations. Also, the nature of the emergency was particularly hazardous for both electricity and water treatment because water can be disastrous for both.

Following physical vulnerability is systemic vulnerability, which in infrastructure, relates to reliance on one single element of a system that exposes end users to the risk of losing service in the case of that one element being damaged or isolated. In this case, each of the infrastructure networks experienced this issue.

Social vulnerabilities include such factors as the misuse of water tanks that were provided by the water company once the water treatment plant had been shut down. This resulted in loss of water and damage to equipment. Additionally, the increased use of water once the closure

was announced as being imminent affected emergency supplies. It also refers to the ability of businesses to respond to the indirect damages because many had no previous experience of flooding of this scale, and emergency services were overwhelmed. The fact that many customers were lost due to being occupied with managing their own situations affected business as did the fact that many thought that particular towns were flooded and avoided shopping there for some time after they were cleared.

Economic vulnerabilities refer to the impact that reduced spending had on business as a result of people being occupied with dealing with the results of the flooding.

13.3.2. Direct Damages

Direct damages are those that directly affected businesses such as the flooding of premises, loss of stock, closure of business due to flooding, and relocation costs.

When it comes to direct vulnerabilities, physical vulnerability is again the most significant as seen in Figure 13.6. This relates mostly to flooding of properties and development of floodplains that led to properties being located in high flood risk areas.

Organizational vulnerabilities are the second most significant vulnerability and include the lack of knowledge with regard to recovery methods. Additionally, many businesses were not prepared for flooding and did not have business continuity plans or emergency action plans in place, which led to increased damage and longer recovery times.

In contrast to indirect damages, economic vulnerabilities are more significant. This relates to the cost of repairs and recovery and to difficulties in accessing insurance and public funding.

Again in contrast to direct damage, systemic vulnerability is less significant. This is due to the nature of business in comparison with the nature of infrastructure networks that depend on many individual elements of the system to operate functionally in order to maintain constant supply of services.

13.4. BUSINESS VULNERABILITY TO NATURAL DISASTERS

The most obvious effects of flooding on businesses is the physical damage to assets including buildings, stock, and transport vehicles.

In the UK, the estimated economic costs of the summer 2007 floods to businesses (buildings, contents, and disruption) was £740 million, comprising 23% of the total costs of the floods, which amounted to about £3.2 billion in 2007 prices, within a possible range of between £2.5 and £3.8 billion. Power and water utilities accounted for about 10% (£0.33 billion) of total costs, and communications (including roads) accounted for about 7% (£0.23 billion). There were between 7,100 and 7,300 business properties damaged by flooding [Chatterton et al., 2010].

Although the physical damage may be obvious, there are difficulties in collecting data about the nature and scale of damage on a micro and macro scale. Reports and papers have repeatedly pointed out that although there is an internationally accepted standard approach for assessing urban flood losses, that is, tracking loss in relation to different types or uses of affected buildings against the depth of inundation, expressed either in monetary value or in percentage loss of the assets including buildings and contents, there remains a large uncertainty about the accuracy and reliability of the estimates [Thieken et al., 2008]. Reasons for this are the limited scale of data collection after events, the inaccuracies of loss data, and other factors such as lack of data on floodwater characteristics, building specifications, and mitigation measures [Smith and Greenaway, 1992; Penning-Rowsell, 1999; Kreibich et al., 2008; Thieken et al., 2005].

The variables relevant to business vulnerability to natural disasters are more extensive than those of physical vulnerability due to the direct and indirect effects of the event. Economic and societal issues play an important part in rendering businesses more or less vulnerable even though they may be affected by the same physical events. A forensic approach offers the ability to identify these vulnerabilities as shown in the previous example.

A summary of potential vulnerabilities is provided by Tierney [2007] and are considered in the following sections.

This study focused on cities in Iowa in the United States (US) following the 1993 floods and concentrated on the effects of lifeline service interruptions on over 1,000 businesses. It concluded that the loss of water supply was the most critical reason for business closure (63.6%), followed by electricity supply (41.7%), with loss of sewer service (34.8%) and customers (34.4%) next. Interruption to telephone service (28.3%), employees unable to get to work (26.3%), and the inability to deliver services and products (25.7%) were next in priority, with evacuation due to threat of flooding (21.4%) and actual flooding (19.9%) being a minority reason. Hence, although physical damage from flooding was an important source of business interruption for those affected by the floodwater, it can be seen that significantly more businesses were closed due to the loss of critical lifeline services leading to loss of customers, employee absence due to travel disruption, and interruptions in the flow of supplies. A study by the UK Royal Institution of Chartered Surveyors (RICS) on the effects of the 2009 flooding on small and medium sized enterprises (SME) in the UK area of Cockermouth confirmed that the impacts extensively spread beyond the flooded area. Most important, secondary effects were travel difficulties for customers, employees, suppliers, and customers directly affected by the floods as well as loss of electricity and other services [Webb et al., 2000].

These results indicate that while increasing resilience of buildings to flooding in businesses at risk will help to ensure continuity of their business operations, a greater stress must be placed on ensuring the continuity of lifeline services, interruptions of which affect far more companies outside the immediate flood area. It should be noted that while most businesses are insured against flooding, few have any insurance against outages of lifeline services.

13.4.1. Business Profile: Categorizing Business Types

Differences in varying impacts of flooding on businesses have been accredited to a large variety of business types and the differences in the vulnerability of businesses to flood damage [Gissing and Blong, 2004; Bingunath Ingirige et al., 2013]. In order to estimate loss, mitigate hazards, and prepare for disasters and subsequent recovery, it is important to understand the nature of business vulnerability and the factors that are significant in measuring the level of vulnerability. According to Tierney [1995], only a small number of studies have focused systematically on this subject.

Very low	florists, churches, garden centers
Very low to low	cafes/takeaway, sports pavilions, consulting rooms, news agents, second hand goods
Low	food, retail outlets, butchers, bakers, service station, pubs
Low to medium	vehicle sales, offices, doctors, surgeries, restaurants, post offices, clubs, hardware
Medium	libraries
Medium to high	printing, clothing, bottle shops
High	chemists, electrical goods, musical instruments
High to very high	cameras, pharmaceuticals, electronics

Figure 13.7 Damage categories according to commercial activities.

The majority of the damages relates to the value of the contents of the buildings rather than the buildings themselves. A review of these was produced by *Smith and Greenaway* [1992], which allocates different damage categories, from very low to very high, to a range of business premises. Examples of the damage categories for different commercial properties are shown in Figure 13.7.

13.4.2. Business Location

Areas vulnerable to natural disasters, whether earthquakes, floods, storms, etc., are often in places of high economic activity, and are therefore attractive places for businesses to locate. The economic and convenience factors usually outweigh considerations about possible natural disasters or of the relative vulnerability of the buildings they occupy [*Tierney*, 2007]. In England, it is estimated that 5.2 million (one in six) properties are at risk of flooding, of which more than one half are at risk from river and coastal flooding and two thirds from surface water flooding [*Penning-Rowsell et al.*, 2014]. It is also noted that different categories of business have different perceptions of risk of flooding, which can affect the resale value of property based on location in areas of low, medium, or high risk of flooding [*Bhattacharya-Mis and Lamond*, 2015].

13.4.3. Business Size

It is the smaller businesses and not-for-profit organizations that are particularly affected by flood events because of their limited resources and lack of geographical spread. This is especially true if they are under stress before the event. Larger firms generally have more resources to draw on in an emergency, and, particularly if they are spread over several sites, they have the potential to shift stock and activities to unaffected areas. They also tend to have a more dispersed clientele beyond the influence of the disaster [*Webb et al.*, 2000; *Dahlhamer and Tierney*, 1998]. Businesses that are part of a chain or franchise can call on resources beyond the effected branch, such as credit and national advertising. Even so, stronger firms, when they have been forced to suspend business for an extended period in the aftermath of the flooding, can lose a vital share of their market that they later find difficult to regain. It is therefore important to understand what "the variables that make the difference between surviving and failure in order to give sensible advice" are [*Alesch et al.*, 2001; *Runyan*, 2006].

Small businesses have an important role in communities by providing much employment and local engagement, while not-for-profit organizations are deeply embedded in the community that has invested much time, effort, and money in maintaining them. The variables relating to the effects of flooding on businesses are numerous and interact in many ways that are difficult to predict and even record. According to *Alesch et al.* [2001], it is not only the physical issues that are important in the recovery process but also the reactions of the customers and the attitude of the business owners. They identify five key factors that can affect the viability of small businesses affected by flooding:

• loss of critical production or service capacity, inventory, and capital assets
• state of the health of the business (e.g., a declining business is particularly vulnerable)
• damage to the customer base due to loss of their resources or moving away (e.g., due to suffering from the same flood event)
• loss of customers to the competitors during shutdown
• shortage of financial strength to overcome disruption
• inflexibility of business owners to adapt to changed market conditions after the flood

13.4.4. Reliance on Infrastructure

In a report produced as part of the FloodProBE project [*Heilemann et al.*, 2013], critical infrastructure is defined as infrastructure that is essential for the functioning of society, whose failure would seriously affect many people. The report offers an assessment of the vulnerability to infrastructure, which includes an analysis of the effects of element failure on a network. The approach adopted accounts for the secondary "knock-on" effects of infrastructure failure and the interdependency of infrastructures. In fact the study focuses more on these secondary impacts than on the direct impacts and offers a relevant analysis of the vulnerability of infrastructure and its impact on businesses.

Rapid recovery of businesses may be hampered more by the loss of critical infrastructure than by direct physical damage. *Tierney* [1997a] found businesses were forced to close by disruption to utility services more often than by direct damage during the 1993 flooding in the US Midwest. In Des Moines, Iowa, 15% of businesses experienced flood damage, and 42% were forced to close due to lack of regular services such as electricity or water. As noted by *Tierney* [1997b] following the Northridge earthquake, businesses located in areas where damage and disruption were widespread had more problems returning to normal operations, noting the influence of larger neighborhood and "ecological" factors. If a business itself is not directly impacted by a disaster, damage to offsite lifelines will still have a negative impact on recovery. Likewise, high tech companies in Silicon Valley lost systems and data when they were hit by a series of electricity failures in January 2001, which cost in excess of $100 million [*Gibb and Buchanan*, 2006]. A study by *Rose et al.* [1997], which involved computer modeling to predict the effects of an earthquake in Memphis, estimated a 7% loss in gross regional production based on a loss of electricity alone.

In a regional context, the most important elements are electricity, water supply, wastewater drainage systems, transportation, and communication systems. Welfare and social systems such as food distribution centers and financial centers as well as emergency services are also of importance for the functioning of normal services during a flood event [*McBain et al.*, 2010]. To quantify the impacts on society, it is important to assess the duration of disruption, the area affected, the number of people affected, and combinations of these, having first analyzed the effect of element failure on the network and then the effect of this on other networks.

Similar problems are faced by businesses subjected to other natural disasters. For example, the effects on small businesses of the Loma Prieta earthquake in the area between Santa Cruz and San Francisco in the US in 1989 were significant not only due to the damage to their buildings but to the disruption of the road transport system. Raised sections of the freeways were damaged causing road closures for a month [*Kroll et al.*, 1991].

Organizations, businesses, and infrastructure systems can be considered to be networks consisting of various actors and information flow. Studies demonstrate that networks consisting of a few key nodes, or hubs, and high connectivity between them, followed by an increasing number of other nodes with decreasing connectivity are less vulnerable to failure due to a deletion of nodes or breaking of links [*Allenby and Fink*, 2005]. Because external shocks to the network can happen at random, the likelihood that a peripheral hub would be affected is small, and the network will not lose its connectedness. However, many networks can have a weakness in that they are vulnerable to a shock that affects the most highly connected hubs. The Internet is one example of this in that Internet traffic is rarely disturbed by disruptions to smaller servers but can be extremely disrupted by a failure of a few key hubs [*Allen et al.*, 2005]. Such perspectives suggest that the more reliant a business is on the satisfactory operation of infrastructure, the more vulnerable it will be to the impacts of climate-related incidents such as flooding.

13.4.5. The Market and Economic Trends

Business vulnerability is also subject to the type of market within which it operates, its diversity, and the competition. Particularly vulnerable are businesses serving a very local market, such as those in the retail sector, because their clients are likely to be similarly affected by the flood incident and might need to curtail discretional spending or suspend purchases during the recovery time. Those businesses in an extremely competitive market could find their clients permanently switching to other companies following the disruption to services or supply of goods caused by flooding [*Tierney*, 2007].

Not all businesses will experience a downturn due to the effects of flooding or other natural disasters. Those in or associated with the construction industry will find a greater demand for their services. Also, those businesses that are able to resume operation more quickly than other competing businesses can benefit from increased sales due to lack of competition, especially if they can serve a local influx of people displaced by the event.

Businesses that are struggling for economic survival are vulnerable to even small disruptions. Also, those that are in a downward cycle of development find it difficult to regain the momentum to continue after an interruption.

13.4.6. Building Ownership

Investment in the premises by leaseholders is likely to be less extensive than that of owners. Even on a full repair lease, the tenants have little incentive to invest beyond the necessary maintenance and renewal, so they are unlikely to spend money on flood mitigation measures at the building or site level. Owners of leasehold properties are also unlikely to invest in measures to protect their leaseholders' property. The only situation where incentives are sufficient for capital investment is when the building owner occupies the building and has sufficient reasons to fear inundation. In addition, building owners are also in a stronger position to obtain loans and other aid based on the property value as collateral [*Dahlhamer and Tierney*, 1998]. In all cases, there is a strong argument to review business continuity plans in the event of an emergency, which may not entail any investment in the buildings.

13.4.7. Building Construction and Planning

In most industrial organizations, the buildings are a small part of the value of the business, but they are a significant element in the investment of many services and especially hospitality sectors. Construction types that are particularly vulnerable are those of traditional construction containing a lot of timber, plaster, and other absorbent materials. These can quickly lose their structural integrity, be difficult to dry out, and are badly effected by pollution. Steel and concrete framed buildings are less vulnerable to structural damage, though the presence of susceptible infill, insulation, and cladding lead to similar problems [*Bramley and Bowker*, 2002].

A basement presents a high risk during flooding, either directly from water pouring in or from seepage, and basements can be difficult to dry out after the event. Single story buildings do not provide the opportunity to store vulnerable items out of reach of the floodwater. The location of switchgear, cabling, and other electrical equipment such as air conditioning machinery, communication installations, etc., can also increase the vulnerability of the company if these are located at ground floor or below.

13.5. CONCLUSION: USING FORENSIC ANALYSIS FOR PLANNING BUSINESS RESILIENCE MEASURES

The use of forensic analysis to uncover the main causal factors and their interconnections of flood events and other natural disasters can provide reliable data for planning business resilience measures and aid predictive models of possible future events. The main feature of this type of analysis is the synthesis of a wide range of data relevant to the events collected and collated in order to identify the main factors that caused the flood and the effects that these had. A multidisciplinary approach is essential for the success of this type of analysis, covering technical, social, and institutional elements. Due to the complexity of the problem, a systems approach is essential to encompass the interplay of the factors, and care must be taken to rate the importance of these in order to produce a hierarchy of causes.

In application of these data to business resilience planning, the individual characteristics of the organization must be taken into account to devise a practical and economical strategy to cope with the threatened disruptions. The outputs of the analysis should be formulated so that businesses can use these to devise resilience measures without needing special expertise in interpreting the data. When the risks are identified, a practical and affordable corporate contingency program should be devised taking into account what is critical for the operation of the business where there is an absence of alternative strategies in the case of a flood. However, *Myers* [2006] asserts that

because there is likely to be a low probability of disruption to business continuity due to natural disaster, a different problem solving process is required to that of other business problems. Lengthy studies cannot be justified; contingency planning requires a specialized planning methodology that minimizes program development costs [*Myers*, 2006]. Whatever the risks, contingency programs should be preventive and forward focused, rather than being limited simply to recovery from incidents.

In order to assist in devising formal plans for response to and recovery from disruption caused by flooding or other events, *Li et al.* [2015] devised an agent-based model of SMEs in order to investigate varying strategies that SMEs may employ when responding to flood events. This models the pre-flood/during flood phase and aftermath phase, based on detailed data collected from two case studies and forensically analyzed using NVivo software to identify patterns and themes and their relationships. They identified the importance of resourcefulness and creative resource utilization within and outside the business as key attributes in underpinning organizational resilience during the disruptive phase of the event, as well as adequate insurance coverage and fast response from the insurance companies to minimize the period of business interruption and rapid full recovery. The agent-based model uses decision tree diagrams to illustrate the processes in the two phases.

Despite the fact that the means are now available to apply appropriate resilience policies in businesses based on reliable evidence, there are still major barriers to implementation [*Jerry Velasquez et al.*, 2015]. The main issue that therefore needs to be resolved now is to combat the apparent lack of concern and interest that most businesses have to instigate appropriate measures to guard against the effects of natural or man-made disasters that might occur sometime in the future.

REFERENCES

Alesch, D. J., J. N. Holly, E. Mittler, and R. Nagy (2001), *Organizations at risk: What happens when small businesses and not-for-profits encounter natural disasters*, Public Entity Risk Institute (PERI).

Allen C. R., L. Gunderson, and A. Johnson (2005), The use of discontinuities and functional groups to assess relative resilience in complex systems. *Ecosystems*, 8, 958–966.

Allenby, B., and J. Fink (2005), Toward inherently secure and resilient societies. *Science*, 309, 1034–1036.

Bhattacharya-Mis, N. and J. Lamond (2015), Risk perception and vulnerability of value: A study in the context of commercial property sector. *International Journal of Strategic Property Management*.

Bingunath Ingirige, D., N. Bhattacharya, J. Lamond, D. Proverbs, and F. Hammond (2013), Development of conceptual framework for understanding vulnerability of commercial property values towards flooding. *International*

Journal of Disaster Resilience in the Built Environment, 4, 334–351.

Bramley, M. E., and P. Bowker, (2002), Improving local flood protection to property. Proceedings of the ICE-Civil Engineering. Thomas Telford, 49–55.

Burton, I. (2010), Forensic disaster investigations in depth: a new case study model. *Environment, 52,* 36–41.

Chatterton, J., C. Viavattene, J. Morris, E. C. Penning-Rowsell, and S M. Tapsell (2010), The costs of the summer 2007 floods in England.

Cross, N. (2001), Design cognition: Results from protocol and other empirical studies of design activity.

Dahlhamer, J. M., and K. J. Tierney (1998), Rebounding from disruptive events: business recovery following the Northridge earthquake. *Sociological Spectrum, 18,* 121–141.

De Groeve, T., K. Poljansek, and D. Ehrlich (2013) Recording Disaster Losses. *Recommendations for a European Approach. European Commission Joint Research Center, Ispra, Italy.*

Gibb, F., and S. Buchanan (2006), A framework for business continuity management. *International Journal of Information Management, 26,* 128–141.

Gissing, A., and R. Blong (2004), Accounting for variability in commercial flood damage estimation. *Australian Geographer, 35,* 209–222.

Heilemann, K., E. Balmand, S. Lhomme, K. De Brujin, N. Linmei, and D. Serre (2013), Identification and analysis of most vulnerable infrastructure in respect to floods. *Floodprobe D2, 1.*

Integrated Research on Disaster Risk (IRDR) (2011), Forensic Investigations of Disasters. Available: www.irdrinternational. org/wp-content/uploads/2012/06/FORIN-REPORT_web. pdf. Last uploaded 5 May 2016.

Jerry Velasquez, D., N. B. Mis, R. Joseph, D. Proverbs, and J. Lamond (2015), Grass-root preparedness against potential flood risk among residential and commercial property holders. *International Journal of Disaster Resilience in the Built Environment, 6,* 44–56.

Kreibich, H., I. Seifert, A. Thieken, B. Merz, B., PROVERBS, D., C. Brebbia, and E. Penning-Rowsell (2008), Flood precaution and coping with floods of companies in Germany. 1st International Conference on Flood Recovery, Innovation and Response (FRIAR), London, UK, 2–3 July. WIT Press, 295–302.

Kroll, C. A., J. D. Landis, Q. SHEN, and S. Stryker (1991), Economic impacts of the loma prieta earthquake: A focus on small business.

Li, C., G. Coates, N. Johnson, and M. C. Guinness (2015), Designing an Agent-Based Model of SMEs to Assess Flood Response Strategies and Resilience. *International Journal of Social, Behavioral, Educational, Economic, Business and Industrial Engineering 9,* 7–12.

McBain, W., D. Wilkes, and M. Retter (2010), *Flood resilience and resistance for critical infrastructure,* CIRIA.

Myers, K. N. (2006), *Business continuity strategies: protecting against unplanned disasters,* John Wiley & Sons.

Penning-Rowsell, E. (1999), Flood hazard assessment, modelling and management: Results from the EUROflood project. *Environments, 27,* 79.

Penning-Rowsell, E., S. Priest, D. Parker, J. Morris, S. Tunstall, C. Viavattene, J. Chatterton, and D. Owen (2014), *Flood and coastal erosion risk management: a manual for economic appraisal,* Routledge.

Rodriguez, H., E. L. Quarantelli and R. R. Dynes (2006), *Handbook of disaster research.* New York: Springer.

Rose, A., J. Benavides, S. E. Chang, P. Szczesniak, and D. Lim (1997), The regional economic impact of an earthquake: Direct and indirect effects of electricity lifeline disruptions. *Journal of Regional Science, 37,* 437–458.

Rosson, M. B., and J. M. Carroll (2009), Scenario based design. *Human-computer interaction. Boca Raton, FL,* 145–162.

Runyan, R. C. (2006), Small business in the face of crisis: Identifying barriers to recovery from a natural disaster. *Journal of Contingencies and Crisis Management, 14,* 12–26.

Smith, D., and M. Greenaway (1992), ANUFLOOD–A Field Guide. *Australian National University, Centre for Resource and Environmental Studies.*

Thieken, A., A. Olschewski, H. Kreibich, S. Kobsch, B. Merz, D. Proverbs, C. Brebbia, and E. Penning-Rowsell (2008), Development and evaluation of FLEMOps-a new Flood Loss Estimation MOdel for the private sector. 1st International Conference on Flood Recovery, Innovation and Response (FRIAR), London, UK, 2–3 July 2008., 2008. WIT Press, 315–324.

Thieken, A. H., M. Müller, H. Kreibich, and B. Merz (2005), Flood damage and influencing factors: New insights from the August 2002 flood in Germany. *Water resources research,* 41.

Tierney, K. (1997a), Impacts of recent disasters on businesses: The 1993 Midwest floods and the 1994 Northridge Earthquake (No. NCEER-SP-0001). *Buffalo: Multidisciplinary Center for Earthquake Engineering and Research, State University of New York at Buffalo.*

Tierney, K. J. (2007), Businesses and disasters: vulnerability, impacts and recovery. *In:* H. Rodriguez, E. L. Quarantelli, and R. R. Dynes (eds.) *Handbook of Disaster Research.* New York: Springer Science + Business Media.

Tierney, K. J. (1995), Impacts of recent US disasters on businesses: the 1993 Midwest Floods and the 1994 Northridge Earthquake.

Tierney, K. J. (1997b), Business impacts of the Northridge earthquake. *Journal of Contingencies and Crisis Management, 5,* 87–97.

Vojinović, Z. (2015), *Flood Risk: The Holistic Perspective: From Integrated to Interactive Planning for Flood Resilience,* IWA Publishing.

Webb, G. R., K. J. Tierney, and J. M. Dahlhamer (2000), Businesses and disasters: Empirical patterns and unanswered questions. *Natural Hazards Review, 1,* 83–90.

Wenzel, F., J. Daniell, B. Khazai, and T. Kunz-Plapp (2012), The CEDIM Forensic Earthquake Analysis Group and the test case of the 2011 Van earthquakes. 15th World Conference on Earthquake Engineering (WCEE).

Wenzel, F., J. Zschau, M. Kunz, J. E. Daniell, B. Khazai, and T. Kunz-Plapp (2013), Near Real-Time Forensic Disaster Analysis. Proceedings of the 10th International ISCRAM Conference, Baden–Baden, Germany.

Part V
Information and Communication Technology Tools

14

Response to Flood Events: The Role of Satellite-based Emergency Mapping and the Experience of the Copernicus Emergency Management Service

Andrea Ajmar[1], Piero Boccardo[2], Marco Broglia[3], Jan Kucera[3], Fabio Giulio-Tonolo[1], and Annett Wania[3]

ABSTRACT

This chapter focuses on the emergency mapping activities generally carried out to support the immediate response to flood events, mainly describing the role of satellite data. An introduction on emergency mapping is provided, presenting the Copernicus Emergency Management Service (EMS), the emergency mapping service provided by the European Union (EU) in the frame of the Copernicus Program. The ongoing relevant international initiatives and their link with the Copernicus EMS are presented.

The focus is on the mapping activities related to the response phase of the emergency management cycle, also known as Rapid Mapping. The main goal of Rapid Mapping activities is described and discussed, i.e., the identification of the affected areas and, when possible, the assessment of damages to the main assets (e.g., infrastructures, agriculture, etc.). A brief analysis of relevant disaster statistics shows the occurrence trend of flood events, explaining why most of the emergency mapping activations are related to this type of disaster. The operational methodologies (and the input data types) generally exploited to identify flooded areas and to assess flood-related damages are described, highlighting advantages and limitations.

Lastly, potential new developments made possible by new technologies (e.g., satellite constellations decreasing the revisiting time and unmanned aerial vehicle [UAV] providing imagery with a higher level of detail) are discussed.

14.1. INTRODUCTION

Disasters and crises affect millions of people annually and cause significant damage worldwide. In particular, floods are among the most catastrophic natural disasters globally in terms of impact on human life and the economy. For example, in the last two decades, they accounted for 55% of people affected and caused a major share of the economic damages [Revilla-Romero et al., 2015]. Although the frequency of geophysical disasters (earthquakes, tsunamis, volcanic eruptions, and mass movements) remained broadly constant in the past 20 years, a sustained rise in climate-related events (mainly floods and storms) pushed total occurrences significantly higher [Centre for Research on the Epidemiology of Disasters, UN International Strategy for Disaster Reduction (CRED, UNISDR) 2015]. Between 1994 and 2013, floods accounted for 43% of all disasters (a total of 6,873

[1] Information Technology for Humanitarian Assistance, Cooperation and Action (ITHACA), Turin, Italy

[2] Interuniversity Department of Science, Design and Land Policies, Politecnico di Torino, Turin, Italy

[3] European Commission Joint Research Centre, Ispra, Italy

Flood Damage Survey and Assessment: New Insights from Research and Practice, Geophysical Monograph 228, First Edition. Edited by Daniela Molinari, Scira Menoni, and Francesco Ballio.
© 2017 American Geophysical Union. Published 2017 by John Wiley & Sons, Inc.

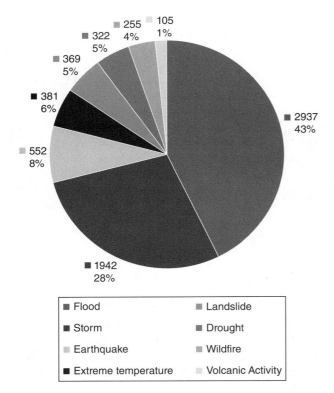

Figure 14.1 Share of occurrence of natural disasters by disaster type (1994–2013). [CRED, UNSIDR, 2015]. *See electronic version for color representation.*

worldwide) registered by the Emergency Events Database EM-DAT [*Guha-Sapir et al.*, 2017]. See Figure 14.1.

Emergency Response (ER) and Disaster Risk Reduction and Management (DRRM) activities need timely information on the event extent, impact, and evolution. A wide spectrum of space/airborne data and ground measurements is available and potentially supports ER and DRRM, offering different options in terms of geometric and radiometric resolution, timeliness, type of sensor (e.g., passive/active), etc. The complete chain from data provision to data processing and information rendering technologies and methods is experiencing a continuous evolution over the last decades allowing several services to evolve from research to the operational domain. One example is the Copernicus EMS, which has emerged from several precursor research projects. It provides, among other services, Rapid Mapping support to emergencies related to flood events. The following paragraphs present the most relevant elements of Rapid Mapping for floods and briefly describe the setup of the Copernicus EMS and a specific example of Rapid Mapping for one recent flood in Europe.

14.2. EMERGENCY MAPPING

Emergency management is defined as "The organization and management of resources and responsibilities for addressing all aspects of emergencies, in particular pre-

paredness, response and initial recovery steps" [*UNISDR*, 2009]. The Emergency Management process is organized into a cycle of four, often overlapping, phases: mitigation, preparedness, response, and recovery [*National Earthquake Hazards Reduction Program (NEHRP)*, 2009]. Mitigation involves actions that are taken to eliminate or reduce the degree of long-term risk to human life and property from hazards. Preparedness is concerned with actions that are taken in advance of an emergency to develop operational capabilities and facilitate an effective response to an emergency. The response phase involves actions that are taken immediately before, during, or directly after an emergency occurs to save lives, minimize damage to property, and enhance the effectiveness of recovery. The recovery phase is characterized by the activity to return life to normal or improved levels [*Cova*, 1999].

Emergency Mapping or EM, defined as "creation of maps, geo-information products and spatial analyses dedicated to providing situational awareness and immediate crisis information for response" [*International Working Group on Satellite-based Emergency Mapping (IWG-SEM)*, 2014], is widely recognized as being among the main and more effective instruments used by organizations and institutions involved in emergency management. Emergency Mapping products are instruments to effectively transfer awareness and knowledge on location and magnitude of a disaster. The specific Emergency Mapping activities carried out to support the response phase in the first hours/days after the disaster are generally referred to as Rapid Mapping. Satellite or aerial imagery is the most exploited source for extracting reference (pre-event) and crisis (post-event) geographic data. This is the reason why Satellite-based Emergency Mapping (SEM) is the solution adopted by most international initiatives to perform emergency-related analysis.

14.2.1. Overview on International Initiatives

The most critical aspect for organizations operating in the field of SEM is the systematic access to appropriate satellite or aerial imagery. For this reason, some of the international initiatives (i.e., the International Charter "Space and Major Disasters,"[1] Sentinel Asia,[2] NASA-SERVIR,[3] and Tomnod[4]) incorporate satellite data providers. Other programs such as Copernicus[5] and organizations (UNOSAT[6]), have signed a special agreement with data providers, to have a structured and privileged access to imagery. Volunteer-based initiatives, based on either

---

[1] *https://www.disasterscharter.org*
[2] *https://sentinel.tksc.jaxa.jp*
[3] *https://www.servirglobal.net/*
[4] *http://www.tomnod.com/*
[5] *http://www.copernicus.eu/*
[6] *http://www.unitar.org/unosat*

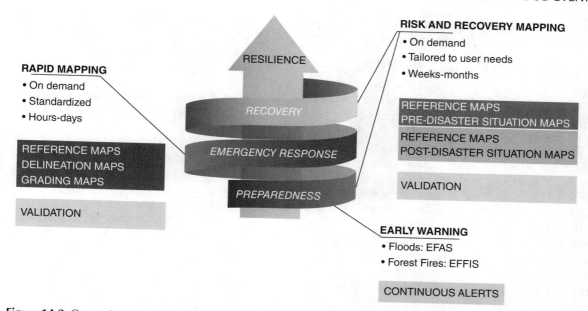

RAPID MAPPING
- On demand
- Standardized
- Hours-days

REFERENCE MAPS
DELINEATION MAPS
GRADING MAPS

VALIDATION

RESILIENCE

RECOVERY

EMERGENCY RESPONSE

PREPAREDNESS

RISK AND RECOVERY MAPPING
- On demand
- Tailored to user needs
- Weeks-months

REFERENCE MAPS
PRE-DISASTER SITUATION MAPS

REFERENCE MAPS
POST-DISASTER SITUATION MAPS

VALIDATION

EARLY WARNING
- Floods: EFAS
- Forest Fires: EFFIS

CONTINUOUS ALERTS

Figure 14.2 Copernicus Emergency Management Service–overview of the two main components, Early Warning and Mapping, with the two mapping modules, and indication of which phases of the disaster management cycle each is supporting. *See electronic version for color representation.*

Geographic Information System (GIS) professionals (e.g., MapAction[7]) or common citizens (e.g., OpenStreetMap[8], MicroMappers[9]), exploit open access remotely sensed data to acquire and update both reference and disaster-specific data. Some of those initiatives are also active in developing and promoting tools for browsing and accessing to openly licensed remote sensing imagery. This is the case, for example, of the OpenAerialMap[10] initiative promoted by the Humanitarian OpenStreetMap Team (HOT[11]). Finally, formal (e.g., UN-SPIDER[12]) and informal (IWG-SEM[13]) initiatives support cooperation and capacity building in the field of space-based information for disaster management and emergency response.

14.2.2. Copernicus Emergency Management Service

The Copernicus EMS is one of the six services offered by Copernicus,[14] the EU's Earth Observation program for global environmental monitoring, disaster management, and security. EMS provides support to all actors involved in the fields of crisis management, humanitarian aid as well as disaster risk reduction, preparedness, and prevention. The service has two main components: Early Warning and Mapping. The Early Warning component currently covers only floods and forest fires, and the Mapping component addresses a wide range of emergency situations resulting from natural or man-made disasters, among them floods. Figure 14.2 depicts the general service overview.

The Copernicus EMS Mapping has been operational since 2012, and it builds on experience gained in GMES and EU 6th and 7th RTD Framework Programme precursor projects (RESPOND, LinkER, BOSS4GMES, PREVIEW, SAFER) and in dedicated thematic workshops, working groups, and other user meetings.

EMS Mapping is an on-demand service that can be triggered through the Emergency Response and Coordination Centre (ERCC) at the European Commission's Directorate-General for European Civil Protection and Humanitarian Aid (DG ECHO). Due to cost restrictions of the service (especially related to the cost of satellite image data), activation of this service is subject to a selective process. The service can be thus requested only by EMS authorized users, which include National Focal Points in EU member states and in countries participating in the European Civil Protection Mechanism (usually institutions involved in civil protection and emergency management), as well as European Commission services (e.g., the ERCC) and the European External Action Service (EEAS). Other users (e.g., international humanitarian aid organizations, the United

---

[7] http://www.mapaction.org/

[8] https://www.openstreetmap.org/

[9] http://micromappers.org/

[10] https://openaerialmap.org/

[11] https://hotosm.org/

[12] http://www.un-spider.org/

[13] www.iwg-sem.org/

[14] Regulation (EU) No. 377/2014 of the European Parliament and of the Council of 3 April 2014, establishing the Copernicus Programme and repealing Regulation (EU) No. 911/2010

Nations, World Bank) must coordinate with and go through the authorized users in order to trigger the service.

EMS Mapping provides geospatial information, maps, and analyses based on satellite imagery before, during, or after a disaster. Depending on the need of the user and for which phase of the disaster management cycle geo-information is required, it is provided in two different modes: Rapid Mapping or Risk and Recovery Mapping. Both differ in terms of map delivery times, type of analysis performed, and level of standardization in the workflow. On the one hand, Rapid Mapping provides geo-information within hours or days, immediately following the event (fast provision). With its availability of 24 hours/365 days, it supports emergency management activities immediately following an emergency event. Both the workflow and products are standardized. On the other hand, Risk and Recovery Mapping operates during work days and working hours only. It is designed for pre- or post-crisis situations and provides geo-information within weeks or months after the service activation (typically one month) in support of recovery, disaster risk reductions, prevention, and preparedness activities. Access to the geo-information and other results generated by both Mapping modes is open and free of charge on the EMS Portal[15] (according to the Copernicus dissemination policy). Under exceptional circumstances, access restrictions may be imposed for security reasons or the protection of third party rights (sensitive activation). For more information, see the EMS User Guide[16] on the EMS portal. The specific service tasks are carried out by service providers (i.e., operating entities) who belong to a consortia of European companies, academia, and public institutions. The current operational setup is supposed to run throughout the current phase of the Copernicus program until 2020.

Within EMS Mapping, service quality and user satisfaction are continuously monitored through dedicated feedback forms, systematic quality check of EMS Rapid Mapping (EMS-RM) maps and during validation exercises.

The validation module of the EMS aims at supporting the continuous improvement of both Mapping modules by checking the outputs of both modules on a sample basis. The first step (product validation) checks the reliability and consistency of the information contents (through comparison with other data sources) as well as the usability of the product. The second step (product evaluation) aims at estimating the added value for and the impact of Copernicus products on the user workflow. Definition, application, analysis, and comparison of alternative methodologies are included as well in the scope of the validation module.

14.3. COPERNICUS EMERGENCY MANAGEMENT SERVICE RAPID MAPPING

Copernicus EMS-RM covers the entire process from the satellite tasking, image acquisition, processing, and analysis of satellite imagery and other geo-spatial raster and vector data sources until the production and delivery of maps and vectors to the user who requested the service and delivery to the public. Since its start in April 2012, Copernicus EMS-RM was activated 144 times in total and 59 times for flood events, which accounts for 41% (status as of 31 October 2015). On average there are 42 activations per year out of which 15 are for flood events.

14.3.1. Most Common Product Categories

EMS-RM products are standardized maps with a set of parameters to choose when requesting the service activation (map type, scale, delivery time, additional information layers to be included).

There are three map types, one pre-event map and two post-event maps, each of which performs specific functions relevant to crisis management. The pre-event or reference maps provide comprehensive knowledge of the territory and exposed assets and population. They are based on satellite imagery and other geo-spatial data acquired prior to the disaster event and are often used for comparative purpose as a baseline for generating the post-event maps. The two post-event map types, known as delineation and grading maps, are produced from post-disaster images. Delineation maps outline the extent of the area affected by an event and its evolution. Grading maps provide an assessment of the impact caused by the disaster in terms of damage grade and of its spatial distribution. They may also provide relevant and up-to-date information on affected population and assets, for example, settlements, transport networks, industry, and utilities.

In the fastest map production mode, the service provides both post-event maps within 12 hours after image data reception and quality acceptance. In addition, a First Available Map (FAM) is delivered within three hours. It is an early information product whose content is as close as possible to the final maps, but which has not undergone extensive quality review and may lack some summary information derived from the analysis. For less time-critical activations, the same maps can be provided in five working days. In such cases, FAMs are accordingly

---

[15] emergency.copernicus.eu/mapping

[16] http://emergency.copernicus.eu/mapping/ems/copernicus-ems-user-guide

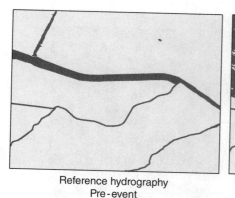
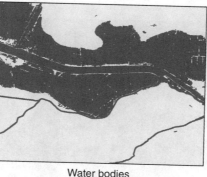
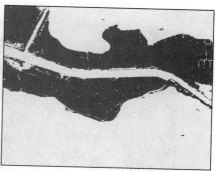

Reference hydrography
Pre-event

Water bodies
Post-event

Flooded areas
Post-event

Figure 14.3 Flooded areas derived from the water bodies extracted from the post-event imagery and the pre-event reference hydrography. *See electronic version for color representation.*

not provided. Reference maps are provided either in nine hours (fastest mode) or five working days.

The final delivery is available in three different formats: geo-referenced raster (GeoTIFF and JPG), printable maps (Geospatial PDF), and vector files of all derived features (Esri shapefile, Google Earth KML or KMZ). The standard sheet format for all raster and printable maps is A1 in three resolutions (100, 200, and 300 dpi).

All maps are by default provided in Universal Transverse Mercator (UTM) cartographic projection using WGS84 geodetic system (EPSG code 4326).

14.3.2. Rapid Mapping of Flood Events

According to the European Union Floods Directive, "[...] 'flood' means the temporary covering by water of land not normally covered by water. This shall include floods from rivers, mountain torrents, Mediterranean ephemeral water courses, and floods from the sea in coastal areas, and may exclude floods from sewerage systems" (Directive 2007/60/EC[17] Chapter 1, Article 2).

Flood events are quite dynamic and, in some cases (i.e., flash floods) can be considered rapid onset disasters. It is therefore fundamental to correctly plan, in time and space, imagery acquisitions. Early warning systems provide adequate information for predicting where and when a flood event may occur. In the EMS, the European Flood Awareness System (EFAS[18]) provides flood early warning information across Europe up to 10 days in advance. Using similar concepts developed for the continental EFAS, a Global Flood Awareness System

(GloFAS[19]) is currently being developed. The information from both is supporting the definition of areas and the scheduling of image acquisitions (including pre-tasking of satellites) in EMS-RM.

In the Rapid Mapping context, the most relevant information that can be extracted from imagery acquired by satellite (or airborne) sensors is the floodwater extent at a given time. In order to derive a flood extent, it is necessary to subtract, from the floodwater extent, all surfaces that are normally covered by water (Figure 14.3); those are usually referred to as normal water bodies and include different natural and artificial surfaces, such as rivers, lakes, reservoirs, etc. The availability of normal water bodies is therefore fundamental to be able to map the flood water extent in areas where reference water extent significantly varies in time, i.e., seasonal variations of braided rivers, a situation as close as possible to the analyzed event allows to derive the event specific flood extent.

The impact of a flood event is estimated from the damages to natural and, mainly, artificial features; therefore, an as complete as possible and up-to-date set of reference features is required. A first and rough identification of the potentially affected features can be obtained by calculating the spatial intersection between the floodwater extent and the reference features. A more complete and accurate damage assessment can be obtained by analyzing high-resolution optical images and by detecting damages, possibly classified on the basis of severity grades. This analysis is normally based on visual interpretation, and its accuracy can be significantly increased by the availability of pre-event imagery acquired with similar characteristics. An example related to an erosion phenomena impacting the road network is shown in Figure 14.4.

[17] *Directive 2007/60/EC of the European Parliament and of the Council of 23 October 2007 on the assessment and management of flood risks.*

[18] *http://www.efas.eu*

[19] *http://www.globalfloods.eu/*

(a)

(b)

Figure 14.4 Example of erosion phenomena related to a flood event and impact on the road network. (a) Pre-event situation (Source: ESRI World Imagery © Digital Globe); (b) Post-event situation (aerial orthoimagery © European Union) with road damage assessment (completely destroyed in red and partially affected in yellow. (Data Source: Copernicus Emergency Management Service © European Union, 2015), [EMSR 138], Flooding and landslides in Emilia Romagna, Italy. *See electronic version for color representation.*

Figure 14.5 shows an example of a damage grading map related to a flood event in southern Italy in November 2013. It shows the result of the damage or impact assessment from the comparison of post- and pre-event imagery, with a focus on the road network, utilities, and the magnitude of flooding.

Products depicting the delineation of the floodwater extent normally include all relevant information about the event, the data sources used for the analysis, and the methods applied to extract the value-added information. In case of flood Rapid Mapping, due to the high dynamics of this type of event, it is fundamental to clearly state the moment in time displayed in the map. For the same reason, a flood delineation map can also depict the evolution in time of the events by synthetizing it in the same product, e.g., by using different colors, the flood extent at different times (example shown in Figure 14.6). Flood impact maps, also known as grading maps, normally include reference features classified according to the level of damages they experienced.

As of 31 October 2015, most of the post-event maps produced in Copernicus EMS-RM service activations for flood events are delineation maps. In fact, 92% of the 534 post-event maps are delineation; 8% are grading maps.

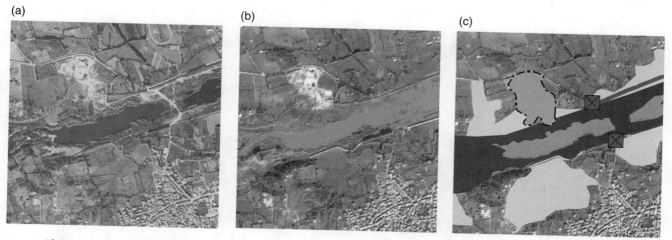

Figure 14.5 Example (detail) of a grading map showing damages on the road network, industrial utilities, and areas affected by floodwater: (a) Pre-event imagery; (b) Post-event imagery; (c) Post-event imagery overlaid with crisis information: road blocks are identified by the crossed orange square; highly affected roads in orange; highly and moderately flooded areas in dark and light blue, respectively; moderately affected quarry in light orange. Data source: Copernicus Emergency Management Service (© European Union, 2013), [EMSR061] Olbia: Grading Map (Detail 1). *See electronic version for color representation.*

14.3.3. Typical Workflow in Copernicus Emergency Management System Rapid Mapping

Figure 14.7 depicts the workflow of Copernicus EMS-RM. The main steps follow:

1. Request for service activation: The authorized user submits the service request form (SRF) to the ERCC, usually via email. The SRF contains background information on the disaster, the approximate area of interest, required map types, production modes, map scales, whether or not monitoring of the post-event maps is required, and required additional topographic features (e.g., hydrology, settlements, transport).

2. Criteria check: The ERCC evaluates the eligibility of the request against a set of criteria in the following categories:

 • Scope: large-scale natural disasters or man-made accidents

 • Impact: exceeding the capacity of the national or regional authorities

 • Urgency: requiring an imminent response

 • Technical feasibility, for example, the extent of area, the feasibility of satellite imagery acquisition, and weather conditions

 • Sensitivity check, for instance, in case the request could harm the security interests of the EU, its member states, or its international partners

 • In case of acceptance, the request is forwarded to the service provider.

3. Product specification and satellite image ordering: The service provider obtains the product specifications and submits a request for satellite image data to the operators of the Rapid Emergency Activation for Copernicus Tasking (REACT, see next step).

4. Image data provision: Copernicus EMS-RM uses satellite image data provided in the frame of the Copernicus program, either by the Sentinels, which are owned by the program, or by other image providers, the so-called Copernicus Contributing Missions (CCM). REACT is the specific mechanism provided by the European Space Agency (ESA) for accessing these data in 24/7 fast mode, either through tasking of new acquisitions or data retrieval from the archive. REACT interacts with image providers on tasking opportunities and image availability and defines accordingly a satellite acquisition and delivery plan.

5. Production: After the reception of the imagery and their quality acceptance, the service provider starts the map production. There might be several iterations between the service provider, REACT, and the user before completion.

6. Product delivery: Final products are delivered to the user via secure file transfer protocol followed by an automatic upload of all products (vector and raster) on the EMS Mapping portal where the public can access them.

Figure 14.8 shows the average duration of steps controlled by the service provider for the handling of reference, delineation, and grading maps from May 2012 to September 2015. Map production and delivery time refer to version 1 (excluding FAM). A large amount of the overall time for these four steps is spent on satellite data ordering, tasking, acquisition, and delivery (66%). Map production and delivery together (steps 5 and 6) cover 25% of the time.

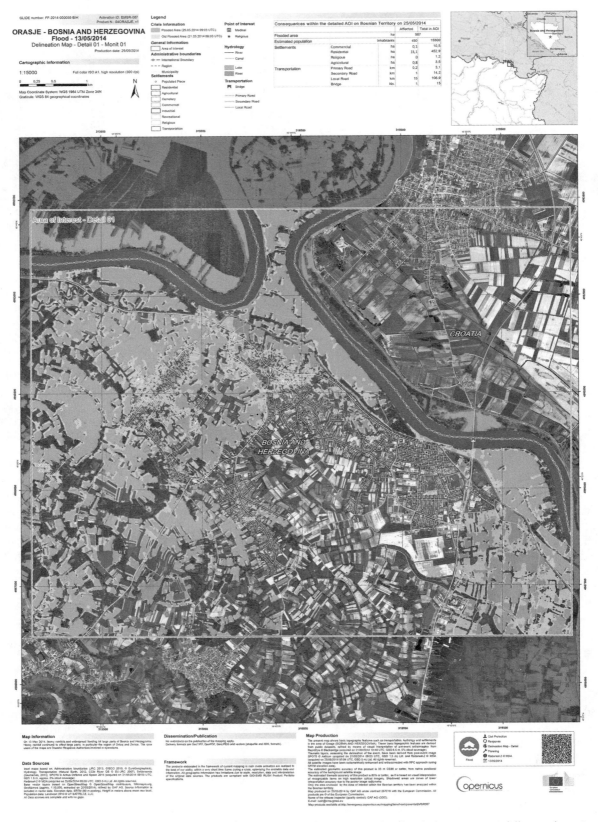

Figure 14.6 Example of a flood delineation monitoring map, where the flood extent at two different dates is displayed with different transparent colors (previous flood extent in light blue, most recent flood extent in dark blue). Data source: Copernicus Emergency Management Service (© European Union, 2014), [EMSR087] Orasje: Delineation Map, Monitoring 1. *See electronic version for color representation.*

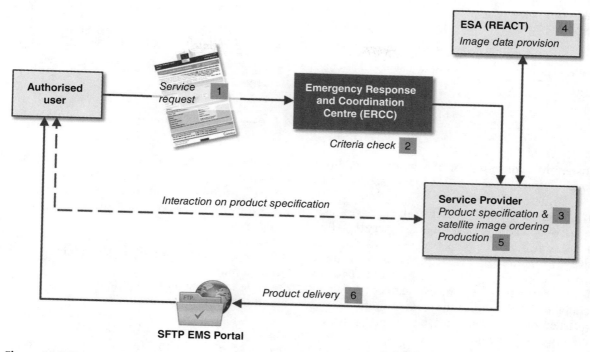

Figure 14.7 Copernicus EMS-RM workflow (simplified). *See electronic version for color representation.*

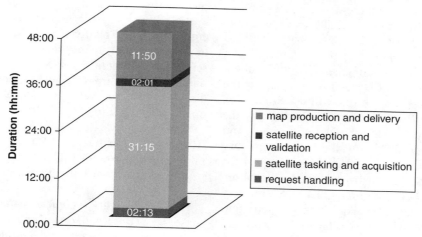

Figure 14.8 Average duration of the four main steps controlled by the EMS-RM service provider (April 2012–September 2015). Note that the value for "map production and delivery time" is the average for version 1. *See electronic version for color representation.*

14.4. FLOOD IMPACT ASSESSMENT: OPERATIONAL APPROACH

This chapter focuses on the description and analysis of common operational approaches generally adopted in the framework of Emergency Mapping activities to extract flood impact information from available post-event imagery. As described in section 14.3.2, the main focus of the post-event analysis is on delineating the affected areas, i.e., the flooded areas, and/or the consequent damages to the main infrastructures. The choice of the technical approach to be adopted is mainly driven by

the required post-event analysis type. The most common approaches are described in section 14.4.2.

14.4.1. Input Data Types

Different input data types (both in terms of sensor type and acquisition platform) can be exploited to extract valuable post-event information in case of flood events. In the framework of Emergency Mapping, the choice cannot be based only on technical requirements (e.g., the level of detail of the analysis or the final product type), since operational constraint such as the timeliness of the

Table 14.1 Imagery Used to Produce Post-Event Maps for Flood Events (as of 31 October 2015); Number of Scenes Per Sensor Type for Delineation and Grading Maps.

Sensor type	Grading + delineation		Delineation		Grading	
	Number of scenes	%	Number of scenes	%	Number of scenes	%
Radar	412	80	410	88	2	4
Optical	102	20	55	12	47	96
Total	514		465		49	

first post-event acquisition or the weather conditions (i.e., cloud coverage) have to be taken into account.

14.4.1.1. Sensor Type.
Both active and passive sensors can be used to acquire post-event imagery for flood impact assessment.

Passive sensors are used to acquire panchromatic or multispectral optical imagery allowing either visual interpretation or semi-automated processing to be carried out. It is a mandatory choice when the focus is on the damages to infrastructures, but it is unusable when the area of interest is covered by clouds.

Active sensors are used to acquire radar (or SAR, Synthetic Aperture Radar) imagery with all-weather and all-light capabilities; that is, with these data it is possible to acquire data even in the presence of cloud coverage or during the night. Radar sensors measure the back-scattered energy that was originally emitted by the sensors itself. An external energy source like the sun is therefore not required. The technology makes use of electromagnetic energy in the microwave part of the spectrum that can penetrate haze, clouds, and light rain (but can be still influenced by the presence of rain cells due to the attenuation-through-rain effect [*Danklmayer et al.*, 2008]. It is therefore the primary data source when the event is still ongoing or when the analysis timeliness may benefit from a night acquisition. On the other hand, a photo-interpretation of SAR imagery is strongly limited by the radar acquisition geometry and the consequent distortion effects (i.e., layover, foreshortening, and radar shadowing).

Table 14.1 shows the share of both sensor types in the production of post-event maps for Copernicus EMS-RM flood activations. Eighty percent of the imagery was acquired by active sensors (radar), and 20% was acquired by passive sensors (optical). Although 88% of the imagery used for producing delineation maps was radar data, grading maps were mostly based on optical data (96%), confirming the aforementioned general considerations.

14.4.1.2. Platforms.
Satellite platforms are nowadays widely exploited to acquire post-event imagery over flood-affected areas [*Boccardo et al.*, 2015; *Ehrlich et al.*, 2009]. The main advantages offered by satellite platforms follow:

• global coverage (with possible limitations on extreme latitudes)
• the possibility to acquire imagery over areas with accessibility issues (a common situation during ongoing flood events)
• the possibility to revisit the same area over time (with different temporal resolutions depending on the satellite orbit parameters)
• large coverage on the ground allowing the ability to monitor events at regional/country level

Aerial platforms can be also exploited, mainly depending on the national capacity (or on international initiatives in case of major events) to quickly trigger aerial surveys over the affected areas while carefully considering the weather conditions. The increase of the imagery spatial resolution is generally the main advantage of employing aerial platforms, in addition to the possibility to derive not only ortho-imagery but also ancillary data sets such as Digital Surface Models (DSM) or three-dimensional features (technically feasible also with satellite data but not yet systematically integrated in an operational workflow).

More recently, Remotely Piloted Aircraft Systems (RPAS) or UAVs, i.e., small aerial platforms without a pilot onboard, are also increasingly adopted to acquire post-event data over limited areas of interest.

14.4.1.3. Imagery Technical Features.
As far as passive sensors are concerned, one of the most relevant technical features is the actual Ground Sample Distance (GSD) of the acquired imagery, i.e., the actual size of a single image cell on the ground, which is intrinsically related to the level of detail of the imagery. It is possible to derive a general relation between GSD and satellite footprint (i.e., the area sensed on the ground). The higher the GSD, the smaller the nominal footprint and the spectral resolution. Therefore, a proper trade-off between level of detail and possibility to cover the entire affected area in the shortest time possible has to be identified during the post-event data tasking phase. Very high resolution (VHR) or very high spatial resolution imagery (GSD of up to 0.3 m for optical sensors as of May 2017) is characterized by a very high level of detail with high positional accuracy (using a standard post-processing algorithm as

the Rational Polynomial Function [RPC] approach) and a high semantic content. Both are crucial features when assessing the impact of a disaster. On the contrary, medium/low spatial resolution imagery is characterized by a larger footprint and a wider spectral resolution, features that may be crucial when monitoring regional-wide disasters (e.g., monsoonal flooding in tropical regions). Furthermore, it has to be highlighted that the actual GSD may be far higher (i.e., lower level of detail) than the nominal one, being strictly related to the sensor off-nadir angle during the acquisition (i.e., the angle between the nadir and the sensor main axis depending on the platform attitude required to frame the requested acquisition area). On the other hand, the agility of modern satellite platforms (potentially leading to high off-nadir angles) allows the nominal revisiting time to be reduced.

Concerning active sensors, in addition to the spatial resolution parameters (for which the aforementioned GSD versus footprint relation generally applies), the most relevant radar technical parameters are the polarization mode and the microwave band that influence the contrast between flooded areas and the surrounding surfaces, depending on the type of flood and impacted areas [*Martinis et al.*, 2015].

The temporal resolution, or revisiting time, is obviously an additional crucial technical parameter in a Rapid Mapping context, especially for floods that may require monitoring the evolution of the event for several days. In general, the relation between nominal revisiting time and spatial resolution classes is positive, i.e., the lower the resolution, the shorter the revisit time (e.g., one day for Moderate Resolution Imaging Spectroradiometer [MODIS] imagery acquired by the same satellite platform). The nominal revisit time can be further reduced by exploiting satellite constellations (e.g., twice a day for MODIS imagery acquired by the Aqua and Terra satellite platforms) and the agility of sensors.

14.4.1.4. Licensing.
The licensing policy of the released imagery is another feature that has to be carefully considered, in terms of both imagery costs (ranging from public domain data generally accessible through the web or commercial imagery that should be purchased through reseller companies) and potential dissemination constraints that may prevent final users to have direct access to raw and processed imagery.

14.4.2. Data Processing Techniques

As detailed in section 14.3.2, the main goals of data processing follow:
• to extract the water body surfaces from the post-event imagery (as well as from a pre-event reference image in case a reference hydrography data set is not available),

using a manual approach (visual interpretation) or a semi-automated approach
• to assess the damages to the main infrastructures, mainly by means of visual interpretation of post-event optical imagery

The main data processing techniques commonly adopted to derive the required crisis information are described in the next paragraphs, depending on the sensor used to acquire the post-event imagery.

14.4.2.1. Optical Data.
Optical data can be exploited for both delineation and damage assessment of flooded areas.

Computer Aided Photo Interpretation (CAPI) allows a skilled image interpreter to identify on the available imagery the areas covered by water, with the additional advantage of also identifying flood traces (and not water) such as mud covered surfaces. Semi-automated procedures exploit the peculiarities of the spectral signature of water (high absorption rates in the infrared bands in the case of clean water) to classify each pixel (or cluster of pixels) as water bodies (with the possibility to differentiate between clean water and turbid water). The most commonly adopted techniques are based on indexes derived from differential band ratios, aiming at making threshold values independent from image acquisition parameters as in the case of a simple but still effective single band thresholding approach. The literature review shows that the general approach to calculate the Normalized Differential Water Index (NDWI) is to use differential ratios based on those bands exalting relative reflectivity differences of water spectral signature [*Chowdary et al.*, 2008].

Other semi-automated approaches described in the literature are based on different types of supervised classifiers, e.g., Support Vector Machines [*Ireleand*, 2015], change detection [*Chaouch*, 2011], Markov random field model [*Chanwimaluang*, 2014], and Decision Tree [*Feng*, 2015].

If the case of damage assessment maps, the main operationally exploited approach is based on CAPI carried out by skilled image interpreters on very high spatial resolution optical imagery. It aims at assessing the damage to the transportation network and possibly identifying other features of interest for the end users, e.g., the presence of erosion areas, landslides, or mudflows induced by the flood. The availability of pre- and post-event imagery is crucial to allow a proper comparison and to improve the thematic accuracy of a damage assessment analysis carried out on a post-event image only. Despite several semi-automated approaches for identifying the most impacted areas that are being developed and piloted, the time required for the preliminary training stage, the strict requirements in terms of acquisition parameters (that should be as similar as possible, also in terms of illumination conditions) as well as the limited thematic accuracy (which requires the operator to review the damage status of the reference features) are currently limiting their adoption in operational services.

14.4.2.2. Synthetic Aperture Radar Data. Imagery acquired by active sensors is exploited to identify water bodies on post-event imagery, to be then compared with the available reference hydrography to extract the flood-affected areas (as described in the previous paragraph). Considering that a visual interpretation approach, despite possible, is very time consuming and in some cases subjective, several SAR-based water classification algorithms can be exploited [*Martinis et al.*, 2015].

Thresholding methods are mainly based on the assumption that calm water can be modeled as a specular reflector, therefore redirecting the incidence microwave signal away from the sensor. This phenomenon leads to water-related pixels in the SAR image to appear darker than other surfaces because no back-scattered energy is measured by the sensor. A threshold, generally defined with a supervised approach by means of visual interpretation, is therefore used to identify all pixels with values below the threshold because those are potentially related to water bodies.

Change detection techniques can be exploited when multi-temporal SAR data with similar acquisition technical parameters are available, comparing the results of before and after independent classifications or integrating before and after data in the change detection algorithm. The latter approach can be based on amplitude data, on the use of coherence measurements derived from multi-pass SAR interferometry [Nico et al., 2000] or on a synergistic use of SAR intensity and coherence. However, the planning of pairs of SAR acquisitions suitable for Interferometric Synthetic Aperture Radar (InSAR) analysis is relatively difficult.

14.4.3. Discussion

Satellite platforms allow limiting the time required to task new acquisitions up to a few hours (depending on the specific combination of parameters like the position of the affected areas, the local time of the event occurrence, the availability of satellite ground uplink stations, possible conflicts leading to satellite unavailability, or other orbit/platform technical constraints). Also aerial and UAV acquisitions can theoretically be triggered in a short time frame but only if structured and operational mechanisms are already in place in the affected region, to limit the time required to deploy the required instrumentations and to correctly cope with local civil aviation regulation requirements.

It is important to highlight that the time span required to task, acquire, and deliver new post-event acquisitions is currently taking most of the total time spent from the initial Emergency Mapping service activation to the delivery of the final outputs to the users, as clearly shown in Figure 14.8. As of September 2015, feedback collected from end users of EMS-RM flood activations shows that users are aware of this issue because only 19 out of the 31 who provided feedback perceived image acquisition timing as good, compared to 7 who think it is medium and 4 who think it is bad. In 1 case, there was no answer.

As far as the choice of the technical features of the input imagery is concerned, it is crucial to consider the time constraints of a Rapid Mapping service. Therefore, a proper trade-off between timeliness and technical requirements should be identified to avoid paradoxical situations where imagery of limited use is acquired in a very short time frame (e.g., SAR imagery for damage assessment at building level) or imagery with optimal technical features is available with huge delays (e.g., days after a flash flood event).

Image tasking should also take weather forecast into account, considering the obvious limitations of optical imagery (and partial limitations of SAR imagery due to atmospheric propagation effects in case of rain cells) and carrying out a proper cost-benefit analysis for tasking acquisitions even with adverse weather forecast.

In terms of data analysis, the well-known general limitations of vertical imagery (perspective deformations, facade occlusions) should be highlighted that may limit especially the CAPI-based damage assessment performed on very high resolution optical imagery.

A completely automated approach, which aims at extracting water bodies from SAR imagery, is generally not operationally exploited because the underlying thresholding concept is complicated by a wealth of other factors that influence the actual backscattering level of the terrain (and therefore leading to non-water surfaces to be erroneously classified as water bodies). Those factors include surface roughness and soil moisture for land areas, the presence of capillary waves for water bodies, or the presence of rain during the acquisition [*Refice et al.*, 2014] as well as the presence of radar shadows due to local morphology. A semi-automated procedure based on an ad hoc choice of data processing parameters is therefore the most common approach in a Rapid Mapping context. Considering the known lower reliability of SAR-based flood analysis in dense urban environments (due to the double-bouncing effect), a final visual interpretation check is required, especially aimed at identifying false positives and increasing the overall accuracy of the derived information.

Overall, Rapid Mapping of flood events and related information provide useful information for emergency management. This is supported by the feedback received from 31 end users of EMS-RM flood activation products. According to 27, EMS-RM products provided a faster overview of the situation, however, this was not valid for 3 end users. Only half (15) of these 31 users think that the accuracy of the analysis is good, whereas 11 think it was medium, and 2 rated it as bad. No answer was provided in three cases.

14.4.4. Potential Future Development

Starting from the main topics discussed in section 14.4.3, this section aims at proposing possible future developments in the different operational steps related to the flood impact assessment. Most of the topics were also addressed in recent works of part of the author group. *Ajmar et al.* [2015] focused on a review of the geomatics role in a generic Rapid Mapping workflow.

As previously highlighted, satellite data tasking, acquisition, and delivery are currently considered to be one of the main bottlenecks in the timeliness of the post-event analyses. In this context, the regular adoption of a laser link download technology, which was recently tested by the ESA[20], is an operational solution that decreases this time span, allowing large volumes of remotely sensed data to be readily available, independent from the location of ground stations and exploiting geostationary telecommunication satellites [*ESA*, 2014]. The increasing availability of constellations of satellites (or virtual constellations composed by different satellite platforms managed by a unique satellite data provider) is also decreasing the satellite revisit time especially for high-resolution sensors. For example, in the case of Planet Labs,[21] the constellation of almost 150 (as of February 2017) satellites allows imaging of the entire globe every single day.

In parallel, a more structured exploitation of RPAS to complement satellite imagery as post-event data source should be envisaged. International initiatives like the Humanitarian UAV Network[22] are trying to bridge humanitarian and UAV communities internationally with the goal of facilitating information sharing, coordination, and operational safety in support of a broad range of humanitarian efforts. In the frame of a pilot study, the role of UAVs in Copernicus EMS is already assessed, using the collected imagery "as an alternative and/or complementary source of post-event imagery in emergency situations and in a rapid response and mapping context."[23]

Concerning the availability of openly licensed imagery, structured initiatives based on distributed systems and on standard web services such as OpenAerialMap are aimed at streamlining and facilitating the discovery and usage of up-to-date post-event image data (including both satellite scenes and RPAS imagery), limiting the need for downloading and processing the data.

In the flood impact extraction phase, it is crucial to automatize as far as possible robust semi-automated procedures for the extraction of post-event crisis information, both for SAR and optical imagery, to reduce the need for CAPI (a time-consuming activity). The goal should be to ensure a thematic accuracy comparable to the one usually achieved by means of visual interpretation techniques, especially in dense urban environments where the currently adopted procedures show strong limitations. In this phase, the possible role of new technologies should be evaluated, allowing to ingest in the operational work-flow high definition video sequences acquired by sensor installed on space platforms, e.g., Urthecast[24] and SkySat[25].

At the same time, the availability of an impressive amount of post-event imagery, especially in case of major flood events, poses new challenges in terms of the effort required to process, store, and analyze it. Therefore, the possibility to integrate collaborative mapping and crowdsourcing initiatives should be investigated (e.g., HOT, TomNod, MicroMappers), evaluating the outcomes in terms of both efficiency and thematic accuracy.

14.5. COPERNICUS EMERGENCY MANAGEMENT SYSTEM CASE STUDY

Between late January and early March 2015, the Ebro Valley (Spain) was affected by two different flood events for which Copernicus EMS-RM provided Rapid Mapping products to the Spanish civil protection (Copernicus EMS internal codes EMSR118, EMSR120). In the following discussion, the example of the second mapping request (i.e., service activation EMSR120) is briefly presented.

The floods in late February and early March were caused by a high degree of saturation of the ground due to almost a month of continuous rain and the snowmelt

---

[20] *http://www.esa.int/Our_Activities/Observing_the_Earth/ Copernicus/Sentinel-1/Laser_link_offers_high-speed_ delivery. Accessed 23 November 2015.*

[21] *www.planet.com*

[22] *http://uaviators.org/*

[23] *http://emergency.copernicus.eu/mapping/ems/ new-phase-brief*

[24] *https://www.urthecast.com/*

[25] *https://youtu.be/OWXN3CXsxTg*

from the Pyrenees. The State Meteorological Agency (AEMET) marked the starting point of the flood relevant for EMSR120 on 26 February. EFAS, the EMS early warning system for floods, also alerted about flood risk in the whole Ebro River and other areas. Copernicus EMS-RM was activated by the Spanish Civil Protection on 27 February (see Table 14.2). The service request covered reference pre-event maps and delineation plus monitoring post-event maps for one overview area (Zaragoza) and three detail areas (Luceni, Ebro Basin, Fuentes de Ebro) in the surroundings of the city of Zaragoza (see Figure 14.9). The flood extent was mapped twice (monitoring the event at two different observation dates) so that in total, eight delineation maps were produced. For both mappings cycles, Radarsat-2 data were used (Multi-Look Fine, Wide Ultra-Fine, 1.5 to 2.5 m resolution).

Table 14.2 shows the timeline for the first flood extent map that was released 17 hours after the service was activated. Figure 14.10 shows one example of the delineation maps produced.

More information and all maps produced in this service activation are available on the EMS Mapping Portal at the following website: http://emergency.copernicus.eu/mapping/list-of-components/EMSR120.

Table 14.2 Timeline for the First Flood Extent Map Released in EMSR120.

Step	Date (dd/mm/yyyy hh:mm UTC)
Event time	26/02/2015
Activation time	27/02/2015 09:02
Submission of data request to REACT	27/02/2015 10:08
Image acquisition (start)	27/02/2015 17:56
Image availability	27/02/2015 22:43
Availability of first available map (v0)	28/02/2015 02:07
Availability of final map (v1)	28/02/2015 07:35

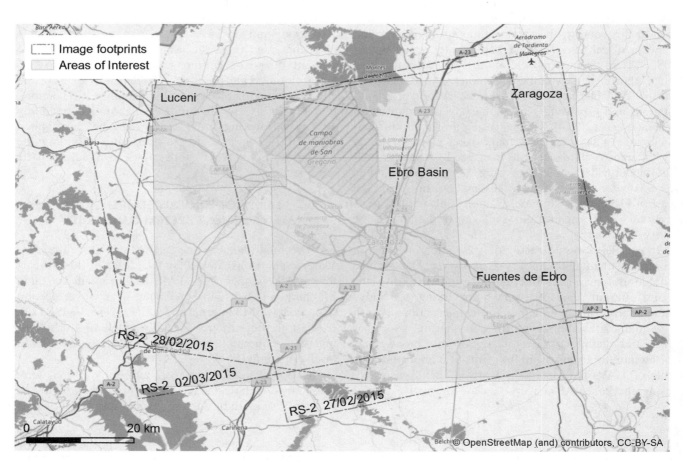

Figure 14.9 Areas of interest and satellite image footprints of Copernicus EMS-RM activation for floods in Spain (EMSR120). *See electronic version for color representation.*

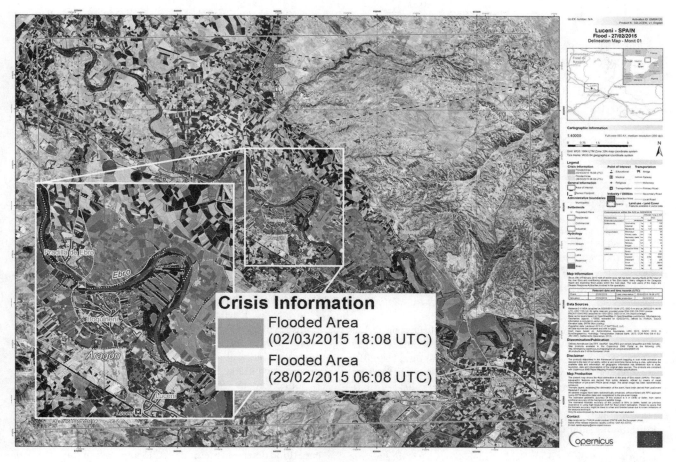

Crisis Information

Flooded Area
(02/03/2015 18:08 UTC)

Flooded Area
(28/02/2015 06:08 UTC)

Figure 14.10 One of the eight delineation maps produced in the Copernicus EMS-RM activation for floods in Spain early 2015 (EMSR120). The map shows the flood extent in the area of Luceni on 28 February (light blue) and 2 March 2015 (darker blue). The zoom, which is not a feature of the standard product, gives an impression of the crisis information and the various reference layers. Data source: Copernicus Emergency Management Service ((©European Union, 2015), [EMSR120] Ebro Basin: Delineation Map [Detail 1, Monitoring 1]). *See electronic version for color representation.*

14.6. CONCLUSIONS

This chapter provided an overview of the Emergency Mapping activities that are operationally carried out to support the immediate response phase to flood events, focusing mainly on the role of satellite data. Benefitting from the operational experience of the Copernicus EMS service, it was possible to describe and showcase the typical post-event information that can be delivered to the users, mainly in the form of emergency maps. The technical approach to assess the impact of flood events in terms of both flooded area identification and assessment of damages to the main infrastructures with a suitable thematic accuracy was described in detail, critically discussing the main advantages as well as the current limitations (section 14.4.3). Potential future developments were proposed in section 14.4.4, mainly exploiting recent tech-

nological developments that need to be further tested and systematically assessed.

As a final remark, the authors would like to highlight that an international and coordinated effort is still required to improve cooperation, communication, and the adoption of professional standards among the global network of satellite-based Emergency Mapping providers, with the paramount goal to make the overall Emergency Mapping effort more effective especially in case of major disasters. This is the main aim of the IWG-SEM, a voluntary group of organizations involved in satellite-based emergency mapping whose vision is "supporting disaster response by improving international cooperation in satellite based emergency mapping." The group is continuously updating a working document on "Emergency Mapping Guidelines" that includes a specific chapter on flood events.

ACKNOWLEDGMENTS

We acknowledge the use of material from Copernicus EMS service providers in particular e-GEOS consortium (e-GEOS, ITHACA, GAF, SIRS, DLR, SERTIT) and Trabajos Catastrales. We thank Iacopo Ceccarini that supported the literature review process related to section 14.4.2, Data Processing Techniques, in the framework of his internship at Ithaca.

REFERENCES

Ajmar, A., P. Boccardo, F. Disabato, and F. Giulio Tonolo (2015), Rapid Mapping: geomatics role and research opportunities. *Rendiconti Lincei, Volume: 26 Issue: 1*, Springer Milan, 63–73. doi: 10.1007/s12210-015-0410-9. http://dx.doi.org/10.1007/s12210-015-0410-9.

Boccardo, P., and F. Giulio Tonolo (2015), Remote Sensing Role in Emergency Mapping for Disaster Response, *Engineering Geology for Society and Territory, Volume 5*, pp. 17–24. doi: 10.1007/978-3-319-09048-1_3.

Centre for Research on the Epidemiology of Disasters, UN International Strategy for Disaster Reduction (CRED, UNISDR) (2015), The human cost of natural disasters 2015: a global perspective, report available at http://reliefweb.int/report/world/human-cost-natural-disasters-2015-global-perspective, 57 pages.

Chanwimaluang T., T. Kasetkasem, I. Kumuzawa, P. Phuhinkong, and P. Rakwatin (2014), A flood mapping algorithm from cloud contaminated MODIS time-series data using a Markov random field model. *Geoscience and remote sensing symposium* (IGARSS), 2507–2510. doi: 10.1109/IGARSS.2014.6946982.

Chaouch, N., S. Hagen, R. Khanbilvardi, S. Medeiros, M. Temimi, and J. Weishampel (2011), A synergetic use of satellite imagery from SAR and optical sensors to improve coastal flood mapping in the Gulf of Mexico. *Hydrological Processes*, 1–12; DOI: 10.1002/hyp.8268.

Chowdary, V. M., R. V. Chandran, N. Neeti, R. V. Bothale, Y. K. Srivastava, P. Ingle, D. Ramakrishnan, D. Dutta, A. Jeyaram, J. R. Sharma, and R. Singh (2008), Assessment of surface and sub-surface waterlogged areas in irrigation command areas of Bihar state using remote sensing and GIS. *Agricultural Water Management, Elsevier, vol. 95* (7), 754–766.

Cova, T. J. (1999), GIS in emergency management. *Geographical Information Systems: Principles, Techniques, Applications, and Management*, 2, 845–858.

Danklmayer, A., B. Döring, M. Schwerdt, and M. Chandra (2008), Analysis of Atmospheric Propagation Effects in TerraSAR-X Images, *Geoscience and Remote Sensing Symposium, 2008. IGARSS 2008*. IEEE International, vol. 2, II-533-II-536, doi: 10.1109/IGARSS.2008.4779046.

Ehrlich, D., H. D. Guo, K. Molch, J. W. Ma, and M. Pesaresi (2009), Identifying damage caused by the 2008 Wenchuan earthquake from VHR remote sensing data, *International Journal of Digital Earth 2*: 4, 309–326. Doi: 10.1080/17538940902767401.

Feng, Q., J. Gong, and J. Liu (2015), Urban flood mapping based on unmanned aerial vehicle remote sensing and random forest classifier – A case of Yuyao, China. *Water*, 7(4), 1437–1455. doi:10.3390/w7041437.

Guha-Sapir, D, R. Below, and Ph. Hoyois (2017), EM-DAT: The CRED/OFDA International Disaster Database, www.emdat.be, Université Catholique de Louvain, Brussels, Belgium.

Ireleand, G., G. P. Petropoulos, and M. Volpi (2015), Examining the capability of Supervised Machine learning classifiers in extracting flooded areas from landsat TM imagery: a case study from a Mediterranean flood. *Remote sensing*, 7(3), 3372–3399; doi:10.3390/rs70303372.

IWG-SEM (2014), Emergency Mapping Guidelines. http://www.un-spider.org/sites/default/files/IWG_SEM_EmergencyMappingGuidelines_A4_v1_March2014.pdf. Retrieved on 13 November 2015.

Martinis, S., C. Kuenzer, and A. Twele (2015), Flood Studies Using Synthetic Aperture Radar Data, *Remote Sensing of Water Resources, Disasters, and Urban Studies*, Remote Sensing Handbook Volume III, 145–173.

National Earthquake Hazards Reduction Program (NEHRP), Introduction to emergency management. http://training.fema.gov/EMIWeb/EarthQuake/NEH0101220.htm Accessed 13 November 2015.

Nico, G., M. Pappalepore, G. Pasquariello, A. Refice, and S. Samarelli (2000), Comparison of SAR amplitude vs. coherence flood detection methods–a GIS application. *International Journal of Remote Sensing* Volume *21*. Issue 8, 1619–1631. doi: 10.1080/014311600209931.

Priolo, A. (2014), Copernicus Emergency Service Overview, Copernicus Emergency Service: Awareness & Demonstration–Warsaw. http://www.copernicus.eu/sites/default/files/documents/User_uptake/Emergency_Events/Warsaw/EMS-Service_Overview.pdf.

Refice, A., D. Capolongo, G. Pasquariello, A. D'Addabbo, F. Bovenga, R. Nutricato, F. P. Lovergine, and L. Pietranera (2014), SAR and InSAR for Flood Monitoring: Examples With COSMO-SkyMed Data, *Selected Topics in Applied Earth Observations and Remote Sensing*, vol.7, no.7, 2711–2722. doi: 10.1109/JSTARS.2014.2305165.

Revilla-Romero, B., F. A. Hirpa, J. Thielen-del Pozo, P. Salamon, R. Brakenridge, F. Pappenberger, and T. De Groeve (2015), On the Use of Global Flood Forecasts and Satellite-Derived Inundation Maps for Flood Monitoring in Data-Sparse Regions, *Remote Sensing*, vol. 7, 15702–15728. doi:10.3390/rs71115702.

UNISDR (2009), Terminology on disaster risk reduction. http://www.unisdr.org/files/7817_UNISDRTerminologyEnglish.pdf. Retrieved on 13 November 2015.

15

Data Collection and Analysis at Local Scale: The Experience within the Poli-RISPOSTA Project

Carolina Arias Munoz[1], Mirjana Mazuran[2], Guido Minucci[3], Danilo Ardagna[4], and Maria Brovelli[1]

ABSTRACT

Information and communication technology (ICT) can play a significant role in Disaster Risk Management (DRM), especially on supporting data collection, storage, and analysis that ultimately is crucial for decision-making, communication, and collaboration. Determining the areas to intervene in case of emergency, and highlighting risk areas, vulnerabilities, and damages in all phases of a disaster event is just one of the many applications of ICT in this field. This chapter discusses the ways in which ICT can positively impact the different phases of risk management. Initially it presents a compilation of best practices in ICT for emergency, disaster, and risk management of different sectors including government agencies, international organizations, academia, non-governmental entities, and the private sector. Then it describes an example of an ICT system in the Poli-stRumenti per la protezione civile a Supporto delle POpolazioni nel poST Alluvione (RISPOSTA) project. Within the Poli-RISPOSTA project, several applications were developed for data management (insertion, update, visualization, etc.) including two components: a mobile application for data gathering and a web portal for data visualization and management. Both components were created to support the different actors in the procedure to perform their tasks before, during, and after a flood event. At the end, some shortcomings and drawbacks on the use of ICT for data collection, storage, and analysis are discussed.

15.1. INTRODUCTION: ROLE OF INFORMATION AND COMMUNICATION TECHNOLOGY IN DISASTER MANAGEMENT

The array of ICT that are well-suited to the disaster risk management is quite wide nowadays and different available technologies are used as tools for communications: radio, cell broadcasting, mobile phones with short message service (SMS) and social media, for collaboration (crisis mapping tools[1]) and for mapping and sensing: satellite imaging, drone-based aerial imaging, and geographic information systems (see for example Chapter 14 in this book).

Information and communication technology is regarded as a key enabler in the process of disaster management [*Santos Reyes and Beard*, 2011]. The relation between ICTs and the four phases (preparedness, mitigation, response, and recovery) of the disaster management cycle has been studied by several researchers [*Dorosamy et al.*,

[1] *Department of Civil and Environmental Engineering, Politecnico di Milano, Milan, Italy*

[2] *Department of Electronics, Information and Bioengineering, Politecnico di Milano, Milan, Italy*

[3] *Department of Architecture and Urban Studies, Politecnico di Milano, Milan, Italy*

[4] *Department of Electronics, Information and Bioengineering, Politecnico di Milano, Milan, Italy*

[1] *http://crisismappers.net/*

Flood Damage Survey and Assessment: New Insights from Research and Practice, Geophysical Monograph 228, First Edition. Edited by Daniela Molinari, Scira Menoni, and Francesco Ballio.

2011; *Wattegama, 2007; Murai, 2006; Nisha de Silva, 2001; Quarantelli, 1997; Fedra and Reitisma, 1990*]. In support of the mitigation phase, ICTs can facilitate the collection and analysis of data and dissemination of risk information. This information can be used to identify risk zones, produce maps, and support settlements and infrastructure development planning as well as to develop mitigation plans. In the preparedness phase of the DRM, ICT tools can help by supporting computerized analysis data to help in forecasting the effects of the impacts or by facilitating the coordination of early warning systems (EWS) and disseminating disaster alerts to people under threats. In the response phase of the cycle, ICTs may help authorities and organizations leading the response in communicate between emergency personnel and the government to determine priorities and to assist those affected. Moreover, ICTs can help in managing the massive load of information related to the emergency, such as victims and relief personnel, available resources, and scientific field measurements [*Li et al., 2014*]. ICTs in disaster recovery and reconstruction can be used to obtain appropriate information required to carry out damage and loss assessments and for decision-making regarding post-disaster planning, project formulation, and implementation.

The role of ICT in the disaster cycle can be seen by the emergence of various disaster management systems such as the DesInventar[2] system, an integrated database in Latin America, and the Sahana free and open source software (FOSS)[3] Disaster Management System (see Chapter 1 in this book); mapping services, for example, the Copernicus Emergency Management Mapping Service, the Global Earth Observation System of Systems (GEOSS) (see Chapter 14 in this book), the International Network of Crisis Mappers); and tools such as EWS and dissemination of warnings using SMS (see *Clothier* [2005]); geographic information system (GIS) [*Wattagama, 2007; Billa et al., 2004; Johnson, 2000*], and web and mobile application such as Ushahidi[4].

As results from the use of different ICTs, according to the disaster phase and from the rising variety of tools, the challenges of ICT are more related to the effective management of technology and its appropriate application than their capacity [*Sagun, 2010*]. With reference to this, flood damage data are currently collected and stored in a manner that is causing problems for an efficient and multipurpose use of data as suggested by *De Groeve et al.* [2013] and *Wirtz et al.* [2014] (Chapter 2 and Chapter 3 in this book). Concerning data collection, data at upper levels can be obtained by applying apposite data aggregation

rules in order to respond to a specific purpose (e.g., strategic or policy making scope), while data at the local scale need to be often collected by surveys on the field. Indeed, during the post-event surveying, information, for instance about water heights, buildings, and infrastructure features and damages, can be collected. This information is then essential for damage assessment and validation of hydraulic models. For this activity, there is the possibility of collecting data using mobile devices such as tablets and smartphones [*Brovelli et al., 2014a; Brovelli, et al., 2014b*]. Concerning this, two systems can be taken as reference since they provide real-time data storage (in a database) and their visualization in terms of maps, supporting this way the field survey/emergency phase (e.g., supporting the coordination of survey teams). On one hand, the Australian damage assessment and reconstruction monitoring system (DARMSys™[5]), which uses a GPS-linked data collection device that allows assessors to collect levels of damage to individual homes and buildings, and then transfer that data in 'real-time' to a central mapping point (see Chapter 12 in this book). On the other hand, the substantial damage estimator (SDE) developed by the US Federal Emergency Management Agency (FEMA) to provide timely substantial damage determinations to residential and non-residential structures so that reconstruction can begin following a disaster. However, the lack of a standardized way to collect spatial data in the aftermath of flood events results in having data collected in different formats, which renders them not immediately usable to perform spatial analysis. At the same time, other important types of data coming from the ground (such as real-time observations, photos, movies from individuals, data collected by different types of sensor) have to be stored and managed in a database supporting the reconstruction of the event, data visualization, and spatial analysis. Whereas, often databases are not in GIS standard compatible formats [*Molinari et al., 2014a*], hence data visualization and spatial analysis cannot be performed. Moreover, as a consequence of the lack of agreed standards to collect and storage damage data [*Hristidis et al. 2010; Guha-Sapir and Below, 2002*], the main issue of the existing database, such as the Emergency Events Database (EM-DAT), NatCat, and Sigma, concerns data comparison and management (refer to Chapter 3 in this book).

This chapter focuses on the information system and tools for collecting, storing, and managing damage data in the aftermath of flood events developed within the Poli-RISPOSTA project, an internal project of Politecnico di Milano, whose aim is to supply tools overcoming the abovementioned limits for supporting civil protection authorities in dealing with flood emergency.

---

[2] *http://www.desinventar.org/*
[3] *Free and Open Source Software*
[4] *http://www.Ushahidi.org*

---

[5] *http://www.qldreconstruction.org.au/about/darmsys*

15.2. POLI-RISPOSTA: A FLOOD DATA MANAGEMENT SYSTEM FOR THE LOCAL/REGIONAL SCALE

Floods constitute a high risk to both rural and urban settlements. The reconstruction process that follows such a tragic event needs to be planned and tailored on an analysis of the damages to infrastructures and properties. Collecting data after floods is necessary not only to verify the request for reimbursements or budget allocations but also to gather knowledge about what factors and artifacts of settlements constitute risk in floods. Although specific expertise is needed to assess damages to complex infrastructures, inspection of residential properties can be more repetitive and need a systematic methodology to address the multitude of sites to be inspected and amount of data to be gathered, organized, and analyzed.

The Poli-RISPOSTA or civil protection tools to support the community affected by floods in the post-event is a project under development at the Politecnico di Milano that aims to build models, tools, and advanced technology solutions for the collection, mapping, and evaluation of post-flood damage data. This project involves the testing of procedures for the integration of an interoperable Web-based system, as well as the development of capabilities to collect and deliver information to mobile devices in the field. This work is expected to support communities in the post-emergency phase, and it will contribute to promote a risk management culture in areas susceptible to flood risks.

Numerous research efforts have been done on flood risk management data collection. However, to date there are no effective methods of estimating post-flood damage, and there is no common methodology that is applied to estimate post-flood damage internationally [*Hammond et al.*, 2011; *Merz et al.*, 2010]. The estimation of economic flood damage to define victim's monetary compensation and intervention priorities has become more important as financial resources have become scarcer. Today flood risk management has become the main approach to flood control policies in Europe [*Begum et al.*, 2007; *European Commission*, 2007].

To measure effectively the direct and indirect flood damages, it is necessary to create a complete picture of the flood event that includes the physical triggers of the event, as well as the exposure and vulnerability of assets and infrastructure (see Chapter 11 in this book). To achieve this task, there is a need for better geographic data collection, storage, analysis, and visualization that will provide reliable flood damage data to scientists and stakeholders in the affected areas. At the national and international levels, there are few experiences on the required interdisciplinary efforts for defining the methods, tools, and procedures to generate the information essential for post-emergency operators and the scientific community.

Creating flood risk management databases is a common practice today. However, this activity is not carried out using standard compatible formats. For example, spatial data are rarely stored in popular GIS formats such as Esri's shape files, and much less in a formal standard formats such as the Open Geospatial Consortium (OGC) Geographic Markup Language (GML). More often, spatial data are stored as text in text files or spreadsheets and even as drawings in paper records. Furthermore, there are incompatibilities regarding scale, level of aggregation of the information, and semantics. The Poli-RISPOSTA project addresses some of these issues by creating an ICT system that allows the post-flood damage data management, from data collection to data visualization, described in the next section.

15.2.1. Poli-RISPOSTA Information and Communication Technology System Architecture

Several tools and solutions have been developed for post-flood monitoring and disaster management [*Abdalla and Niall*, 2009; *Raltman et al.* 2011; *Karnatak et al.* 2012]. For the Poli-RISPOSTA project, architecture for a Web-based system was created, using FOSS for geospatial data processing and sharing over the Web, following the system requirements described in *Molinari et al.* [2014a]. Figure 15.1 illustrates the proposed elements and information flows in the architecture. A big part of the architecture is based on GeoNode[6] Open Source Geospatial Content Management System, an open source platform that facilitates the creation, sharing, and collaborative use of geospatial data. GeoNode core is based on Django web framework[7] with few more dependencies necessary for the communication with the geospatial servers (such as GeoServer and pyCSW). GeoNode aims to provide and create a spatial data infrastructure by implementation of GIS and cartographic tools [*Balbo*, 2013], and it has been mainly used to deploy platforms for disaster risk management [*Balbo et al.*, 2014; *Gutierrez et al.*, 2007].

In the backend, the Database Management Systems PostgreSQL with the spatial extension PostGIS is used to store geospatial data, using a *data model* explained in detail in the next section. Vector and raster layers are converted to PostGIS spatial tables. After the geospatial layers are entered into PostgreSQL-PostGIS, they can be queried, managed, and subjected to any of more than 100 geo-processing functions available in PostGIS (see PosGIS manual http://postgis.net/stuff/postgis-2.1.pdf).

For the publishing and discovery of geospatial metadata GeoNode uses pycsw,[8] which provides a standard catalogue and search interface based on Catalogue Service CSW

---

[6] *http://geonode.org/, http://docs.geonode.org/en/master/*
[7] *https://www.djangoproject.com/*
[8] *http://pycsw.org/*

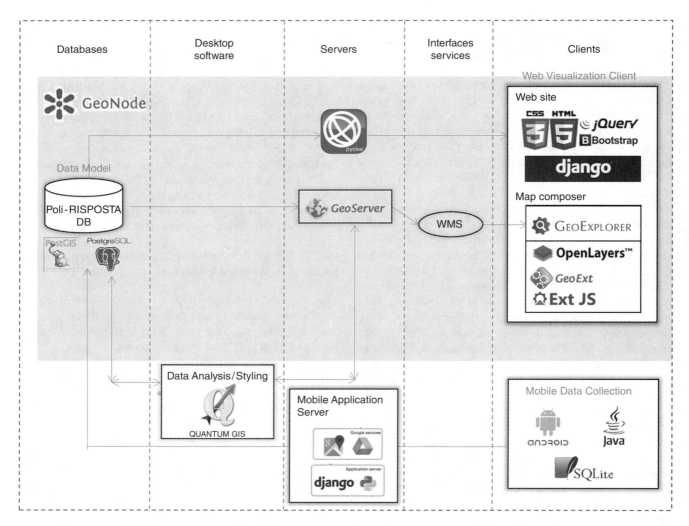

Figure 15.1 Poli-RISPOSTA general architecture. Its tree main features Data Model, Web Visualization Client, and Mobile Data Collection, which are explained in detail in sections 15.2.2, 15.2.3, and 15.2.4, respectively.

OGC[9] standard. It is used to create, update, search, and display metadata records when they are accessed in GeoNode. Since pycsw is embedded in GeoNode, layers published within GeoNode are automatically published to pycsw and discoverable via CSW. No additional configuration is needed to publish layers, maps, or documents to pycsw.

GeoServer[10] is used as the GIS server, creating and managing several types of OGC Web services such as Web Map Service (WMS), Web Feature Service (WFS), and Web Coverage Service (WCS). Each service is appropriate for different types of data and functionality. The Poli-RISPOSTA database is added as a data source, also called datastore, in Geoserver, in such a way that all geodata from the database can be visualized on the *Web Visualization Client* using the WMS.[11] GeoNode is always associated with the GeoServer catalog. An ad hoc security module allows the two modules to strictly interact and share security and permissions rules.

The *Web Visualization Client* can be divided in the *Website* and the *Map Composer*. For the *Website*, GeoNode counts with a Django site that allows the user to easily manage the following content: maps, layers, documents, users, and groups. It includes tools to handle user administration, user profiles, and helper libraries to interact with GeoServer and pycsw in a programmatic and integrated way. It is based on jQuery[12] and Bootstrap[13] frameworks. A set of jQuery JavaScript plugins is used to

---

[9] http://www.opengeospatial.org/
[10] http://geoserver.org/

[11] *Until now, GeoNode only uses WMS to comunicate with Geonode Client Geoexplorer.*
[12] https://jquery.com/
[13] http://getbootstrap.com/

drive the user interface, and Bootstrap cascading style sheets (CSS) is used to style the pages. It is also designed in such a way that any type of project can fit, but because of the requirements of Poli-RISPOSTA project, we decided to change and personalize the site interface. This new interface is described in section 15.2.3.

The *Map Composer* is based on the GeoExplorer[14] web application. It allows creation and publication of web maps with OGC and other web-based GIS services and provides the GIS functions that are a main part of GeoNode. It is built on top of GeoExt[15] and uses OpenLayers[16] and GXP,[17] ExtJS for component-based user interface (UI) construction and data access, and OpenLayers for interactive mapping and other geospatial operations and GeoExt for integrating ExtJS with OpenLayers. GXP provides higher-level components and developing facilities for GeoExt, such as a grid for displaying symbolizers, layer properties panel and styles dialog for displaying histograms of a feature attribute distribution. To create the layers styling (color, labeling, stroke, among others) in the Poli-RISPOTA project, Styled Layer Descriptor or .sld extension files (see Styled Layer Descriptor OGC standard) were created using the FOSS QGIS desktop.[18]

One of the distinct characteristics of this architecture is the inclusion of mobile technologies for data collection in the field that have been proven to be an effective way to make surveys by common users. The Poli-RISPOSTA damage data collection is made through an Android mobile application, created specifically for the Poli-RISPOSTA project without making use of or relying on any existing application. Technologies like Django, Java, SQlite, as well as external services like Google maps and Google drive are used to get access to map services and remote storage services.

All information collected reside on the Poli-RISPOSTA database, the same one used to create maps and graphs, and to store user information. For now, the collected information is only for residential buildings, but in the future, it could be expanded also for other types of infrastructures. In section 15.2.4, more detail on the mobile application is given in terms of requirements and design.

The data flow can be described (see Figure 15.1) as follows: first, damage data are collected in the field by civil protection volunteers and other trained users using the *Mobile Data Collection* application by filling specific damage data forms. Data are transmitted in real time from the mobile device to the Poli-RISPOSTA database, updating the building's layer's attributes already present in the database. Then data are transmitted to the *Web Visualization Client* through the WMS standard geo-web service, to be used for map creation and/or editing. This flux demonstrates how the different components are integrated, creating a single operative and simple ICT system, from data collection and storage to visualization.

15.2.2. Poli-RISPOSTA Data Model

The problem of gathering and analyzing data collected during and after a flooding event is multifaceted. We have several heterogeneous sources of information that need to be structured and integrated into a single database containing all the information we need to proceed with the objective of damage understanding and measurement.

To this aim, starting from the surveys eliciting all the questions to be collected for flooded residential buildings (see Chapter 6 in this book), we have designed a PostGIS database able to contain the required information. Figure 15.2 shows the Entity Relationship diagram of such database that stores all the data related to the procedure, including the surveyor teams, the exact locations of the buildings inserted, and the main characteristics of the flood within the buildings (for example, water depth, presence of contaminants or sediments). In the database, the damage that occurred to each unit surveyed, whether a single house, an outbuilding, a dwelling in a condominium, or the common parts of the latter, is recorded. Damage attributes, not shown in the figure, range from damage to structures and content, through the physical disruption of equipment, to the days required for cleanup. The database was designed in a database management system (DBMS) that supports GIS functionalities; hence, geo-referenced information can be easily imported and processed.

After designing the database according to the requirements expressed by the surveys, we have migrated all the data that were collected on paper to the database itself. During this migration phase, we have faced several challenges related to data interpretation and cleaning. Some writings were hard to understand, but mostly we noticed several inconsistencies in the collected answers due to missing data, redundancies, and similar values expressed in different ways (such as abbreviations, different formats, etc.). A lot of effort has been put into the process of data cleaning in order to provide a uniform view of the information inside the database.

All the data we collect using the surveys are geo-referenced in the sense that they are related to a certain place, that is, a building. On these data we can perform either classic SQL queries, that is, ones that do not involve geographical information (e.g., number of flooded buildings, average

---

[14] *http://suite.opengeo.org/opengeo-docs/geoexplorer/*
[15] *http://geoext.org/*
[16] *http://openlayers.org/*
[17] *https://github.com/boundlessgeo/gxp*
[18] *http://www.qgis.org/it/site/*

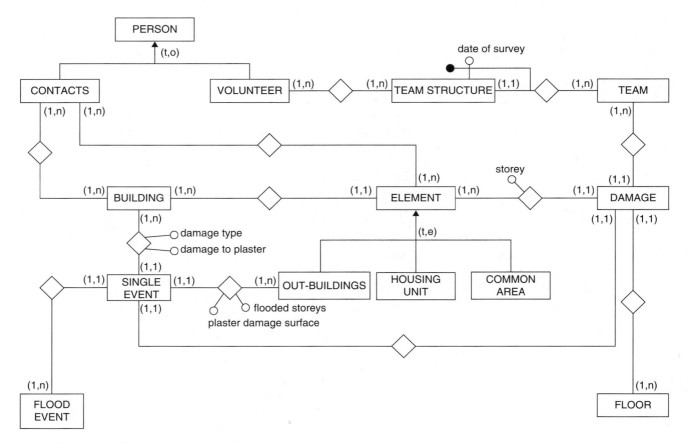

Figure 15.2 Database Entity Relationship diagram.

age of flooded buildings, cost of infrastructure divided by city and so on) but also spatial SQL queries, that is, ones that involve geographical information (e.g., geometry of buildings near a certain place, area with the highest number of flooded buildings, and so on). Moreover, apart from the information described by the surveys, a very important role is played by geo-referenced data; thus, we designed the database adding the GIS extension of the PostgreSQL DBMS to be able to easily import any shape file at any time. Once a shape file is imported into the database, it can be related to the information we collect through the surveys by using spatial SQL queries. For example, one of the most important shape files that we use during the Poli-RISPOSTA procedure is the one containing PAI zones (which are fluvial bands as identified by an hydrogeological plan) (Law 183/89), which is an external shape file that we do not collect but receive from the outside (i.e., a basin authority). However, as soon as we import it into the database, we can relate the information it contains with our own data in order to find, for example, all of the flooded buildings that are part (or not part) of a given PAI zone.

Therefore, the designed database allows us to (i) store all the data we collect using the surveys related to residen-

tial buildings, (ii) import any kind of geo-referenced data in the form of shape files, and (iii) query the data using both standard and spatial SQL in order to retrieve any subset of the stored data. However, we do not require all users to import data or write SQL queries because these tasks require some deeper knowledge that not all users are (or should be) aware of. Therefore, as we will see in the following sections, SQL queries are actually hidden to final users that can perform them by selecting some features from both a web portal and a mobile application. On the other hand, advanced users can still directly access and browse the database.

15.2.3. Poli-RISPOSTA Web Portal

The web portal represents an integrated access point to the data. It is a multi-owners data management environment where each user, according to his role and security restrictions, can access and manipulate portions of data acquired during the Poli-RISPOSTA procedure (see Chapter 6 in this book). The portal described here is a first demo, which can be downloaded at http://www.polirisposta.polimi.it/about/risultati-ottenuti/ (WP4 link). The main functionalities of the web portal (see Figure 15.3) follow:

Figure 15.3 Web Portal Functionalities. Eight basic functionalities were defined for the Poli-RISPOSTA project: layer and documents upload, flooded area definition, damage report, direct damage survey, layers, maps, creation of graphs and tables, and archive.

- ① *Layer and documents upload*: Layers represents raster or vector Geospatial content. GeoNode allows users to upload Shapefiles and raster data in their original projections using a web form. Users can browse and search for geospatial data. Geospatial content (e.g., layers of cadastre, flooded area, survey points) in different formats (e.g., vector *.shp, raster * geoTIFFs). For the specific case of the Poli-RISPOSTA web client, only shape file upload is supported for now, but the upload of .sld extension files was added as a functionality, so the layer can have its predefined style when it is uploaded into GeoServer. The upload functionality is also available for documents (e.g., law or directive associated to a certain layer and survey reports).

- ② *Flooded area definition*: Similar to *Upload Layer and Documents*, this section is used to exclusively load the layers representing the flooding areas.

- ③ *Damage report*: This section is intended to be a communication tool between the stakeholders and the Poli-RISPOSTA system. Here, the damages of infrastructures and vital utilities (rail roads, electricity lines, industrial buildings, public goods, public space, among others) different from residential buildings can be reported by anyone through a form, leaving the contact information.

- ④ *Direct damage survey*: In this section, users can view the tasks that need to be fulfilled and that are assigned to the survey teams. Teams can fill out the forms they would fill out on the field using the mobile application (refer to the Poli-RISPOSTA mobile application). They can use this functionality in case the mobile application does not work on the field.

- ⑤ *Layers*: By clicking the Layers link, users get a predefined list of all layers that need to be published, divided into the following categories: Base layers, Residential Sector layers, and Industrial Sector layers. If a layer is missing, a warning message is displayed to advise users to upload the missing layer.

- ⑥ *Maps (see Figure 15.4)*: Maps were identified to provide a complete flooding event scenario (see Chapter 11 in this book) regarding damage, exposure, vulnerability, and the physical event. The symbolization of the geographic features were set up, allowing the comparison between different event maps and minimizing the user effort of map creation.

In GeoNode, Maps are a set of layers and their styles. Users can create and share interactive web maps by combining layers from the local Geoserver instance, as well as WMS layers from other servers, web service layers such as Google or OpenStreetMap, ArcGIS REST Service (REST) layers or Tiled Map Service (TMS). All the layers are automatically reprojected to web mercator for maps display, making it possible to use different popular base

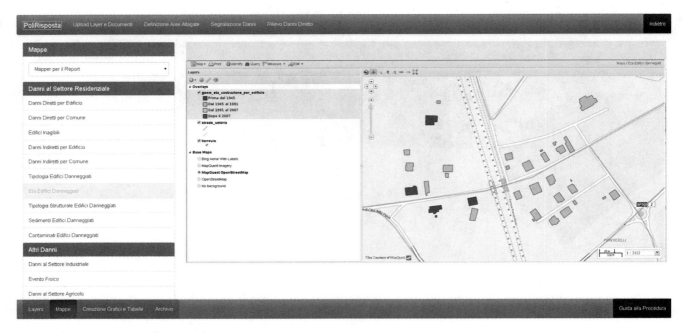

Figure 15.4 Maps functionality.

layers, like OpenStreetMap, Google Satellite, or Bing layers. By clicking the Map link, users get a pre-defined list of all maps. The map functionality includes cartography tools for styling and creating maps graphically in the same way of traditional desktop GIS applications, including editing. Users can gain enhanced interactivity with GIS-specific tools such as querying and measuring. Maps of the flooded area, damage to buildings, points of survey, and economic value of buildings can be visualized and shared using the web portal.

When the layer style has not been uploaded using the Upload layer and Document functionality, layers can be styled using the Manage layers styles tool on the map application. This tool works very similar to the exact same tool on a typical desktop GIS. Figure 15.4 shows an example of a map construction year of the buildings.

- ⑦ *Creation of graphs and tables (see Figure 15.5)*: The Graphs and Tables section allows displaying in a tabular and graphical manner predefined queries related to the damage report. Whenever the output of a query is a multi-dimensional matrix (e.g., damage at each building within a given area), results can be visualized through charts with pre-defined structure. Users can choose from a variety of pre-defined queries, such as total direct damage by building, total direct damage by municipality, duration of buildings unavailability in days, direct damages by flooded level, among others. Graphs are updated in real time with the consolidated data from the surveys.

- ⑧ *Archives*: Here users can find a chronological order list of all the documents uploaded (using the upload layers and documents functionality).

15.2.4. Poli-RISPOSTA Mobile Application for Data Collection

The Poli-RISPOSTA mobile application [*Brondolin and Ardagna.,* 2015] is an Android application developed to support the data gathering step of the Poli-RISPOSTA procedure performed after a flood event, and its aim is to digitalize the current paper-based approach [*Molinari et al.,* 2014b] (see Chapter 6 in this book). The proposed system aims to provide a consistent approach for collecting and storing data and to develop a long-term application solution by providing an environment for analyzing, managing, accessing, and reusing information, objects, and data.

15.2.4.1. Requirements. The mobile application has been developed by strongly keeping in mind the requirements that have emerged by observing the paper-based method currently adopted for data collection. In this manual system, data are collected through interviews and questionnaires. The data collectors go door to door to the victims' houses and ask various questions related to the flood event and the damages it caused as well as other relevant questions and then record the answers on paper. The volunteers have to map the victim's tale in the fields of the form, and this becomes a difficult task when the victim jumps from one topic to another without following the form's structure. (See Chapter 6 in this book.) The Poli-RISPOSTA mobile application addresses this behavior and implements a fast way to move from one question to another. Another important requirement of

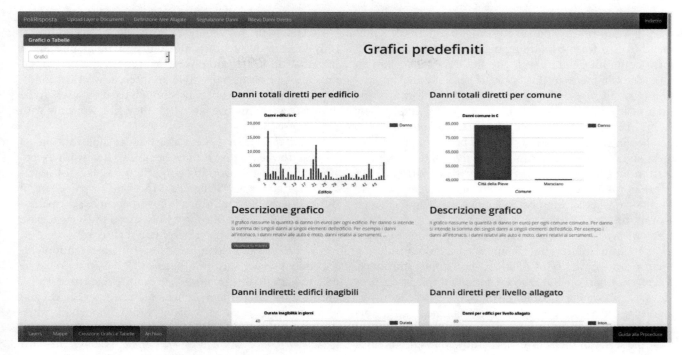

Figure 15.5 Graph and Tables functionality.

the application is to mitigate problems related to misplacement and misfiling of data and to avoid loss of documents, attachments, and photos. Moreover, it is important to reduce time loss for exchanging data between volunteers at different layers of the organization. These issues are solved by means of an infrastructure that takes into account data collection, storage, and analysis. In other words, the Poli-RISPOSTA mobile application is a part of the solution, and it is supported by a back-end Web portal. The users of the application have to be registered to the system, and they will be assigned a team group, in which they will work on the field to collect data. We assume they are knowledgeable about the floods and related concepts but do not have a deep knowledge of computer science and in particular of software. Therefore, the application has to be as easy to use as possible, and it must be intuitive to the user. In order reach these objectives, the application needs to implement the following high-level operations:

• Collect data in a fast and intuitive way.
• Collect information that is easy to analyze afterward.
• Allow the users to search through questions.
• Check the correctness of the data at runtime.
• Let the user jump from a question to another in a fast way.
• Allow the users to make effective team collaboration.
• Help users to efficiently collect data on the field by means of maps.

These features can be further detailed and classified as functional and non-functional requirements of the system.

Regarding the functional requirements, we can identify the following features offered by the Poli-RISPOSTA mobile application:

• *Login.* The application must allow users to log into the system to correctly perform data collection operations but also provide them with off-line operations, i.e., some operations that can be performed even if the tablet is not connected to the Internet.

• *Map management.* The map is a fundamental feature of the application. Users must be able to visualize it, scroll it, zoom in and out, and, if desired, move to the Google maps application. Moreover, they must be able to see their position on the map and the position and information about all the tasks they have to perform, i.e., all the surveys they have to fill in. Moreover, as they are part of a team, they should also be always aware of the position and progress of other teams since they are all collaborating to reach similar objectives.

• *To Do list.* Users' main objective is to collect data by performing different tasks, i.e., going to flooded buildings and filling in surveys about the corresponding flooding event. Therefore, it is important that the application shows users the set of tasks they have to perform, let them create a new task from scratch when needed, and remove tasks that are not necessary anymore. Tasks are stored locally on the device so they can be accessible in the absence of an Internet connection; however, users must be able to send them to the server as soon as they are completed.

• *Task management*. Each task represents a survey that users have to fill in, and surveys might contain different forms (e.g., a form containing data about the building, a form containing data about the flooding event, and so on). (See Chapter 6 in this book.) Therefore, the application must allow users to visualize the list of forms inside each task and to possibly modify them by adding new forms or removing old ones. Of course, each form has to be filled in so the user should always be aware of the progress status of each task through some qualitative information on the completeness of the various forms. When a task is completed then the application must allow users to finalize it and to indicate whether it is suitable to be sent to the server or not.

• *Form editing*. One of the main features is to allow users to input information in an easy way by considering there is a huge variety of data that users must introduce. Therefore, the Poli-RISPOSTA mobile application provides users with the capability to support different answer data types such as numerical values, sentences, gps information, dates and times, istat codes, and both single and multiple selections. Users are also allowed to add pictures both by taking one with the tablet's camera and uploading one from the gallery and to add short notes related to each question in the form. The application provides strong flexibility features by allowing users to duplicate questions and sections to express different information with the same form structure. Moreover, at each moment during the data input procedure, the application shows if a single question is filled in a correct way and if the whole form is fully filled in.

• *Search capabilities*. Users must be able to quickly find the questions they want to fill in, thus, the application provides them with the capability to search for questions both inside tasks and inside forms. The search must be performed on the form structure, not on the gathered values, in order to help users find the correct question when necessary and, when this happens, users must be quickly redirected to the desired request.

• *Local data management*. Users must be aware about the information they hold on the tablet, that is, they must know which tasks are assigned to them and which forms are currently stored on the tablet. When new tasks are assigned to them, they must be able to interact with the server and download both the tasks and the forms.

• *Server coordination*. The system must update the server about its position on a regular basis and must be able, after user inputs, to send the selected tasks and forms, with all the information they contain, to the server.

As far as non-functional requirements are concerned, we can identify several categories that are important in the development of the Poli-RISPOSTA mobile application.

• *Usability*. The system must be able to guarantee fast responsiveness on different devices, in particular by implementing the capability to adapt to different screen sizes. Due to the particular scenario, the project addresses only tablets, due to their flexibility of managing much information in the same view. Moreover, the user interface must be clear and visible in different weather conditions and must be responsive enough to allow users to be at least as productive as when they are working with paper-based forms.

• *Reliability*. The system must be reliable during its work in the sense that it must be able to deal with exceptions and difficult working conditions like lack of a network signal or GPS signal. Moreover, even when in difficult working conditions it must guarantee the persistence of all the data filled in by users and thus has to save information as soon as possible in a secure way.

• *Security*. The system must store the information collected in a secure way. In other words, the server must be able to acquire information using authenticated sessions, and it must keep them protected from unexpected and unauthorized accesses. This is quite important when dealing with sensitive data.

• *Performance and energy efficiency*. The system must be able to operate for several hours, thus the mobile application should work with simple user interface effects and without consuming too much of the device's battery. The main focus is to provide an application that is responsive and performs its internal processes in a fast and transparent way by reducing memory usage and sensor usage as much as possible

• *Maintainability*. The architecture of the application must be as general as possible. In particular, the method used to generate different survey visualizations must be general in such a way that every kind of questionnaire specified in a correct way can be processed and displayed.

15.2.4.2. Design. To achieve the requirements specified in the previous section, a complex design is needed that is able to take into account several aspects of the system. The chosen system architecture can be classified as the classical three-tier client-server architecture. On the server side, the system has the storage layer, on which all the final information is saved. Moreover, there is an application layer useful to control communication between client and server. The application layer is also present, with a more complex architecture on the client side. The client is also responsible for the presentation of the data, which are stored locally for the time needed to perform the required operations. In Figure 15.6, we can see an overview of the entire system.

The server side of the system provides the application program interface (API) required for forms download, the tasks, and information about survey area and groups on the field. It allows also the ability to upload the completed tasks and information about the current position.

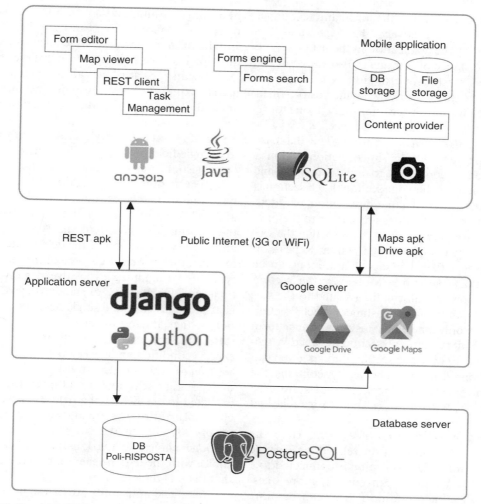

Figure 15.6 Graph and Tables functionality.

External services like Google maps and Google drive are used, respectively, to get access from the tablet to map services and remote storage services for picture uploads.

On the client side, starting from the information coming from the server, users can perform different operations. In Figure 15.6, we have listed some macro-functionalities offered by the application. Users can view the map and get information on their survey area while looking at other team positions. They also have the possibility, after downloading the assigned tasks, to see where they have to go in order to fulfill the assignments. Moreover, users can download the tasks assigned to their group, and after that, have the possibility to access the task management functionalities. In this view, they can add new forms or modify the pre-filled ones. When a form is selected, the form engine processes the selected form model and opens the form editor, in which it is possible to fill the required questions. If the user does not find a question or a field, he can search for it through the search functionality.

In order to manage the search functionalities and to be able to handle forms in the most general way, the mobile application relies on an internal database used to store form models, instances, and tasks. This design choice makes the application able to persist data into this structure independently with regard to the used form. Most of the solutions currently available on the market, such as Open Data Kit, use the database only to store data and not form models. In order to do so, they create one table for each form model inside the system. This approach works in case of data analysis with forms with fixed data structure, but it does not fit to a mobile application, which focuses only on data collection and not on data analysis. Moreover, with our database schema, it is no longer needed to delete the old data or the entire table in case of update of the form structure. In particular, the database allow the application to store information about the following:

• *Form and form instance entities.* Each form is characterized by an id, name, text name, and description. Name is a unique string that identifies the form, and text name and description are, respectively, the name and the description shown by the application in the list of forms. Moreover, every instance of a frame is also stored in the database, that is, all the information related to the instances created by the user or downloaded as pre-filled by the server. Each form instance refers only to one form, but a form can have more than one instance.

• *Section and section instance entities*. The database stores information related to a single section within a form. One section is connected to only a form, while a form can contain more than one section. Each section is described by an id, name, text name, description, required, and repeat. The first four features are similar to the ones in the form entity. Required is a flag that says that the section is required to be completed by the user. Repeat, on the other hand , is a flag that says if a section can be repeated more than one time. Moreover, the section instance entity stores the information related to the sections created within a form instance. Each section instance refers to only one form instance, but a form instance can have more than one section instance. A section instance refers to only one section, but a section can have more section instances. The entity contains the attributes id, and index. The index attributes are used as weak key in order to address and store replicated sections.

• *Question and question instance entities*. Here the application stores the information related to a single question within a form. One question is connected to a single section, but a section can contain more questions. Moreover, for each question there can be more resources attached (such as images, audio, or video resources). The attributes of the entity are the same as the one presented for the section entity. When an instance of a question is created, the application stores the information related to the questions generated within a form instance. Each question instance refers to only one section instance, but a section instance may have more question instances. Each question instance refers to only one question in the model, but a question may have several question instances. The attributes are id and index. As for section instance, the index attribute is used as weak key in order to store the repeated question instance belonging to the same form instance and the same form model.

• *Resource and resource instance entities*. The resource entity stores the information related to an image or audio or video within a form. Each resource is related only to a question, but a question may have more than one resource attached. The attributes of the entity are id, name, path, and type. Name is a unique string identifier, path is the identifier of the resource on the local storage, and type holds information about the type of the resource, like image, audio, and so on. On the other hand, the resource instance entity stores the attachments for the specific form instance. An attachment is related to only a question instance, but a question instance may have more attachments. The attributes are id, type, and path. Type is used to express whether a resource instance is an image or an audiotape. Actually the application supports only images as attachment.

• *Field*. The field entity stores the information related to a single field of the form. Each field is related to only one question, but a question may have more than one field. The attributes of the entity are id, name, text name, type, required, repeat, cond, condition statement, op1, and op2. We want to focus on the type attribute. It stores the information related to the type of field, and this affects the visualization of the field in the UI and the way in which the answers are stored in the database. The data types are string, integer (positive, negative, or standard), float (positive, negative, or standard), gps, date, date time, istat code, measure, and single selection or multiple selection. Other interesting attributes are cond, condition statement, op1, and op2. They are the attributes used to express conditioning between fields.

• *Choice*. The choice entity stores the choices for fields of type single selection or multiple selection. One choice is related to one field, but a field can have more than one choice. The attributes are id, text, and other. The attribute *other* is a flag used to say whether or not that choice is an *other* choice. In the UI, the user can fill the other choice with the data he/she wants if there is not a choice that fits that data.

• *Note*. The note entity stores the notes written by the user during the survey activity. A note refers only to one question instance, but a question instance may have more than one note. The attributes are id and text.

• *Answer*. The answer entity stores the answers prompted by the user during the survey activity. Each answer is related to only one question instance and to only one field. A field can have more answers related, and also one question instance may have more than one answer. The attributes are id, prefilled, applicable, type, filled, answer string, answer integer, and answer float. The prefilled attributes say if the answer was prefilled and thus downloaded from the server. The applicable and the filled attributes say, respectively, if the answer (and so the related field) is meaningful and if the answer was completed. The type attribute has the same rules as the one in the field table. The answer attributes are used to store the answer given by the user. Depending on the type attribute, they are used in different ways.

• *Task and task module*. The task entity stores the information related to the task assigned to a group. Each time a user downloads a task from the server, its basic information is stored in this entity and in task module.

Figure 15.7 Mobile map functionality.

Each task can contain more than one form instance, but one form instance is related to only one task. Each task can have more than one task module, but one task module is related to only one task. The attributes are id, name, description, group id, generated, user defined, and sent. The attributes name description identifies the task from the user point of view. The attribute group id refers to the group that the task is assigned to. Generated is a flag that says if the task is already generated in terms of form and form instances. User defined is a flag that states if the task comes from the server or if the user has created it. The sent flag is used to store whether the task was sent back to the server or not. On the other hand, the task module entity stores the information related to the elements that must compose a task. Each task module is related to only one task and specifies which form model is related to the task. A task can have more than one task module, and one form instance can be related to more than one task module. The attributes are id and n instances. This attribute states that the maximum number of the selected form instance can be placed inside a task. If the value is 0, there is no limit on the number.

15.2.4.3. User Interface. In this section, we explore the user interface able to provide the functionalities previously introduced. The first interface displayed by the application after launch contains, on the top right corner, a wheel icon, which opens the settings view where the user

can insert the login data with the address of the server to connect. The rest of the screen is comprised of three parts shown as the horizontal scrolling tabs ToDo, Map, and Downloads:

• *ToDo* allows the user to control the local tasks. Each time the user opens the application, the system shows this view where users can select a task by clicking on it in the list; create a new task; or delete a task or upload it to the server. In this latter case, the system checks whether all fields have been answered and, if not, asks for a send confirmation. Moreover, for each task, the user can add, edit, and delete the forms contained in it.

• *Map* helps the user with geo-localization and group coordination (shown in Figure 15.7). Here users see a map displaying his/her position inside the survey area. The map shows the gps position of each group and of each task with a brief description. Moreover, for each task, the user can click on its description and the system will redirect him/her to the related forms. Each time the map is loaded, the application asks to the server for updated data of positions and area. The system triggers periodically a position update, which is sent to the server to inform the other groups about the user's position. Each gps position inside the map can be passed to the Google maps application in order to leverage its navigation facilities.

• *Downloads* is deputed to communicate with the server. Here we find the list of local form models, the list

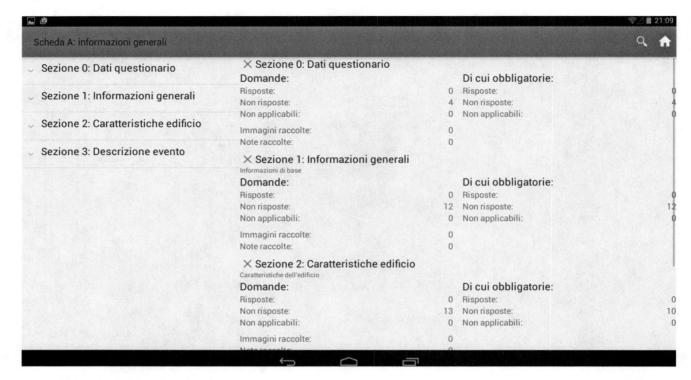

Figure 15.8 Forms functionality.

of remote form models, and the list of remote tasks. A long-click on each element will show the option menu and allow users to either download remote elements or delete local ones.

A functionality that can be reached from the ToDo view is the addition of a new task. The user has to choose first the name and a brief description of the task then the forms that can be used in the task and their maximum amount. After a new task has been created, it is shown in a grid with its forms, and for each of them some information is provided such as name and description of the form, the date, and the progress in terms of answered questions. The user can add a new form anytime and can then proceed to edit it with the necessary information. When all forms have been completed then the whole task can be sent to the server.

When a form is selected, all sections and questions related to it are shown, as in Figure 15.8, each with their title and the possibility to compile the corresponding answer or add notes or multimedia objects. For each form statistics about its fulfillment are provided. When an answer is added to a question in the form, the system checks its correctness with respect to its domain and highlights it with a green check if it is correct or it shows a red cross if it is not correct. These completeness signals are spread through the whole form and task that are completed only when all answers have a green check. Within a given form, users can duplicate sections and questions, if

the form supports it. This is useful when there is several homogeneous pieces of information that we want to collect without limiting their amount. Similarly, duplicated sections and questions can also be deleted.

Another fundamental feature offered by the mobile application is the possibility to search through the forms in order to find a given question or section of field. This is very important because usually the flow of narration actuated by the people being surveyed is different than the flow of the survey, and the volunteers performing the surveys need to be able to move from one question to another in a fast way. Thus, the search functionality comes into play when the user does not know the exact position or name of the needed question. To use this functionality, the user simply needs to insert some search keywords and the system will provide the list of questions and sections of fields that comply with the request. By selecting one of the results, the user will automatically be redirected to the chosen element inside the form.

15.2.4.4. Software Architecture. The mobile application is split into different packages, typically one for each activity. Each package collects the view code and the methods responsible to manage and store in the model the data coming from the view. The model is in a separate package, with model classes, database definition, and content provider. Particularly important is the form engine. This part of the application is able to take the user input and to manage it, starting from the storage in the database, moving to the correctness and completeness check and the statistics computation. Given the event-based paradigm of the Android applications, the form engine starts each computation after an input event. When the user selects a form, the form engine loads it from the database and shows it in the view. When a question is selected, the form engine reads the content of the question from the form model and creates the right interface to display dynamically the field views. Each field is generated starting from a view XML description that is customized depending on the field type and text. The views are inflated in the content fragment, which contains the name, the description of the question, the list of fields grouped in a linear layout, the list of images, and the list of notes.

When the user answers a question, the system stores the answer and checks if the content of the answer is correct for the data type. If the answer is not applicable, the content is saved in any case, but it is not further considered. When the user leaves the question to move to another or closes the activity, the system computes the new statistics related to the form instance.

When it is needed and if the form model allows it, the user can replicate a question or a section. In this case, the system looks at the form model and replicates the selected item creating a new instance of it. This allows a flexible data collection with regard to paper-based one, in which the number of eventual replicas is fixed. This approach allows us to specify a form model in terms of rows in a database, and the mobile app is in charge of rendering the model in the proper way. The administrator of the system is only in charge to define the form model in terms of form structure and type of the fields. He/she can specify if a question or a section is mandatory and/or repeatable. The form model defines the table attributes, while mapping the questions to that attributes without taking into account the structure of the survey. Our approach tries to manage the structure from a logical and graphical point of view. This means that the form specification takes into account a different way to show information, which is more flexible for our case study. Moreover, with our approach, when the form structure changes, the data inside the database are not affected by the change in the structure.

Another important consideration is that the Android operating system (OS) runs on resources with limited performance and power consumption. To maintain the UI responsive in each condition, the most time consuming operations must be done in background. Moreover, the networking operations must be done in the background due to OS limitations. Given this environment, we decided to move all the complex database operations and the networking activity on several async tasks, a particular

implementation of the thread library. With this class, the UI is constantly updated with the information coming from background threads. Moreover, the code under the onPreExecute and onPostExecute methods are executed in the UI thread, warning the user that something is happening. External APIs, like Google maps and Google drive, are asynchronous by design. For the first one, we use it in an asynchronous way, and the second is used synchronously inside a background thread.

15.3. REMARKS ABOUT THE POLI-RISPOSTA INFORMATION AND COMMUNICATION TECHNOLOGY SYSTEM

Within the Poli-RISPOSTA project, tools and advanced technical solutions have been developed working with stakeholders and by involving experts from several fields such as ICT, geomatics, engineering, urban planning, economists, etc., in order to support civil protection authorities in dealing with flood emergency. Such tools aim to ensure an effective solution for collecting, storing, analyzing, and representing a multitude of data as well as a flexible way to support future changes or updates. The development of the database and the software for data management has been carried out contemporaneously as they unfold complementary activities characterized by continuous interaction and feedbacks.

One important result of the Poli-RISPOSTA project has been the integration between different ICT technologies, which allowed overcoming, on one hand, the lack of integration among tools used during the emergency and recovery phases. On the other hand, the projects achieved to develop tools that integrate different data formats and support data visualization and spatial analysis, and to use a tablet application for field surveys, which responds to a standardized data collection procedure, called RISPOSTA [Ballio et al., 2015].

15.4. CONCLUSIONS

This chapter discussed the information system and tools for collecting, storing, and managing damage data in the aftermath of flood events developed within the Poli-RISPOSTA project (stRumenti per la protezione civile a Supporto delle POpolazioni nel poST Alluvione), a Data model, web portal, and mobile application.

This is just an example on how ICTs in disaster recovery and reconstruction can be used to obtain appropriate information required to carry out damage and loss assessments and for decision-making regarding post-disaster planning, project formulation, and implementation. The latest developments in OS for the collection and maintenance of data, as well as for sharing these data (and geospatial data in general) over the Internet, in conjunction with the latest capabilities of FOSS for the development of web-based applications for collecting, processing, and distributing data are greatly facilitating the integration of systems like Poli-RISPOSTA. This type of system would have been extremely complicated to integrate even just a few years ago. These technologies have lowered the knowledge and effort necessary to develop these systems. This is extremely valuable given that many scientists, managers, agencies, and governments around the world do not have access to ample and sophisticated resources for the creation of their Web-based information systems for disaster management data or any other application.

As an improvement of the web portal, users should be able to give and access information not only according to their role but also according to the timeline of the procedure. This will guide the user to provide and obtain information in a quick and efficient manner, without being overwhelmed with links and functionalities that are not all relevant in a specific time of the procedure. Hence, functionalities should be extended so, at each step of the procedure, the web portal provides users with the specific knowledge required to perform each action (e.g., flooded areas, location of buildings), to remind them about tasks they have to perform (e.g., gather data), and to alert them when new information is required or made available (e.g., data gathered during the field survey), according to their role in the procedure.

ACKNOWLEDGMENTS AND DATA

The PoliRISPOSTA project was funded by Poli-SOCIAL funding scheme of Politecnico di Milano. The authors acknowledge with gratitude all those involved in the Poli-RISPOSTA and in particular Simone Corti, Francesco Zoffoli, and Rolando Brondolin for the development of the web portal and mobile application.

REFERENCES

Abdalla, R., and K. Niall (2009), WebGIS-based flood emergency management scenario. Adv. Geogr. Inf. Syst. Web Serv. 2009. GEOWS'09. Int. Conf., 7–12.

Balbo, S., P. Boccardo, S. Dalmasso, and P. Pasquali (2014), A Public Platform for Geospatial Data Sharing for Disaster Risk Management. ISPRS - Int. Arch. Photogramm. Remote Sens. Spat. Inf. Sci., XL-5/W3 (February), 189–195.

Ballio, F., D. Molinari, G. Minucci, et al. (2015), The RISPOSTA procedure for the collection, storage and analysis of high quality, consistent and reliable damage data in the aftermath of floods. J. Flood Risk Manag.

Begum, S., M. J. F. Stive, and J. W. Hall (2007), Flood Risk Management in Europe: Innovation in Policy and Practice, Physica-Verlag.

Billa, L., S. Mansor, and A. Rodzi Mahmud (2004), Spatial information technology in flood early warning systems: an overview of theory, application and latest developments in Malaysia. Disaster Prev. Manag. An Int. J., 13 (5), 356–363.

Brondolin, R., and D. Ardagna (2015), Poli-RISPOSTA Mobile App, Technical Report.

Brovelli, M. A., L. Dotti, M. Minghini, et al. (2014) Volunteered Geographic Information for water management: a prototype architecture. HIC 2014 – 11th Int. Conf. Hydroinformatics "Informatics Environ. Data Model Integr. a Heterog. Hydro World."

Brovelli, M. A., M. Minghini, and G. Zamboni (2014), Public Participation GIS: a FOSS architecture enabling field-data collection. Int. J. Digit. Earth, 1–19.

Clothier, J. (2005), Dutch trial SMS disaster alert system. CNN. Retrieved 6 June 2010, from http://edition.cnn.com/2005/TECH/11/09/dutch.disaster.warning/index.html.

De Groeve, T., K. Poljansek, and D. Ehrlich (2013), Recording Disaster Losses. Recomm. a Eur. Approach. Eur. Comm. Jt. Res. Center, Ispra, Italy.

Dorasamy, M., and M. Raman (2011), Information systems to support disaster planning and response: problem diagnosis and research gap analysis. Proc. 8th Int. ISCRAM Conf. Port.

European Commission (2007), Directive 2007/60/EC of the European Parliament and of the Council of 23 October 2007 on the assessment and management of flood risks (Text with EEA relevance).

Fedra, K., and R. F. Reitsma (1990), Decision support and geographical information systems, Springer.

Guha-Sapir, D., and R. Below (2002), The quality and accuracy of disaster data: a comparative analyses of three global data sets, World Bank, Disaster Management Facility, ProVention Consortium.

Gutierrez, F. V., M. A. Manso, D. H. Lang, et al. (2007), Designing Web-Enabled Services to Provide Damage Estimation Maps Caused by Natural Hazards.

Hammond, M., S. Djordjevic, D. Butler, and A. Chen (2011), Flood Impact Assessment Literature Review.

Hristidis, V., S.-C. Chen, T. Li, et al. (2010), Survey of data management and analysis in disaster situations. J. Syst. Softw., 83 (10), 1701–1714.

Johnson, R. (2000), GIS technology for disasters and emergency management. An ESRI white Pap.

Karnatak, H. C., R. Shukla, V. K. Sharma, et al. (2012), Spatial mashup technology and real time data integration in geo-web application using open source GIS–a case study for disaster management. Geocarto Int., 27 (6), 499–514.

Li, J., Q. Li, C. Liu, et al. (2014), Community-based collaborative information system for emergency management. Comput. Oper. Res., 42, 116–124.

Merz, B., H. Kreibich, R. Schwarze, and A. Thieken, (2010), Review article "Assessment of economic flood damage." Nat. Hazards Earth Syst. Sci., 10 (8), 1697–1724.

Molinari, D., M. Mazuran, C. Arias, et al. (2014a), Implementing tools to meet the Floods Directive requirements: a "procedure" to collect, store and manage damage data in the aftermath of flood events.

Molinari, D., S. Menoni, G. T. Aronica, et al. (2014b), Ex post damage assessment: an Italian experience. Nat. Hazards Earth Syst. Sci., 14 (4), 901–916.

Murai, S. (2006), Monitoring of disasters using remote sensing GIS and GPS. Int. Symp. Disaster Prev., 9–11.

Nisha de Silva, F. (2001), Providing spatial decision support for evacuation planning: a challenge in integrating technologies. Disaster Prev. Manag. An Int. J., *10* (1), 11–20.

Quarantelli, E. L. (1997), Problematical aspects of the information/communication revolution for disaster planning and research: ten non-technical issues and questions. Disaster Prev. Manag. An Int. J., *6* (2), 94–106.

Raltman, M. T., K. R. Rahaman, and A. Sadie (2011), Real-Time Web-Based GIS And Remote Sensing For Flood Management.

Sagun, A. (2010), Advanced ICTs for Disaster Management and Threat Detection: Collaborative and Distributed Frameworks, chapter Efficient Deployment of ICT Tools in Disaster Management Process. IGI Glob., 95–107.

Santos-Reyes, J., and A. N. Beard (2012), Information communication technology and a systemic disaster management system model. Dev. Distrib. Syst. from Des. to Appl. Maint., *294*.

Wattegama, C. (2007), ICT for Disaster Management, Asia-Pacific Development Information Programme" e-Primers for the Information Economy. Soc. Polity.

Wirtz, A., W. Kron, P. Löw, and M. Steuer (2014), The need for data: natural disasters and the challenges of database management. Nat. hazards, *70* (1), 135–157.

Conclusions

Daniela Molinari¹, Scira Menoni², and Francesco Ballio¹

1. THE HISTORY OF FLOOD DAMAGE DATA COLLECTION AND MANAGEMENT

The fact that damage data available in official records are not as reliable and complete as would be wished for is a complaint that has already been brought up by *Hoyt and Langbein* [1955] in their book Floods. In their book, they analyzed damage data related to floods in the United States (US) in the period between the years 1900 and 1949. Reasons for limited reliability were identified in the fragmentation among offices and bureaus each responsible for some information but which were uninterested in getting a comprehensive overview. Differences in the criteria used for entering data along with a limited number of organizations and bureaus covering the entire country caused problems.

Those reasons can be summarized as lack of an institution that is made responsible for collecting and managing the data on disaster losses across sectors in a continuous manner in such a way to feed a national level database. Similar considerations can be found in much more recent reports and studies that seem to have increased in the last two decades, perhaps as knowledge on hazard factors has increased and both researchers and practitioners turned their interest to exposure and vulnerability. Analysis of the latter requires that damage be much better known than it actually is, with respect to both physical and systemic aspects that connote damage.

In 1999, *Hubert and Ledoux* conducted research and held a workshop to trace the history of pre-disaster damage assessment and post-disaster damage and impact estimation. In their study, they compared the situation of France, Germany, and the United Kingdom (UK) in Europe and the US. They suggested that in the US efforts were much more focused on the development of risk modeling computerized tools than on creating a systematized post-disaster losses data collection. HAZUS is the most relevant result of such an effort; it is a computerized model that supports planners and decision makers in assessing the risk in a given area using data and information on hazards, assets, and geography of places that are largely available at the federal and national levels. HAZUS is certainly grounded on knowledge regarding impacts of different types of hazards, but it is not necessarily the type of systematic and cross-sectoral survey of direct and indirect losses that is required to overcome the problems highlighted by Hoyt and Langbein a couple of decades earlier. This is not to say that such problems were totally disregarded; in fact, in 1999, the US National Research Council mandated a report to the established Committee on Assessing the Costs of Natural Disasters, with the aim of providing recommendations on how to improve knowledge and the informational base on disaster-related losses. The committee considered as a major obstacle the absence of "an official disaster cost accounting system in the United States" and suggested that "one agency of the Federal Government should be made responsible for compiling a comprehensive database containing the losses of natural disasters." However, results in this domain have not been that compelling as stated by Gall in her contribution (Chapter 4). The Spatial Hazard Events and Losses Database (SHELDUS), the only one reporting damage due to a variety of hazards in the US, which was established in 2004, is an academic initiative funded by

¹*Department of Civil and Environmental Engineering, Politecnico di Milano, Milan, Italy*

²*Department of Architecture and Urban Studies, Politecnico di Milano, Milano, Italy*

Flood Damage Survey and Assessment: New Insights from Research and Practice, Geophysical Monograph 228,
First Edition. Edited by Daniela Molinari, Scira Menoni, and Francesco Ballio.
© 2017 American Geophysical Union. Published 2017 by John Wiley & Sons, Inc.

research grants and does not result from the coordination of institutions responsible for disaster management.

In Europe more attention has been devoted to post-disaster damage data collection at least by some institutions, but with limited results. For example, in 1976 the French Ministry of Environment led an initiative to establish a national disaster losses database. However, no resources were committed nor mechanisms put in place for the active involvement of the local institutions that actually are the first data collectors, leading to the abandonment of the initiative.

The Joint Research Centre (JRC) of the European Commission established a first initiative on the issue of gathering knowledge and lessons learned from past disasters in the Nedies project [2001–2003] consisting of workshops conducted with responsible organizations in the field of civil protection and disaster risk prevention. In a report summarizing the results of one of those workshops, with the rather intriguing title "In search of a common methodology on damage estimation: from a European perspective," *Van der Veen et al.* [2003] point at the fact that "Only a limited number of countries try to record in an informative manner the damage of previous disasters. This is done at local or national level, but there is no institutionalization of the procedure."

Comprehensive and comparative assessments of global databases to our knowledge are more recent. Grasso and Dilley [*United Nations Development Programme (UNDP)*, 2013] compiled an overview of existing global databases, highlighting strengths and weakness of each. In general, they found all of them to be very weak in reporting economic losses.

Margottini et al. [2011] carried out a rather extensive analysis of international databases, comparing the figures provided for the same events and showing how discrepancies can be an order of magnitude between one source and the other. They also examined databases provided by reinsurance companies such as Swiss-Re and Munich-Re, finding that apart from being biased toward insured risks for obvious reasons, they heavily rely on information provided by the companies they insure in their turn. As previously mentionened also by *Hubert and Ledoux* [1999], insurance companies often do not keep a separate recording system for natural hazards, and the reporting quality largely depends on the expert who surveyed the affected items, ranging from a rough estimate to very detailed description of damage and associated likely repair costs. Finally, *Margottini et al.* [2011] also analyzed the mismatches between national and global databases, due also to the significant differences in coverage, duration of the recorded period, and quality of recorded data as shown also by Ehrlich et al. in Chapter 2, who analyzed the situation of national losses databases across European countries.

An issue that arises from the analysis of the past relates to the consistency across databases horizontally, among data provided by different institutions, and vertically, across scales. Consistency has to do with the way aggregation of data at larger spatial levels is achieved; intrinsic data quality depends on the accuracy with which data are collected and managed locally and regionally, where damage actually occurs. If data are not collected in a consistent and standardized way locally where events actually occur and produce damage on the ground, there is no hope for better data at larger scales, where figures are the result of aggregation and summation of locally constructed data.

A second important issue relates to the motivation behind data collection and management. Until very recently, it seems that the main motivation was identifying needs that arose during the emergency and sustaining recovery and reconstruction. Among such needs, financial compensation of losses is certainly prominent for most institutions, including insurance companies. Accounting is another relevant motivation, though data that feed national accounting systems are those required for recovery management and compensation and display all the limitations that have been mentioned above. Global accounting relies heavily on data provided by national governments and in case of lack of such data turn to secondary sources, often newspapers. This type of accounting has been used mainly to identify trends of events and associated costs. In order to advance in the field of post-disaster damage data collection, new rationale and revised understanding of traditional motivation are therefore essential.

2. CURRENT MOTIVATION AND RATIONALE FOR DAMAGE DATA COLLECTION AND ASSESSMENT

"Understanding risk" is one of the pillars of the Sendai Framework for DRR and it is explicitly stated that it implies enhanced data at global, national and regional level about the different damages provoked by disasters' impact. In other words, "understanding risk" implies enhanced knowledge of hazards and exposed built and natural environments, digging into the losses that show in action the combination of hazards with vulnerable systems.

Serje (Chapter 1) introduces the reader to the targets set by the Sendai Framework for Disaster Risk Reduction (DRR) and that will have to be measured by quantitative indicators. This is a rather relevant departure from the previous Hyogo Framework for Action. In the latter indicators of performance were highly qualitative making it very hard to assess to what extent a country has accomplished the prescribed goals. Despite evident difficulties, the Sendai Framework is setting a very clear path towards increased capacity to measure damage so as to be able to say on the basis of numbers if mitigation policies have been effective or not in reducing the risk.

It may be held that the requirements of the Sendai Framework for DRR are opening the door for a new generation of losses accountings that go beyond identifying trends toward more decisively supporting programs and policies. In more recent years, the link between accounting and programming different kinds of initiatives and expenses related to risks has been strengthened also as a response to the financial crisis and to a general shrinking of public welfare expenditure.

One very important purpose that such accounting should serve is the comparison between costs sustained to repair and reconstruct after disasters and costs of mitigation to avoid and reduce damages and losses. In this regard, the *Organization for Economic Co-operation and Development (OECD)* [2014] has published a very interesting report, suggesting that the two sets of costs should be compared in a consistent way to assess the validity of choices made so far and to make decisions for the future. In both accounting spheres though, the OECD report highlights problems and challenges that have to be met. On the one hand, there is a lack of consistent and reliable damage data; on the other hand, there is a lack of comprehensive figures of public expenditure for mitigation, that is spread across a very large number of vertically and horizontally different governmental organizations and agencies.

A similar strand can be found in the contribution of Moon (see Chapter 12) describing the mandate of the Queensland Reconstruction Authority responding to a governmental quest of proving and providing effectiveness to the public expense in the field of natural hazards mitigation [*Australian Government Report*, 2014]. In the latter report, a quite significant change of policy is reconized, shifting part of the responsibility for mitigation to individual subjects while leaving to the state the burden to respond and mitigate only large disasters.

Even though in a less radical sense than the one hypothesized in the Australian Government Report, still the direction is clear also in Europe that the provision of help and post-disaster support is expected to be outbalanced by states' efforts to reduce damage over time, and part of this responsibility has to be shifted to local governments and to citizens themselves. For now, the reports that have been produced so far in the last decade or so, responding to increasingly strict national and European requirements for compensation, have increased the capacity at least to make rough estimates of how much will be needed to face emergencies similar to the ones that have occurred in the last decade. The interesting audit made by the former Head of the Italian Civil Protection to the Senate in December 2014 [*Gabrielli*, 2014] can be quoted as an example, but there are similar instances in France

[*Direction Territoriale Méditerranée du Cerema*, 2014] and in the UK [*Pitt Report*, 2007]. The same Flood Directive, an important piece of legislation at the European level, is conceived as an iterative effort to increasingly base flood risk management plans on evidence of how much has been lost and how much can be saved.

In the US, more pragmatic attempts to assess the cost-benefit ratio of mitigation measures in order to boost political initiative have been conducted, leading to the figure of 1 to 4 [*Federal Emergency Management Agency (FEMA)*, 1997] or 1 to 6 [*US Army Corps of Engineers (USACE)*, 2009] depending on the considered source. Yet, as Gall in Chapter 4 warns, such figures are actually based on not so reliable damage estimates that do not constitute such a solid basis as one would wish. So even though politically these proclamtions sound really good, in reality they lack the evidence base that would be required to sustain not the general decision to commit resources for mitigation, but to define exactly how much, and especially toward what sectors and according to which priorities.

In order to improve accounting at different levels of government and governance, including the global one, different steps need to be followed. Partly the latter imply new and better procedures and practices of damage data collection and management, and partly they relate to the actual uses such data may serve. In this regard, the contributions by Ehrlich et al. in Chapter 2 stress the need of better data for forensic post-disaster investigation and to improve current risk modeling capacity. The first part of the endeavor, that is, innovating damage data management practices, requires coupling administrative procedures and technological instruments already in place for assessing post-disaster resources and compensation needs with further data surveys needed to serve the other purposes that are recognized as essential nowadays.

The second part of the endeavor requires that the scientific and practitioners communities agree upon advanced types of uses of damage and loss data that justify whatever extra burden is put on already stressed organizations for enhanced data collection and management. Somehow the two components of the problem, motivation rationale on the one hand and innovative practices, need to go hand in hand. One requires the other and vice versa as shown by Berni et al. in Chapter 6.

2.1. Innovative Good Practices of Damage Data Collection and Management

In this book, a number of good and innovative practices for post flood damage data collection and management can be found, at different spatial scales.

At a global level, in Chapter 1, Serje discusses the DesInventar database that covers several countries in Latin America and Asia. Originated by an academic initiative, it provides a good example of how information can also be managed in developing countries to serve at least very basic level requirements of the Sendai Framework of DRR.

At a national level, the effort of maintaining and managing the HOWAS 21 database in Germany is certainly very relevant. It is an academic endeavor carried out by the German Research Centre for Geosciences GFZ and works on a community-based concept, according to which those who use the data are also required to feed it with data as discussed by Kreibich et al. in Chapter 5. In Chapter 7, Thieken et al. describe the methodology followed to poll data into the system, regarding damage to households and economic activities.

At the state level, in Chapter 12, Moon discusses the experience gained by the Queensland Reconstruction Authority that provides an important model of how coordinated damage data collection after the series of floods that affected the Australian State can be structured and sustained to support a more resilient recovery and reconstruction. Such enhanced damage data management system is fundamental also to monitor the advancement of recovery over time, comparing the needs risen by the event with the response public and private organizations provide. The Queensland Reconstruction Authority show how a more systematic, systematized, and technology supported way of managing data can be actually used to guarantee transparency and accountability of governmental action during recovery but also, in the longer run, in implementing mitigation measures.

At the regional level, the effort of collecting damage data across sectors and at different temporal thresholds, considering the spatial scales relevant to explain the damage due to the two floods of 2012 and 2013 that affected the Umbria region in Italy, results from a true partnership among governmental and scientific institutions (Chapter 6 and Chapter 11). An important aspect of this practice is the achieved integration between requirements set at the national level for compensation and the requirements to be fulfilled in order to use the data also for forensic and risk modeling objectives.

Finally, King and Gurtner in Chapter 8 describe a number of initiatives of damage data collection aimed at eliciting social drivers of flood risks, looking at vulnerabilities created by European settlers in an environment that was not familiar to them. This experience introduces the use of damage data for forensic analysis of disasters.

2.2. The Different Roles of Post-Disaster Forensic Investigation

Forensic investigation actually covers a rather wide range of practices. Originally as suggested by Szoenyi et al. in Chapter 10, it regards court practices, and the increasing demand for engineering and geological expertise to serve the judge or the two opponents in trials related to disasters triggered by natural hazards. Such increasing recourse to what has been labeled as "forensic engineering and geology" [*Slosson and Shuirman*, 1992] testify to an important cultural change: disasters are not seen anymore as acts of God but as the result of human action that provoked directly or indirectly losses, or inaction in cases where correct intervention could have saved lives and goods. For wrong action or failure of initiative, it is deemed that responsible individuals or institutions should be pursued. This the origin of the word, in its strict sense; however the term "forensic" when applied to disaster damage and losses analysis refers to a much wider set of practices that we are starting to explore now. The Integrated Research on Disaster Risk (IRDR) [*Burton*, 2010; *Oliver Smith et al.*, 2016] in particular has launched a sort of research campaign aimed at understanding the "root causes" of disasters, investigating not only the more visible vulnerabilities and the more evident features of the combination between hazards and human exposure but also the underlying social, political, and economic drivers that led to the more visible pattern of risks in a given territory. While the IRDR initiative is grounded on the strong interest for social and political drivers, similarly to the experience brought by King and Gurtner in Chapter 8, the same idea of "root causes" relates to the body of techniques and practices of incidents investigation particularly in the field of hazardous plants and industries [for an overview, see *Livingston et al.*, 2001]. Although the first examples of disaster forensic analysis carried out following the IRDR approach display a mainly narrative form, the application of techniques used in the industrial domain requires a more standardized and formalized way of reporting disasters, in order to be able to identify key parameters and indicators to be fed with real values each time a disaster occurs. The current capacity of existing databases both at the global and the national level is still too limited in this regard, as discussed by Ehrlich et al. in Chapter 2. At regional and local levels, such values can be found, but due to the lack of standardization and formalization, only individual cases can be investigated, because we lack the capacity to consolidate statistics. This is in part understandable because even though plants and industries are relatively closed systems that can be studied and analyzed in situ, natural disasters affecting an entire region happen in a totally open system and are rather complex. In this case, although it is relatively easy to point at some general drivers such as poverty or governmental failures in passing regulations and laws toward safety or in making individuals and firms comply with existing norms, it is more difficult to determine how exactly those general drivers translate in a specific place into exposure and vulnerability and because such root causes may be operating over a very long time. Nevertheless, despite of all those difficulties, interesting practices aimed at tracking the nexus between some decisions and practices to the damage as it manifested in a concrete case are provided in the book.

In Chapter 10, Szoenyi et al. describe the Post-Event Review Capability (PERC) methodology aimed at providing insight on why and how damage occurred as a result of floods. The methodology requires one to follow a checklist of questions to be discussed with a wide range of stakeholders with different responsibility in the field of disaster management and prevention. The aim is to learn the main drivers and causes of damage to be able to reduce exposure and vulnerabilities in similar situations. Understanding each individual event may mean different things, ranging from a shared description of how the event unfolded and what can be considered the main factors explaining the damage to more formalized efforts to elicit the contribution of each component of the risk equation, namely hazard, exposure, vulnerability, and resilience. The objective in this case is to analyze how the event unfolded in order to identify the explanatory variables of the observed damage [Burton, 2010], elicit the root causes behind the explanatory variables, and be able to define a causality chain between the explanatory variables and the damage.

Our capacity to carry out those three activities of course varies depending significantly on the level of detail and the reach of the data and information related to each event. At the very least, the forensic investigation is aimed at finding out those components of the risk function that can be held responsible for the damage and possibly rank them. It is a matter of defining if the damage is due to excessive severity of the event (the hazard component), to extensive exposure of people and goods in a dangerous zones (exposure), to overly high levels of physical, social, economic, or systemic vulnerabilities of exposed systems. The ranking can be done as of now only in semi-qualitative terms, adopting a scoring system as proposed by Dolan et al. in Chapter 13 on the forensic investigation of damages due to the 2007 flood on businesses in the UK.

As a second step, as proposed by King and Gurtner in Chapter 8, for example, one may want to discover what the reasons are behind the identified levels of hazard, exposure, and vulnerabilities in a given place. Such reasons encompass legislation to mitigate floods, rules and norms enforcement capacity (or lack of), cultural factors, history of places, etc. It is clearly more difficult to connect in a rigorous way such root causes to the damage that has been observed. A wider investigation than the already challenging one to collect and analyze damage data would be required. Developing a full causality chain linking observed damage to explanatory variables and to root causes is still an objective for the future.

A first step toward this direction, though, would be to define a structure of damage analysis reporting that comprises all the elements that are needed to reconstruct such causality. Such damage analysis framework should include, apart from pure damage data, information regarding land use and urban plans, emergency planning, and legislation and administrative arrangements that constitute the environments in which damage has occurred. Menoni et al. in Chapter 11 have proposed such structure to report the 2012 and 2013 floods in the Umbria region of Italy. A standardized index is proposed, according to which damage is analyzed sector by sector in all relevant components (objects and systems), across spatial and temporal scales that are relevant to understand the development and the full consequences of the event.

The effort and time required for the types of reporting and analyses that are proposed in the above mentioned experiences (PERC, reporting in the Umbria region, the assessment of post-disaster situations in Australia, the analysis of post-flood business interruption in UK firms) are justified as long as they serve a number of relevant purposes. The latter provide substantive grounds for decision making during recovery and reconstruction to achieve a more resilient condition in flood-prone areas; support enhanced accounting capacity; and offer new information and better insights into real disaster scenarios to improve risk modeling capabilities.

The first of the three objectives is perhaps the most evident. In order to recover and reconstruct in a more resilient way, the components that contributed more to the damage must be known at the finest possible level of detail as well as the drivers behind such components. Decision makers need to know if they have to invest in structural measures to avoid the repetition of floods in the affected areas of similar extent in the future or instead convince planners and communities to limit exposure in hazardous zones, partially relocate, or increase assets' and people's resilience.

In terms of accounting, forensic investigation may guide toward the identification of priorities for action, addressing resources to counteract hazards, exposure, and vulnerabilities, depending on how those factors are shaping the risk condition in the areas at stake. In fact, it makes a big difference if one has to turn toward structural measures or instead act on the exposure of communities and the vulnerability of their assets. In this regard, the study compiled by Pielke [2014] building on the data and conclusions of the IPCC 2012 Report is an important example. As Mohleji and Pielke [2014] already did for the US case, at the global level he shows that the increase of damage due to natural disasters that are climate related can be explained as the effect of demographic growth and the consequent increased exposure in the most hazardous zones. This has of course important implications for policies.

As for improving risk modeling capacity, on the one hand, some contributions in this book show that enhanced damage data collection and analysis can provide a more realistic and comprehensive perspective on hazards. In fact, in many cases (if not in most), events are not simple and linear as usually represented in handbooks but derive from enchained and combined natural phenomena. Floods do not occur only as a result of peak discharge in catchments due to excessive rainfall but also as a secondary occurrence of hurricanes, storm surges, and landslides. Classification methods used thus far for recording such events have often failed to provide such complexity, misrepresenting the reality that proves to be much more multifaceted than disciplinary boundaries permit to recognize.

On the other hand, damage data are essential to understand vulnerability and resilience in a number of field related to natural hazards. Interest regarding vulnerability and vulnerability aspects is not limited anymore to only seismic hazard, because clearly in the latter structural intervention on existing structures and building codes for new ones is key to reducing the risk but should encompass virtually all hazards. In the case of floods, there has been in the more recent years an increasing effort to develop flood damage functions for different residential buildings, lifelines, agriculture, and industries, to find correlations between flood relevant parameters and expected damage, given the differences in structures, material, and layout of exposed assets. In order to develop damage curves, one needs to have a much more refined idea of how different structures and buildings react to being inundated. Survey of actual flooded homes and structures provides the necessary empirical support to the flood damage models as discussed by Kreibich et al. in Chapter 5 and Thieken et al. in Chapter 7.

However, interest in vulnerability is not limited to physical aspects because events such as flood also reveal other types of vulnerabilities, such as systemic, social, economic, and related damages that are indirect, secondary, and systemic that affect not only individual objects but entire systems, including social and economic ones. In some instances, ripple effects and interconnection among affected and nonaffected systems have been more important than what was actually inundated. Floods can be very costly in terms of the total dollar amount needed for repairs and a return to normalcy but also as far as indirect losses for the economy are concerned.

There is still much to do in order to develop the capacity to track and then store information regarding systemic, indirect, and secondary damages in a systematic and comprehensive way. Disaster reporting is still too focused on tangible physical damage to objects, mainly buildings, while consequences on lifelines and infrastructures, as well as on productive systems are underreported or reported only in anecdotal fashion, failing to provide data and systematic information. More needs to be done and for all kinds of disasters, including floods, if we want to get a more precise and measurable assessment of different types of impacts and damages and to understand how the latter are connected.

Scale aspects are crucial when dealing with systemic and indirect damage. Depending on the level of analysis, the secondary and indirect effects of a flood can be considered to be a damage or not. For a region, the relocation of firms to another region is a damage to its economic system, differently from how the same relocation would be considered at a national level.

However, also scale matters considering direct damage to objects and assets. For example, Kreibich et al. in Chapter 5 discuss how synthetic methods for assessing flood risk are by order of magnitude mismatching real data. Gall in Chapter 4 suggests that while the HAZUS tool can be effective for county level estimates, it lacks consistency when applied at the local level.

3. RECOMMENDATIONS FOR THE FUTURE FOR ENABLING TECHNOLOGIES AND INTEGRATING SECTORS AND STAKEHOLDERS IN A NEW GENERATION OF POST-FLOOD DAMAGE INFORMATION SYSTEMS

The current status as depicted in the previous section is connoted by important changes in the way the effort of managing post-disaster data is conceived and perceived by public administrations at different levels and by various scientific institutions. Such changes are likely to push forward toward hopefully a more systematic and structured way of dealing with these data that deserve some reflection. For the future, four main challenges require attention and should constitute the subject of future research and experimentation. Ranging from "theoretical" to more operational challenges, the following can be considered: the search for coherence between economic and "engineering" damage assessments; the full exploitation of available technologies; the control of data quality; and last but not least, the organizational environment that may transform the technological potential into an everyday practice as suggested by Rudari et al. in Chapter 3.

3.1. An Open Question: How Does One Integrate Different Perspectives on Damage and Losses?

The convergence between physical damage estimates and economic damage representations is still a big, open question. Often the distinction that is made is between an "engineering" view of the damage (the physical damage) and the "economic damage," represented as figures of costs associated with repairs and reconstruction. However, the situation is actually more complex than this simple distinction would suggest. There is certainly a description of damage in terms of the physical disruption of an artifact, be it a lifeline, a building, or a plant. Such description is generally provided by engineers to develop damage/vulnerability curves. Associating with physical damage and repair needs actual costs, requires a specific economic expertise appraising the prices of components, materials, and workforces in the construction industry. The description of the physical damage and its representation in terms of costs of repair is generally made at the local level, at the individual asset. Then of course there is the need to provide estimates at larger scale, to assess the expenditure that is likely to be needed at the municipal and regional scale for the event.

There is also a geographic and territorial perspective on damage that looks beyond individual artifacts to assess the effects of damages on larger systems, which can be services, lifelines, or communities. Such damage can be also viewed in economic terms but has been also represented as a description, in few cases as quantitative estimates of the systemic consequences and disruptions a major event may cause, due to the complexity of modern assets, territories, and built and natural environments. In fact not all those consequences can be actually translated into economic terms.

What is generally referred to as economic damage is instead related to the losses to the economy, which depends also on the total physical damage, but much more on what has been damaged (strategic industries, economic assets, critical infrastructures) and how relevant what has been damaged is for a given economy and for its functioning. Even when talking about economic damage, a distinction can be made between damage to economic sectors (tourism, manufacturing, services, others), which depends on the damaged businesses, and to the economy as a whole, measured for example as a percentage of loss on the gross domestic product (GDP).

3.2. Full Exploitation of Available and Frontier Technologies

In this book, the importance of more recent technologies for damage data collection, storage, and query is evidenced. Ajmar et al. in Chapter 14 show how satellite imagery is used to produce first damage reconnaissance maps and then damage maps at increasing speed and accuracy. In Chapter 9, Roberts and Doyle discuss the role social media have and may play in reporting damage by crowdsourcing. Contrarily to some naïve claims of how the latter can revolutionize damage data collection efforts

by making citizens actual partners of official organization, their article points instead at the real novelty and the challenges ahead. One thing in fact is the spontaneous testimony of a tragedy one is living directly or indirectly, and another is to channel digital volunteerism toward usable and useful components of crisis management.

Recent innovation particularly connected to the web potential (crowdsourcing, crowdmapping, cloud computing) is paving the way for easing the task of collecting and inputting data into databases and making the latter easier to connect. Yet the current situation is rather patchy and fragmented. Different tools exist and pursue their own path of development, without a unified strategy or design for what concern specific domains such as damage data collection and treatment may take. In their contribution, Mazuran et al. in Chapter 15 highlight the many criticalities encountered in the effort to integrate different technologies and to develop an information system able not only to store and retrieve relevant data but also provide different types of representations including maps when this is necessary.

What is certainly needed and is already the object of current research is (for example issued by the European Commission) the integration of technologies and the creation of information systems customized to the needs of users. The latter makes up a rather large set of stakeholders, including those who will need to carry out the surveys, different authorities in charge of reporting damage to the assets they are responsible for, and those who subsequently carry out the analysis of data to produce damage reports. Finally, such a system should be open to a variety of uses of the stored data such as supporting enhancement of risk modeling capacity to forensic investigation as discussed above.

3.3. Defining the Requirements to Collect Good Quality Data

Ehrlich et al. in Chapter 2, Rudari et al. in Chapter 3, Gall in Chapter 4, Kreibich et al. in Chapter 5, and Thieken et al. in Chapter 7 all stress the extreme importance of data quality and the suitability of information systems designed to manage data. The requirements of stakeholders who will actually input and use flood damage data should be carefully investigated and fully embedded in the design of information systems. Such requirements depend also on what stakeholders expect in terms of data quality, particularly if we talk about regulators such as the Australian Government or the European Commission. The following aspects are of particular relevance: consistency, accessibility, and infrastructure [*Australian Government*, 2014].

Consistency is necessary if we wish to use the data to compare situations regionally, nationally, or globally.

This can be achieved by centralizing the collection or by adhering to agreed upon formats and standards. This is an effort that has been going on for some years, at the international level in particular, with the IRDR and the International Strategy for Disaster Reduction (ISDR) in the front line. Consistency is strictly linked also to the softer issue of commitment of governments and administrations to keep and maintain them.

Accessibility is even more complex. It refers first to where data can be found, or put another way, it is the true source of a given datum. When data are official, the thread leading to the organization responsible for it needs to be followed up to the end. This is the reason why all relevant data providers should be engaged in the endeavor, including novel ones such as statistical offices that have remained at the margin of the process until very recently.

When the initial source of the data is found, as suggested by Roberts and Doyle in Chapter 9, there are two aspects related to openness that need to be considered. The first is technical openness, that is the extent to which the format and the place where the data is stored is easy to access or not. In fact, the data can be in paper form and therefore it may be transmitted as a pdf or in a fax. Or it may be in digital form but one that is not compatible with others or stored in a server or computer that is not reachable through the web. The second type of openness is legal. Data can be perfectly digitalized and stored in an easy-to-access place but one needs special permission to get them.

Both types of openness are relevant for the objective of improving post-flood damage data. The official source of data, the legal status, and the technical accessibility are connected to each other. Real interoperability requires that technical formats are somehow compatible and that institutions are willing to share the data. Interoperability is not only a technical issue. Data availability does not immediately coincide with their usability in different types of analysis. Here again the issue of designing a performing information system comes back to the front, suggesting that the design needs to take into account the multiple forms, types, and kinds of data that different authorities, organizations, and people actually collect.

3.4. Integrating a Wide Range of Public and Private Stakeholders in Post-Flood Damage Data Collection and Management

The efforts and the burden to significantly improve the methods and the practice of post-disaster damage data collection needs to be rewarded. In this book, it is suggested that the possibility to use data for multiple purposes and by a variety of stakeholders can represent a great reward.

Apart from the most obvious compensation and identification of needs, improved accounting, enhanced forensic investigation serving the objectives of feeding risk assessment with real figures, and deeper understanding of the dynamic of a flood and of supporting more resilient recovery and reconstruction can be considered of use for a wide range of stakeholders. The latter include public administrations at different levels of government, decision and policy makers, service providers, insurance companies, and at the very bottom, the same affected or at-risk communities. Showing the potential benefit that enhanced damage databases can bring to them, in terms of better understanding of risk situations and being in a better position to decide about future land use developments and investments, is perhaps not so difficult, even though still to be done. This book has the goal of offering this wide range of stakeholders an overview of what is currently done and what the frontiers are that could bring to a new generation of post-flood damage databases to support the variety of goals mentioned above.

What is perhaps more difficult to achieve is the commitment of many stakeholders, particularly but not only those pertaining to the private sector to agree to exchanging and sharing damage data with others. Information Technology (IT) actually permits a protected way of doing this, yet the challenge is a change of mindset.

Certainly the recent stress on open data, the same fact that data are increasingly difficult to hide in an overly wired and connected world, will help to overcome present barriers. However, technical integration is not enough and will not be fully achieved unless a change at the administrative and institutional levels does not take place. First, there is the issue of commitment to the program of better collecting and managing losses data. This is considered by Serje in Chapter 1 as the major issue, much more than financial costs associated with maintaining a damage database that are quite negligible. Long-term commitment despite changes in governments and legislation, despite differing views on priorities and programs for mitigation, is harder to achieve, particularly in some countries.

Second there is the issue of how to organize the task of collecting data. Several contributions in this book highlight the relevance of a shift of practices in this sense. For examples, Gall in Chapter 4 suggests that "best estimation practices need to be developed, shared, and trained on." Moon in Chapter 12 points at the work of the Queensland Reconstruction Authority as a first pillar of a longer term work that should be further supported. In Chapter 2, Ehrlich et al. refer to a data coordinator and a data curator that must be established nationally to collect data from different sources, to guarantee that a comprehensive picture of the direct, indirect, and overall impacts of disasters are registered and maintained with acceptable quality. In Chapter 6, Berni et al. and Menoni et al. in

Chapter 11 hold that such a data coordinator should exist also at lower levels of government, on the regional, provincial, or county level, depending on the magnitude of events and on responsibilities allocated to the different authorities in charge of managing emergency and reconstruction. The Umbria region made clear that without such a figure (that can be one person or an office), data are likely to be lost among the many stakeholders who collect them for different purposes and will be impossible to retrieve once the main objective for which data were collected (i.e., compensation) has been achieved.

3.5. Will Things Actually Change or Will the Past Repeat Itself?

Considerations made in the previous sections, for example, regarding the need to bridge between physical and economic perspectives on damage and losses are not new. The US Committee established in 1999 (quoted at the beginning of the current conclusions section) recognized the need to define models describing potential ripple effects to the economy on the basis of more empirical evidence of the indirect effects of disasters on businesses and services. It seems somehow that what we contemplate for the future has been already suggested in the past. Therefore, the question of whether or not things will change and why they should change right now seems legitimate because for a long time at given time intervals researchers, professional, or institutions brought up the issue of improving damage and losses databases but no actual improvement occurred.

We would not have done this effort and we would have not involved so many colleagues from different countries and different institutions if we were not convinced that there is room for being optimistic regarding the possibility to advance.

Let's try to explain what can be considered different today from past situations. First, there seems to be more commitment and interest on the side of official national and international institutions on the issue of improved databases. There seems to be agreement about that, but without a convincing accounting mechanism, it will be impossible to evaluate the effectiveness of mitigation measures. In a period characterized by scarce resources and financial crisis, the wise use of public money, be it given by governments or donors, is a priority.

Second, technologies seem to provide stronger support than in the past, and we are able to better tailor information systems to our needs. Computer technologies have become pervasive and are a common tool for virtually any worker today, at least in administration and service settings. The step toward using such technologies instead of paper for collecting and managing disaster data is today really small. If we will be able to design usable and effective information systems and databases, the possibility that they will be used as an ordinary tool is greater than in the past.

Last but not least, we may also say that the perception of knowledge needs has changed across the various stakeholders groups. At one time, researchers did not meet often with practitioners or public officers, and decision makers did not participate to scientific research projects. The more mixed environment that has been shaped also by funding agencies wishing to improve the channels of research and knowledge dissemination has fostered the mutual exchange of expertise and experience among academics, professionals, and workers of public and private agencies dealing with risks. Many recording initiatives that have been described in this book started as an initiative of an academic institution (for instance, Sheldus in the US, Howas in Germany). In the future we may expect that more decisive forms of institutional and private/public partnerships will be permitted to co-develop and co-design enhanced information systems and have them run by a mixed body of institutions (academic, governmental, private) sharing the objective of being able to rely on better damage and losses data for different purposes and with different mandates.

REFERENCES

Australian Government, Productivity Commission, Natural Disaster Funding Arrangement (2014), Inquiry report, vol. 1, 2, No. 74, 17 December.

Burton, I. (2010), Forensic disaster investigations in depth: a new case study model, in Environment, 52(5): 36–41.

Direction Territoriale Méditerranée du Cerema (2014), Retour d'expérience sur les inondations du département du Var les 18 et 19 janvier 2014 Volet 2 – «Conséquences et examen des dommages», 2014. Accessed 10 February 2016, available at: http://observatoire-regional-risques-paca.fr/sites/default/files/rapport_rex83_2014_dommages_sept14_0.pdf.

Federal Emergency Management Agency (FEMA), (1997), Report on Costs and Benefits of natural hazard mitigation, April.

Gabrielli, F. (2014), Former Head of the Italian National Department of Civil Protection, Audit at the XIII Committee of the Italian Senate on the "Situation of national emergencies connected to hydrogeological risks due to recent meteorological events," 5 March, 2pm.

Hubert, G., and B. Ledoux (1999), Le coût du risqué. L'évaluation des impacts socio-économiques des inondations, Presses de l'école nationale Ponts et chaussées, Paris.

Jongman, B., H. Kreibich, H. Apel, J. I. Barredo, P. D. Bates, L. Feyen, A. Gericke, J. Neal, C. J. H. Aerts, and P. J. Ward (2012), Comparative flood damage model assessment: towards a European approach, Nat. Hazards Earth Syst. Sci., 12: 3733–3752.

Livingston, A. D, G. Jackson, and K. Priestley (2001), Root causes analysis: Literature review, Contract Research Report, Prepared by WS Atkins Consultants Ltd for the Health and Safety Executive 325.

Margottini C., G. Delmonaco, and F. Ferrara (2011), Impact and Losses of Natural and Na-Tech Disasters in Europe. S. Menoni, C. Margottini (eds.). Inside Risk. Strategies for sustainable risk mitigation. Springer.

Mohleji, S., and R. Pielke Jr. (2014), Reconciliation of trends in global and regional economic losses from weather events: 1980–2008. Natural Hazards Review.

Oliver-Smith, A., Alcántara-Ayala, I., Burton, I., and Lavell, A. (2016), Forensic Investigations of Disasters (FORIN): a conceptual framework and guide to research (IRDR FORIN Publication No. 2). Beijing: Integrated Research on Disaster Risk. 56 pp.

Organization for Economic Co-operation and Development (OECD) (2014), Improving the evidence base on the costs of disasters: Towards an OECD framework for accounting risk management expenditures and losses of disasters. 4th meeting of the OECD High Level Risk Forum, 11–12 November. Document: GOV/PGC/HLRF (2014)8.

Pielke, R. Jr. (2014), The Rightful Place of Science: Disasters & Climate Change, Consortium for Science, Policy and Outcomes, Arizona State University.

Pitt, M. (2008), The Pitt review: learning lessons from the 2007 floods, 2008. Accessed: 10 February 2016, available at: http://archive.cabinetoffice.gov.uk/pittreview/thepittreview/final_report.html.

Slosson, J., and G. Shuirman (1992), Forensic Engineering: Environmental Case Histories for Civil Engineers and Geologists, Academic Press.

United Nations (UN) (2015), Sendai Framework for Disaster Risk Reduction 2015, 2015–2030, A/CONF.224/CRP.1. Retrieved at: http://www.unisdr.org/we/coordinate/hfa-post2015.

United Nations Development Programme (UNDP) (2013), Bureau for Crisis Prevention and Recovery, A comparative review of country-level and regional disaster loss and damage databases.

US Army Corps of Engineers (USACE), (2009), Value to the Nation: flood risk management. Alexandria, VA.

Van der Veen, A., A. L. Vetere Arellano, and J-P. Nordvick (cur.) (2003), In search of a common methodology on damage estimation. In "Workshop Proceedings," European Commission – DG Joint Research Centre, European Communities, Italia, **EUR 20997 EN**.

INDEX

Flood Damage Survey and Assessment: New Insights from Research and Practice, Geophysical Monograph 228,
First Edition. Edited by Daniela Molinari, Scira Menoni, and Francesco Ballio.
© 2017 American Geophysical Union. Published 2017 by John Wiley & Sons, Inc.